Springer Series in
SOLID-STATE SCIENCES

Series Editors:
M. Cardona P. Fulde K. von Klitzing R. Merlin H.-J. Queisser H. Störmer

The Springer Series in Solid-State Sciences consists of fundamental scientific books prepared by leading researchers in the field. They strive to communicate, in a systematic and comprehensive way, the basic principles as well as new developments in theoretical and experimental solid-state physics.

Yu. G. Naidyuk   I.K. Yanson

Point-Contact Spectroscopy

Springer

Yu. G. Naidyuk
I.K. Yanson
B. Verkin Institute for Low Temperature Physics and Engineering
National Academy of Sciences of Ukraine
47 Lenin Avenue
61103, Kharkiv, Ukraine
E-mail: naidyuk@ilt.kharkov.ua, yanson@ilt.kharkov.ua

Series Editors:

Professor Dr., Dres. h.c. Manuel Cardona
Professor Dr., Dres. h.c. Peter Fulde*
Professor Dr., Dres. h.c. Klaus von Klitzing
Professor Dr., Dres. h.c. Hans-Joachim Queisser

Max-Planck-Institut für Festkörperforschung, Heisenbergstrasse 1, 70569 Stuttgart, Germany
* Max-Planck-Institut für Physik komplexer Systeme, Nöthnitzer Strasse 38
  01187 Dresden, Germany

Professor Dr. Roberto Merlin

Department of Physics, 5000 East University, University of Michigan
Ann Arbor, MI 48109-1120, USA

Professor Dr. Horst Störmer

Dept. Phys. and Dept. Appl. Physics, Columbia University, New York, NY 10027 and
Bell Labs, Lucent Technologies, Murray Hill, NJ 07974, USA

Cover concept: eStudio Calamar Steinen

Library of Congress Cataloging-in-Publication Data is available upon request

ISBN 0-387-21235-3   Printed on acid-free paper.

Printed in the United States of America.   (SBA)

9 8 7 6 5 4 3 2 1   SPIN 10943526

springeronline.com

To the staff of Point-Contact-Spectroscopy Department of B. Verkin Institute for Low Temperature Physics and Engineering, National Academy of Sciences of Ukraine.

Preface

The main goal of solid-state physics is investigation of the properties of the matter including the mechanical, electrical, optical, magnetic, and so on with the aim of developing new materials with defined characteristics. Nowadays, it consists of the synthesis of superconductors with high critical temperature or fabrication of new heterostructures on the base of semiconductors, in creation of layered, amorphous, organic, or nanofabricated structures and many others. To do all of these, the various methods of investigation are developed during the past. Because it is impossible to find an universal method to investigate a variety of materials, which are either conducting or insulating, crystalline or amorphous, thin-layered or bulk, magnetic or segnetoelectric, and so on, various kind of spectroscopies, like optical, neutron, electron, tunnel and so on, are widely used in solid-state physics. Recently, a new type of spectroscopy, namely, the Point-Contact Spectroscopy (PCS), was designed for study of the conduction–electron interaction mechanism with a whole class of elementary excitations in the solids. In PCS, a small constriction, about a few nanometers large, between two conductors plays a role of a spectrometer. Namely, because of inelastic scattering of accelerated electrons, the $I - V$ characteristic of such a tiny metallic contact is nonlinear versus an applied voltage and its second derivative surprisingly turns out to be proportional to the electron–quasiparticle–interaction spectrum. This property of point contact carries spectroscopic investigations of different excitations in solids, first of all the phonons, as well as magnons, magnetic, and some nonmagnetic impurities with internal degrees of freedom, crystal-field-induced electron levels, and many other processes influenced by the electrical transport. To provide an analogy, the PCS can be compared with the optical absorption spectroscopy, where frequency-dependent absorption of light gives characteristic energy of excitations in insulating materials. Hence, the great power of absorption spectroscopy is peculiar to PCS, which measures the imaginary part of conduction electron self-energy. This can be compared with its "antipode", the Tunneling Spectroscopy, where net conductivity is proportional to the electron density of quasiparticle states and, consequently, measures the real part of the conduction electron self-energy in solids. PCS can be considered in a much wider sense. Suppose we have a constriction between any "massive" banks that size is much smaller than the inelastic mean free path of any carriers (charge, magnetic moments, energy, mass, etc.). The flow of these carriers versus the generalized forces applied to the banks contains the spectral information similar to PCS. As an example of such generalized PCS, one can consider the flow of ballistic phonons between the small constriction connecting the banks with the applied difference in temperatures. Such a theory and experiment are successfully elaborating in a recent time for dielectrics. Another well-known example is the Knudsen regime of mass flow through the small orifice to the vacuum versus the gas pressure in the source. If there are some seldom impurities with the internal degree of freedom present in vacuum, then such a process can be considered as a PCS of these excitations.

The goal of the book is to introduce the readers to PCS, the method that has already been widely used since the beginning of 1970s. Created at the Institute for Low Temperature Physic and Engineering, National Academy of Sciences of Ukraine, and placed in Kharkiv, PCS was soon recognized in the physical laboratories of Europe, Japan, and North America. Among them there are Grenoble High Magnetic Field Laboratory (France), Leiden and Nijmegen Universities (The Netherlands), the Physical Institutes of Cologne, Darmstadt and Karlsruhe (Germany), Institute of Experimental Physics in Kosice (Slovakia), Universities in Cornell, Maine (USA), and Madrid (Spain). In the book we put more attention on the experimental aspects of PCS and present above 200 figures. The theoretical part is restricted essentially in presenting the main relations without sometimes superfluous details. This is because to provide the theory of PCS in full measure, a separate book is necessary. We are sure that the method of PCS because of its simplicity and direct information should gain more places where the properties of new materials are studied, especially in the mainstream of mesoscopic or nanoscale physics.

The authors are grateful to all who assisted in preparing the book. We acknowledge all authors whose results we used for preparing the figures and whose names are written in the figure captures. Special thanks are addressed to A. G. M. Jansen and P. Wyder for the discussion of the book contents and practical comments. We wish to acknowledge very enlightening discussions and long-time collaboration with I. O. Kulik, A. N. Omelyanchouk, R. I. Shekhter, A. A. Lysykh, O. I. Shklyarevskii, and Yu. A. Kolesnichenko. We are thankful to K. Gloos, G. Goll, J. M. van Ruitenbeek, E. Scheer for giving us the figures in the final form, and to O. E. Kvitnitskaya for the help in digitizing of numerous curves and preparing of many pictures. Y. N. thanks P. Wyder for the opportunity to stay in Grenoble HMFL, where the draft of the book was written. I. Y. thanks the Solid State Division of Kamerlingh Onnes Laboratory guided by L. J. de Jongh for long-term mutual collaboration with J. M. van Ruitenbeek and his group. The authors are indebted to H. von Löhneysen, Alexander von Humboldt Stiftung, DFG and MPI für Festkörperforschung (all from Germany) for support of the PCS method during the last few decades.

Kharkiv,   *Yu. G. Naidyuk*
December 2003   *I. K. Yanson*

Contents

1 Introduction

With the miniaturization of electronic circuits, the number of transistors per integrated circuit continues to follow Moore's law by doubling every 18 months. The demand for reduced size and therefore increased speed of operation is a continuous challenge for modern device technology on the nanometer scale. It can be foreseen that further reduction will finally lead to basic changes in the principles of device operation. The physical phenomena playing a role at reduced dimensions attract, therefore, a lot of attention in fundamental research.

It is evident that the investigation of submicron electrical contacts is very important from a practical point of view if one considers size reduction and electronic connections. Our interest will be mainly concerned with the fundamental understanding of the microscopic transport of electrical current in a small metallic constriction. A small contact means here a metallic constriction with size d smaller than the mean scattering length l of the electrons. For such a situation $(l > d)$, the charge carriers transport electrical current through a contact ballistically with minimal scattering with impurities. Like the situation encountered in mesoscopic physics, considering the conservation of energy in a scattering process, the above given condition can be relaxed into an inelastic diffusion length larger than the contact dimension.

In the middle of the 1960s, Sharvin (1965) from the Institute of Physical Problems in Moscow realized the importance of ballistic electron transport in small metallic constrictions of dimensions smaller than the electron mean free path. He proposed to use these metallic contacts as injectors of ballistic nonequilibrium electrons into a metal by applying a voltage across the contact. Using a longitudinal magnetic field, the electrons injected into a single crystal plate can be focused according the shape of the Fermi surface on the opposite side of the plate. Another detector point contact, placed within a mean free path from the injector contact, was used for the detection of the focused electrons. These electron-focusing experiments contain information on the Fermi surface of the metal under study. In a different geometry with a transverse magnetic field, the emitted electrons can be focused back on a laterally displaced spot on the same side of the crystal. Such experiments were made by Tsoi (1974) at the Institute of Solid State Physics in Chernogolovka

(Russia) and opened the possibility of investigating the specular reflection of electrons at the sample boundary from the inside.

Sharvins ideas stimulated further investigations with point contacts. In 1972, Bogatina and Yanson, from the Institute for Low Temperature Physics and Engineering in Kharkiv (Ukraine), called attention to pronounced nonlinearities in the $I - V$ characteristic of a metallic short-circuit in the insulating barrier of a tunnel junction. In subsequent experiments with this type of metallic contacts, Yanson (1974) has shown that point contacts can be used for energy-resolved spectroscopy of the electronic scattering inside a metal. In his first measurements of the nonlinear current-voltage characteristics, he could directly measure the energy dependence of the electron–phonon interaction. From that time, the epoch of point-contact spectroscopy (PCS) began. In order to understand the essential mechanism of energy-resolved spectroscopy in ballistic contacts, one should realize that for an applied voltage V, the electrons are accelerated corresponding to a characteristic increase of energy eV upon passing through a contact without inelastic scattering. The applied voltage tunes the characteristic energy eV of the electron system. The accelerated electrons can be scattered by, for instance, phonons that lead to a decrease in the probability for the electrons to pass through the constriction. As a result, the current I has a negative correction at specific bias voltages corresponding to the characteristic phonon energies. The nonlinear $I - V$ curves can be directly analyzed in terms of energy-dependent information on the electron–phonon scattering mechanism.

PCS has some relation with the classic tunneling spectroscopy encountered in structures with an insulating barrier between metals [Wolf 1985)]. However, the underlying mechanisms leading to spectroscopy in the metallic constriction and tunnel junction are different, and one should not confuse these two spectroscopic methods. PCS deals with metallic constrictions or conducting bridges between two metals with the characteristic size d smaller than the electron mean free path l. In this case, the inelastic scattering of accelerated electrons with energy increase eV can be probed by measuring the voltage-dependent resistance change of a contact. These scattering processes occur inside the metallic contact volume. The fulfillment of the ballistic regime $l \gg d$ is the main condition for energy-resolved spectroscopy. In tunneling spectroscopy, the energy difference of the electrons between the left and the right side of the junction is achieved by applying voltage over a thin dielectric or vacuum gap. Here energy-resolved spectroscopy is determined by quantum mechanical tunneling processes, which depend on energy caused by structures in the electronic density of states and by the interaction of electrons with quasiparticle states (phonons, local modes, etc.) in the barrier or at the interface of the electrodes. Whereas with metallic point contacts one probes scattering processes inside the contact volume, tunneling spectroscopy is a more surface-sensitive technique because of its dependence on the overlap of electronic wave functions from both sides of the junction.

For clean metals, the ballistic condition $l \gg d$ is satisfied at low temperatures for contacts with a dimension of about 10 – 100 nm. Current densities of $10^9 - 10^{10}$ A/cm^2 can be achieved in these ballistic point contacts without substantial heating. Bulk conductors would evaporate under such high current densities. These dramatic heating effects do not play a role in ballistic contacts because the heat will be spread out over a region far away from the contact at a distance l. The typical point-contact current densities are considerable larger than the value, as high as 10^6 A/cm^2 encountered today in small-scale devices.

Nowadays PCS is widely used to study the interaction of conduction electrons with elementary excitations or quasiparticles in conducting solids. It is a comparatively simple, easily available, and highly informative method. In this method, the contact plays the role of an analytical instrument, a certain kind of spectroscope. Within the last two decades, many new results have been obtained in this rapidly developed field, which were not available using other experimental techniques.

In this book, we would like to describe the theoretical background of PCS and to cover in a complete overview the experimental work done in this field concerning different aspects of solid-state-physics research. We intend to review most of the publications in the field. A detailed tabulation of the electron–phonon interaction spectra measured by means of point contacts can be found for many pure metals in the atlas written by Khotkevich and Yanson (1995). With regard to other topics, we recommend the following research-level reviews: The first review on point-contact spectroscopy in metals was published by Jansen et al. (1980). A detailed analysis of point contact electron–phonon interaction functions was done for 26 pure metals by Yanson (1983). The principles of point-contact spectroscopy in systems with a short elastic mean free path along with experimental data for many compounds have been presented by Yanson and Shklyarevskii (1986). The possibilities of point-contact techniques and a number of new experiments were surveyed by Duif et al. (1989). The study by means of point contacts of heavy-fermion systems in the normal and superconducting state was treated by Naidyuk and Yanson (1998).

References

Duif A., Jansen A. G. M. and Wyder P. (1989) J. Phys.: Condens. Matter **1** 3157.

Jansen A. G. M., van Gelder A. P. and Wyder P. (1980) J. Phys. C **13** 6073.

Khotkevich A. V. and Yanson I. K. (1995) *Atlas of Point Contact Spectra of Electron–Phonon Interaction in Metals* Kluwer Academic Publisher, Boston.

Naidyuk Yu. G. and Yanson I. K. (1998) J. Phys.: Condens. Matter **10** 8905.

Sharvin Yu. V. (1965) Sov. Phys. - JETP **21** 655.

Tsoi V. S. (1974) JETP Lett. **19** 70.

Wolf E. L. (1985) *Principles of Electron Tunneling Spectroscopy* Oxford University Press, Inc. New York.

Yanson I. K. (1983) Sov. J. Low Temp. Phys. **9** 343.

Yanson I. K. (1991) Sov. J. Low Temp. Phys. **17** 143.

Yanson I. K. and Shklyarevskii O. I. (1986) Sov. J. Low Temp. Phys. **12** 509.

2 Metallic point contacts as a physical tool

Already more than 100 years ago Drude developed a theory for the electrical and thermal conduction of metals based on the classic kinetic theory of gases. Drude considered a metal to be a lattice-reservoir filled by a gas of electrons, which move freely between the scattering events on the ions of the lattice. Considerable improvements in the quantitative estimates were made later by Sommerfeld, which account for the quantum mechanical Pauli exclusion principle of the electrons, what leads to the Fermi–Dirac energy distribution of the electrons with occupied electron states up to the Fermi energy. At zero temperature, the Fermi surface separates in wave vector $\mathbf{k}$-space the occupied electronic states from the empty ones. For free electrons, the Fermi surface consists of a sphere with radius k_{F}. In real crystalline metals, the shape of the Fermi surface is much more complicated and determined by the number of conduction electrons per atom as well as the symmetry of the lattice. Because the electron states close to the Fermi surface are responsible for the main characteristic properties of metals, *viz.* electrical and thermal conductivity, superconductivity, thermoelectric effect, and so on, knowledge of the Fermi surface has an important meaning for the understanding of the electronic transport. Many experimental methods exist for the investigation of the Fermi surface. The most important ones are known as the de Haas–van Alphen and the Shubnikov–de Haas effects, where oscillations of thermodynamic and transport properties of metals in a magnetic field are measured to get detailed information on the shape of the Fermi surface. In this chapter, we briefly describe milestone experiments connected with the main topic of this book. These achievements gave undoubtedly the necessary impact to the further development of point-contact spectroscopy.

2.1 Electron focusing

Sharvin (1965) theoretically examined for the first time the resistance of a *ballistic* contact where the mean free path l of the electrons is larger than the size d of the constriction. Here one can model the contact as a circular orifice of diameter d in an insulating plane between two metallic parts (Fig. 2.1). Sharvin's description of the contact resistance was as follows. In the voltage-biased ballistic contact, the speed increment δv for an electron

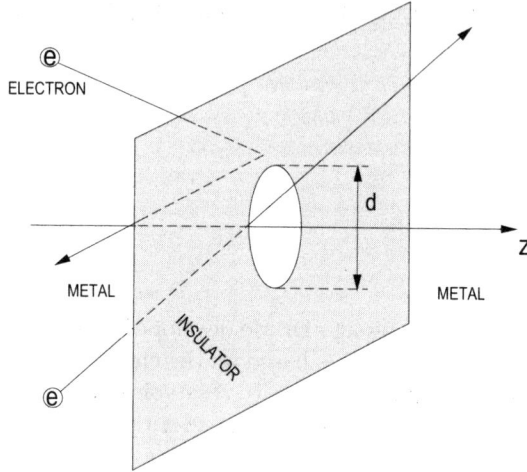

Fig. 2.1. Model of a ballistic point contact presents an orifice with diameter d in an insulating plane between two metallic banks along with trajectories of electrons.

passing the orifice is proportional to the applied voltage: $\delta v = eV/p_F$, where p_F is the Fermi momentum. This results in a current I through the contact: $I \simeq e(\pi d^2/4)\, n\, \delta v = (\pi d^2/4)\,(n\, e^2/p_F)\, V$, where n is the electron density. Using the Drude formula for the resistivity $\rho = p_F/ne^2 l$, one finds an expression for the resistance of a contact: $R \simeq 4\rho l/\pi d^2$. The integration over all possible angles gives a numerical factor $4/3$ leading to the Sharvin resistance of a circular contact:

$$R_{\mathrm{Sh}} = \frac{16\rho l}{3\pi d^2}. \tag{2.1}$$

The Sharvin resistance depends only on the contact dimension d and is independent of the material purity, whereas $\rho l = p_F/ne^2$ is a material constant for a specific metal. It is of interest to compare the ballistic resistance with the resistance of a contact in the opposite limit $d \gg l$. Already Maxwell (1904) has calculated the resistance of a "dirty" metallic contact and found that it depends on the resistivity and the contact diameter in a simple way:

$$R_{\mathrm{M}} = \frac{\rho}{d}. \tag{2.2}$$

Apart from a geometrical constant, this expression can be easily deduced by considering the resistance of a sample of length d and diameter d.

Realizing the mechanism of ballistic electrical transport in small metallic constrictions at low temperatures, Sharvin proposed to use contacts placed on the opposite sides of a thin metallic slab for a study of the Fermi surface. One of them is used as an injector of electrons and the other as a detector of electrons. By applying the appropriate longitudinal magnetic field, injected electrons can be focused onto the opposite electrode [see Fig. 2.2(a)]. The

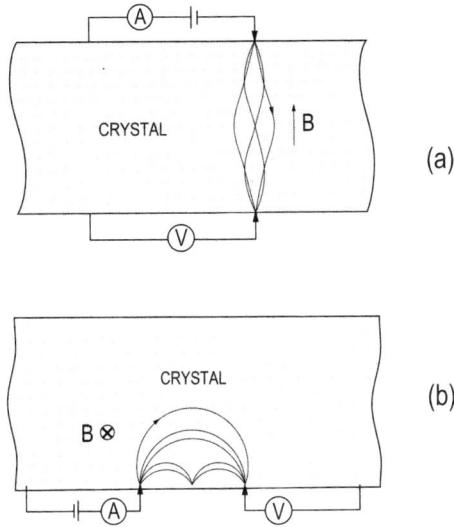

Fig. 2.2. (a) Longitudinal electron focusing in a magnetic field according to Sharvin (1965). (b) Transverse electron focusing in a magnetic field according to Tsoi (1974).

idea is based on the ballistic electron flow without collisions between the two contacts. In this case, the trajectory of the electrons in a magnetic field is determined by the geometry of the Fermi surface. Deflected by a magnetic field of appropriate focusing strength, the injected electrons cause an increased concentration of charge carriers and therefore an increased potential at the second contact. In the experiments of Sharvin, it is not necessary that the flow of electrons through the contact be ballistic or energy conserving; only the trajectory between the contact periphery of injection and detector contacts should behave ballistically without collisions. Collisions between emitter and collector would strongly decrease the amount of focused electrons.

In 1974, Tsoi further developed Sharvin's method of focusing by placing the emitter and collector contacts on the same side of the metallic slab and applying a transverse magnetic field parallel to the surface [Fig. 2.2(b)]. This method was found to be simpler compared with the Sharvin one, where it was difficult to place two contacts exactly opposite to each other on a thin metal slab. For increasing magnetic field, or decreasing cyclotron radius, the direct focusing of electrons is followed by electron focusing after specular reflection from the inside of the metal surface. In Fig. 2.2(b), we have shown electron trajectories close to the field for direct focusing and for focusing after one internal reflection. Transverse focusing provides not only information on the Fermi surface, but also the probability of specular reflection from the crystal surface. Moreover, other scattering boundaries such as the normal metal-superconductor interface can be investigated. These examples already show that metallic point contacts are a powerful tool in transport experiments where the *ballistic* regime of electron transport with minimal electron scattering plays a decisive role.

2.2 Normal-metal contacts

During the investigation of small constrictions between two metallic films separated by a thin (a few nanometers) dielectric layer, in the beginning of the 1970s, Yanson paid attention to their nonlinear current-voltage $(I - V)$ characteristics at helium temperature. Further detailed study showed that these deviations from Ohm's law occurred at an energy eV corresponding to the phonon frequencies of the metals under study. Moreover, Yanson (1974) found that the measured second derivative $d^2V/dI^2(V)$ of the $I - V$ curve resembles the Eliashberg function $\alpha^2 F(\omega)$ of the electron–phonon interaction. Roughly speaking, the dimensionless $\alpha^2 F(\omega)$ function is the convoluted product of the phonon density of states $F(\omega)$ and the matrix element squared α^2 of the electron–phonon interaction. As shown in Fig. 2.3, the maxima of

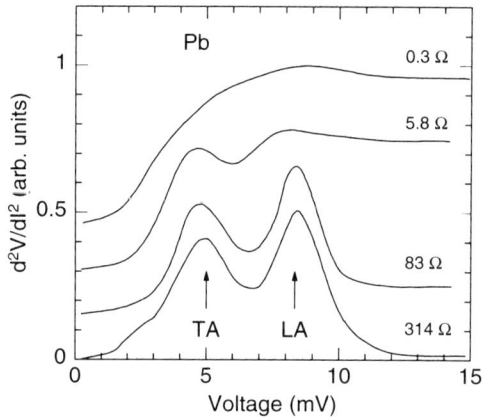

Fig. 2.3. Second derivative of $I - V$ curves for a Pb tunnel junction with micro-constriction at helium temperature in the process of successive decreasing of the resistance of the constriction from 314 Ω (bottom curve) to 0.3 Ω (top curve). The positions of longitudinal acoustic (LA) and transverse acoustic (TA) phonon peaks in Pb are indicated by arrows. The curves are offset vertically for clarity. Data taken from Yanson (1974).

the $d^2V/dI^2(V)$ point-contact spectra for a Pb constriction coincide with the main phonon peaks of Pb – transverse and longitudinal. Why $d^2V/dI^2(V)$ directly reflects $\alpha^2 F(\omega)$ is easily to see by including in the Sharvin formula for the contact resistance a contribution (2.2) caused by scattering of electrons in the constriction, such that

$$R \simeq \frac{16\rho l}{3\pi d^2} + \frac{\rho}{d}. \tag{2.3}$$

This expression of the contact resistance is an extrapolation between the two limiting expressions of a clean contact ($l > d$) and of a dirty contact ($l < d$) [see Eqs. (2.1) and (2.2)]. The second term of (2.3) describes the corrections to the ballistic Sharvin expression caused by scattering, which will be small in the limit $l > d$. We can rewrite (2.3) as an expansion in d/l

$$R \simeq R_{\text{Sh}} \left(1 + \frac{3\pi d}{16\, l}\right) = R_{\text{Sh}} \left(1 + \frac{3\pi d}{16 v_{\text{F}} \tau}\right). \tag{2.4}$$

An energy-dependent scattering time $\tau(eV) = l(eV)/v_{\text{F}}$ leads to a voltage-dependent contact resistance $R(V)$. Using a golden rule argument for the emission of phonons by a nonequilibrium electron at an energy eV above the Fermi energy, $\tau_{\text{el-ph}}(eV)$ can be related to the electron–phonon interaction function $\alpha^2 F(\epsilon)$, according to Grimvall (1981),

$$\tau_{\text{el-ph}}^{-1}(eV) = \frac{2\pi}{\hbar} \int_0^{eV} \alpha^2(\epsilon) F(\epsilon)\, \mathrm{d}\epsilon. \tag{2.5}$$

Then the derivative of (2.4) yields

$$\frac{\mathrm{d}R}{\mathrm{d}V} \left(\propto \frac{\mathrm{d}^2 V}{\mathrm{d}I^2}\right) \simeq R_{\text{Sh}} \frac{3\pi^2\, ed}{8\hbar v_{\text{F}}} \alpha^2 F(eV). \tag{2.6}$$

This estimation shows directly the proportionality between the second derivative of the $I - V$ curves measured by Yanson (1974) and the electron–phonon interaction function. Using a picture of ballistic electron trajectories, electrons passing through the constriction gain the excess energy eV and can relax by the spontaneous emission of phonons with maximum energy eV. Some of the electrons are reflected back from the contact because of these electron–phonon scattering processes, leading to a backflow correction to the current through the contact. Finally, this results in an $I - V$ characteristic with nonlinearities at voltages corresponding to the characteristic phonon energies.

A more comprehensive theory by Kulik et al. (see Section 3.2) based on the solution of the Boltzmann equation confirmed the phenomenological formula (2.6) and became the founding issue for PCS. The voltage V applied to the contact defines the energy scale eV for the scattering processes. For the nonequilibrium situation of an applied voltage across a ballistic contact, the Fermi surface splits into two parts with a difference in their maximum energy given by the bias-voltage energy eV (see Section 3.1). This energy step in the distribution of available electron states close to the contact is, figuratively speaking, the energy-resolving probe in point-contact spectroscopy.

The use of point contacts provides spectroscopy with all kinds of quasi-particle excitations in conductors. Along with the traditional investigation of the electron– phonon interaction, other types of interactions that result in an energy-dependent relaxation time $\tau(\epsilon)$ can be studied by PCS: electron–magnon interaction, Kondo scattering, crystal-field levels of the rare-earth ions, and so on.

2.3 Superconducting contacts

Point contacts with superconductors were used for the investigation of the superconducting phenomena as well as for many applications. It is difficult to cover all aspects of superconducting contacts, but from a spectroscopic viewpoint, one mainly deals with such phenomena as the Josephson effects and Andreev reflection. For the Josephson effects in superconductor–superconductor contacts, we refer to the books by Kulik and Yanson (1970), Barone and Paterno (1982), and Likharev (1985). Here we will mostly concentrate on the Andreev reflection phenomenon for the electrical transport across a normal-metal superconductor interface. The Andreev reflection describes the process of charge transfer between single particle charge carriers in normal metals and Cooper pairs in superconductors.

An electron at the Fermi level cannot cross an interface with a superconductor because of the existence of an energy gap with forbidden quasiparticle states. In 1964, Andreev predicted that the electronic transfer across the interface involves the simultaneous reflection of a hole leaving a Cooper pair into the superconductor for the charge transport. It results in a doubling of net current through the contact between normal metal and superconductor with a resistance twice smaller compared with a contact fully in the normal state. The extra current is known as the excess current, and it is a characteristic feature of the $I - V$ curve of a contact between normal metal and superconductor. For excitation energies above the forbidden energy gap, quasiparticles are allowed in a superconductor, so that direct transport via single-particle excitations is again possible across the interface. This change in charge transfer results in nonlinear $I - V$ curves around the voltage corresponding to the gap energy. We show this in detail in Section 3.7.1 by taking into account the specific electron/hole character of the quasiparticles in a superconductor close to the energy gap. Actually, spectroscopy of the energy gap by point contacts began after the breaking work of Blonder et al. (1982) (BTK). They used energy-dependent transmission coefficients obtained from a solution of the Bogolubov equations at the boundary to calculate the current through the ballistic N-S contact. In spite of the one dimensionality of this model, their calculations perfectly describe experimental data. As seen from Fig. 2.4, one can directly observe the gap feature in the measured dV/dI curve and in principle estimate the gap value from the voltage position of the minimum[1]

---

[1] A few years before Artemenko et al. (1979) noticed that at low temperatures, conductivity of diffusive S-c-N contact has maximum at $V = \Delta$.

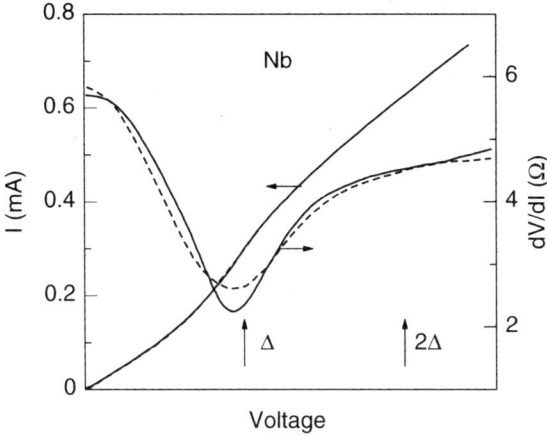

Fig. 2.4. $I - V$ curve and its derivative for a Nb-Cu point contact (solid lines) compared with the calculation by virtue of the BTK theory (dashed line) using Δ = 1.47 meV, $Z = 0.65$ and $T/T_c = 0.138$, where T_c is the critical temperature of Nb. Data taken from Blonder and Tinkham (1983).

in dV/dI without the need of a full calculation according to the BTK theory. The latter includes also some phenomenological transmission coefficient Z of the N–S interface, which in a simple way describes a continuous transition from a metallic contact to a tunnel junction. In brief, with point contacts, it is possible to investigate the energy gap in superconductors as well as with the tunneling measurements. Over the last two decades, point contacts have been widely applied for the investigation of Δ in high-T_c materials, heavy fermion systems, borocarbides, organic and other superconductors.

2.4 Tunneling phenomena

Quantum mechanical tunneling in physics is connected with a long-time known phenomena of the emission of α-particles from radioactive nuclei, the ionization of atoms, or the cold emission of electrons in a strong electric field. In solid-state physics, tunneling is mainly connected with artificial structures that contain a potential barrier on an atomic length scale. Starting from the Esaki (1958) tunnel diode developed in 1958 and the metal-insulator-metal structure produced by Giaever (1960) a few years later, the scanning tunneling microscope initially developed in 1982 by Binning and Rohrer is nowadays transformed into a routine device with the possibility of precise control (better as 0.1Å) of the vacuum gap between the electrodes. A schematic model of the potential distribution near a tunnel junction between two conductors is shown in Fig. 2.5(a) with different transmission channels. By applying to a tunnel contact with nonzero tunneling transmission, a small bias voltage

(a)

(b)

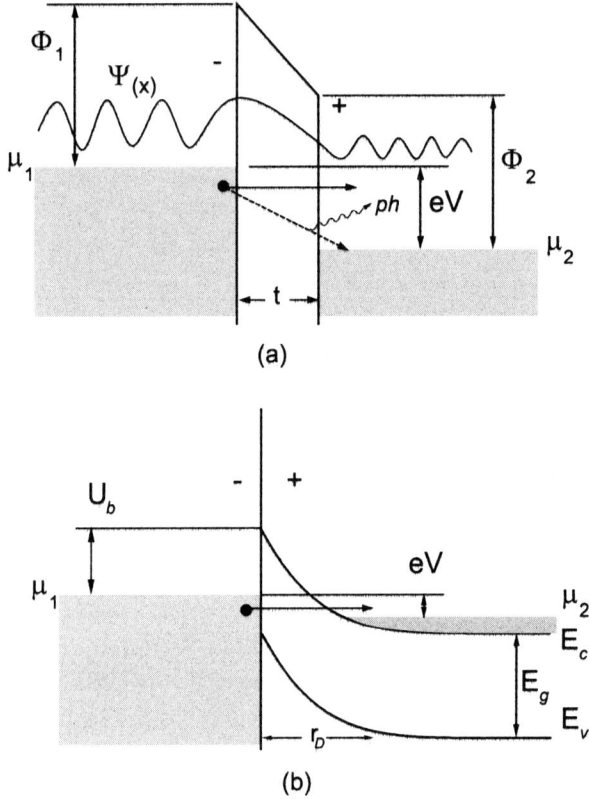

Fig. 2.5. (a) Scheme of a voltage V biased tunnel contact between two metals 1 and 2. Elastic and inelastic tunneling processes are shown by solid and dashed arrows, respectively. Ψ is the wave function, Φ is the workfunction, μ is the chemical potential, and t is the barrier width. ph marks excitations like phonons. (b) Scheme of voltage V biased Schottky barrier between a metal and a degenerate n-type semiconductor with a gap E_g between the bottom E_c of the conduction band and the top E_v of the valence band. r_d marks the Debye charge screening length in semiconductor, and U_b is the height of the barrier.

compared with the workfunction of the metal, the tunnel current will be proportional to the applied voltage and is defined by the junction resistance

$$R = R_0 \exp\left(\frac{2t}{\hbar}\sqrt{2m\Phi}\right), \tag{2.7}$$

where R_0 is a prefactor that depends on various parameters, t is the barrier width, m is the electron mass, and Φ is the metal workfunction. This exponential dependence of the resistance on the width of dielectric gap, e. g., increase of the gap with 1Å , changes the resistance about one order of magnitude, is used in a scanning tunnel microscope as an extremely sensitive method to measure surface corrugations with a subangström resolution.

One of the very fruitful applications of tunneling spectroscopy was the observation of the superconducting energy gap Δ by Giaever (1960). The current of a voltage-biased $S_1 I S_2$ tunnel junction depends on the quasiparticle density of states (DOS) $N_i(\epsilon)$ in the superconducting electrodes S_i $(i = 1, 2)$ such that [Wolf (1985)]

$$I(V) \propto \int_{-\infty}^{\infty} N_1(\epsilon)N_2(\epsilon + eV)(f(\epsilon) - f(\epsilon + eV))\mathrm{d}\epsilon, \qquad (2.8)$$

where $f(\epsilon)$ is the Fermi–Dirac distribution function. For superconductors with a gap Δ in DOS ($N_i^S(\epsilon)=0$ at $e|V| \leq \Delta_i$), the tunnel current is zero for a bias voltage below $(\Delta_1 + \Delta_2)/e$ and increases to the normal-state current at higher bias voltages. In the case of SIN junction, the derivative of (2.8) with respect to the applied voltage assuming a constant density of states for the normal-metal electrode N results in

$$\mathrm{d}I/\mathrm{d}V \propto N_\mathrm{S}(\epsilon)$$

for $T \rightarrow 0$. Hence, the tunnel junction presents a tool for the spectroscopy of the quasiparticle density of states in superconductors, which gives direct information on the superconducting energy gap [Giaever (1960)]. Moreover, the electron–phonon interaction (EPI) modifies the DOS of the superconductor, what results in small variations of the tunnel current that can be clearly seen in the $\mathrm{d}^2I/\mathrm{d}V^2\,(V)$ curves of junctions of superconductors with a strong EPI (Fig. 2.6). As an example for Pb, Fig. 2.6 reveals distinct features in the second derivative reflecting the EPI in the metal. However, unlike in PCS, there is not a direct proportionality between $\mathrm{d}^2I/\mathrm{d}V^2\,(V)$ and the EPI function α^2F. The EPI function can be reconstructed from the experimental data by a self-consistent iteration procedure based on the solution of the Eliashberg equation for the gap function [McMillan and Rowell (1965)].

Tunnel contacts between normal metals can also be used for spectroscopy. Electrons with a well-defined excess energy up to eV may interact with the quasiparticle excitations in the electrodes and/or in the barrier as well as with the molecules absorbed at the interface of the tunnel junction [see Jaklevic and Lambe (1966)]. The energy resolved detection of these excitations, mostly vibrational modes, is known as inelastic electron tunneling spectroscopy. Therefore, the vibrational frequencies of molecules or barrier oxides, the energies of plasmons or spin waves, the Kondo scattering, and a variety of other excitations have been measured by tunneling spectroscopy [Wolf (1985)]. Here we should emphasize that tunneling spectroscopy of the inelastic scattering of electrons leads to an opposite sign of the correction to the current compared with PCS. The tunneling current correction increases the net current because inelastic scattering opens additional channels for the

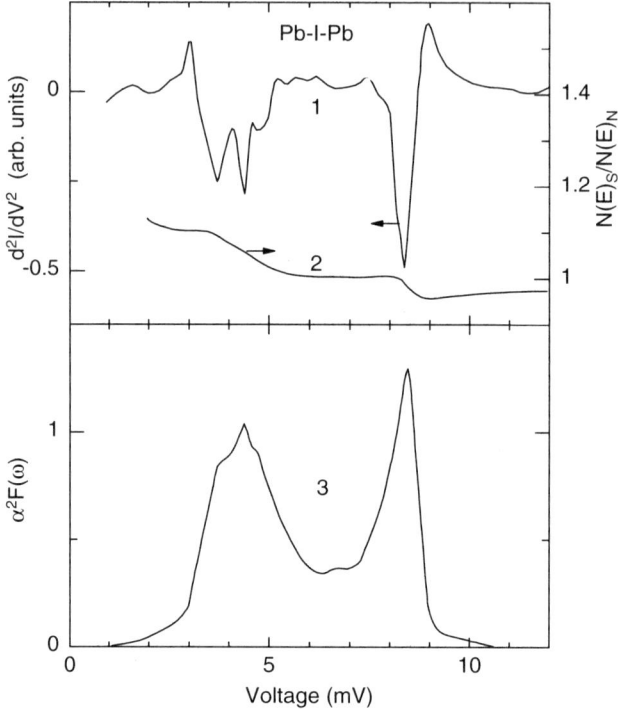

Fig. 2.6. Normalized second $\mathrm{d}^2 I/\mathrm{d}V^2$ (V) derivative (curve 1) of Pb-I-Pb tunnel contact (I stands for insulator) with Pb in the superconducting state at 0.8 K. Curve 2 is the experimentally obtained normalized density of states in the superconducting state. Curve 3 represents the EPI function derived from the above experimental data [compare with point contact spectrum presented in Fig. 2.3 (bottom curve)]. Data taken from McMillan and Rowell (1965).

electron tunneling across the insulating barrier [see Fig. 2.5(a)]. The tunneling current depends strongly on the DOS. As we will show further, DOS effects can be neglected in a first approximation for the transport in ballistic point contacts.

Finally, we present an example of a tunnel junction known as the Schottky barrier [Fig. 2.5(b)]. This barrier originates in the contacts between a metal and a semiconductor if the difference between workfunction of metal and semiconductor is positive $\Phi_m - \Phi_s = U > 0$. One should keep in mind that this kind of barrier caused by a charge redistribution at the interface could also play a role in the investigation of metallic contacts with conductors of a low (at least on the surface) carrier concentration.

References

Artemenko S. N., Volkov A. F. and Zaitsev A. V. (1979) Solid State Communs. **30** 771.

Barone A. and Paterno G. (1982) *Physics and Applications of the Josephson Effect* Wiley-Interscience, New York.

Blonder G. E. and Tinkham M. (1983) Phys. Rev. **27** 112.

Blonder G. E., Tinkham M. and Klapwijk T. M. (1982) Phys. Rev. **25** 4515.

Esaki L. (1958) Phys. Rev. **109** 603.

Giaever I. (1960) Phys. Rev. Lett. **5** 147, ibid. 464.

Grimvall G. (1981) *The Electron–phonon Interaction in Metals* North-Holland Publ. Co., Amst., N.-Y., Oxf.

Jaklevic R. C., and Lambe J. (1966) Phys. Rev. Lett. **17** 1139.

Kulik I. O. and Yanson I. K. (1970) *The Josephson Effect in Superconductive Tunneling Structures* (Nauka, Moskow) [English trans. by Israel Program for Scientific Translation Ltd; 1972, Keter Press, Jerusalem].

Likharev K. K. (1985) *Introduction to the Dymamics of Josephson Junctions* Moscow: Nauka (in Russian).

Maxwell J. C. (1904) *A Treatise of Electricity and Magnetism* Clarendon, Oxford.

McMillan W. L. and Rowell J. M. (1965) Phys. Rev. Lett. **14** 108.

Sharvin Yu. V. (1965) Sov. Phys. - JETP **21** 655.

Tsoi V. S. (1974) JETP Lett. **19** 70.

Wolf E. L. (1985) *Principles of Electron Tunneling Spectroscopy* Oxford University Press, Inc. New York.

Yanson I. K. (1974) Sov. Phys. - JETP **39** 506.

3 Fundamentals of PCS theory

In this chapter, the highlights of the canonical theory of the point-contact spectroscopy are presented. Some theoretical considerations will be introduced with more details by discussion of the experimental results in the other chapters as well.

3.1 Ballistic regime

The theory of PCS has been developed in the pioneering paper of Kulik et al. (1977). It was based on the solution of the Boltzmann equation for the point-contact geometry. The most used theoretical models represent the point contact as an orifice with a diameter d in the untransparent for the electrons partition between two metallic half-spaces (see Fig. 3.1(a) and Fig. 2.1). Often one exploits a one-dimensional approach, that is, the model of a long metallic channel with the length $L \gg d$ [Fig. 3.1(b)]. In some cases, contact with a shape of the three-dimensional rotating hyperboloid is used [Fig. 3.1(c)]. However, the shape of the contact does not influence critically on the physics of the processes in the constriction and does not modify quantitatively the final formula.

The electron distribution function for the ballistic point contact derived by Kulik et al. (1977) is shown in Fig. 3.2. If the voltage V is applied to the contact, the Fermi surface will split into two parts with a maximal energy difference defined by the bias energy eV between electrons arriving from the right and the left regions, correspondingly. It becomes important that the energy separation between any two states of electrons at the Fermi surface is equal to eV or zero. This is the circumstance that enables us to use the point contacts as a specific energy "probe" to determine the energies of other quasiparticles.

A current density j at the orifice with a normal parallel to the z-axis is given [see, e.g., Jansen et al. (1980)] by

$$j_z = 2e \sum_{\mathbf{k}} (\mathbf{v_k})_z f_{\mathbf{k}}(\epsilon), \tag{3.1}$$

with $f_{\mathbf{k}}(\epsilon)$ being the nonequilibrium Fermi–Dirac distribution function (Fig. 3.2) reduced in the contact center to $f_{z=\pm 0}(\epsilon) = \left(\exp(\frac{\epsilon - \epsilon_F \pm eV/2}{k_B T}) + 1 \right)^{-1}$ and $\mathbf{v_k}$

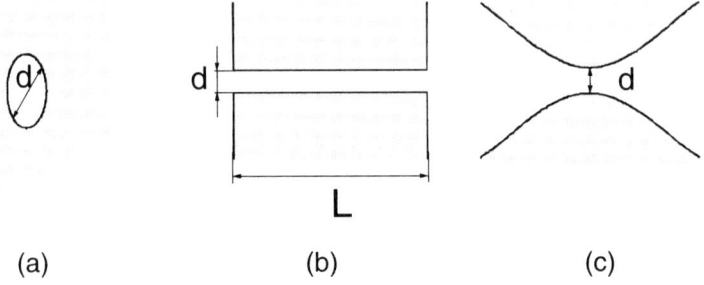

Fig. 3.1. Models of the point contacts: (a) circular orifice with a diameter d in opaque screen, (b) long channel with a length $L \gg d$, and (c) single-cavity three-dimensional hyperboloid of revolution.

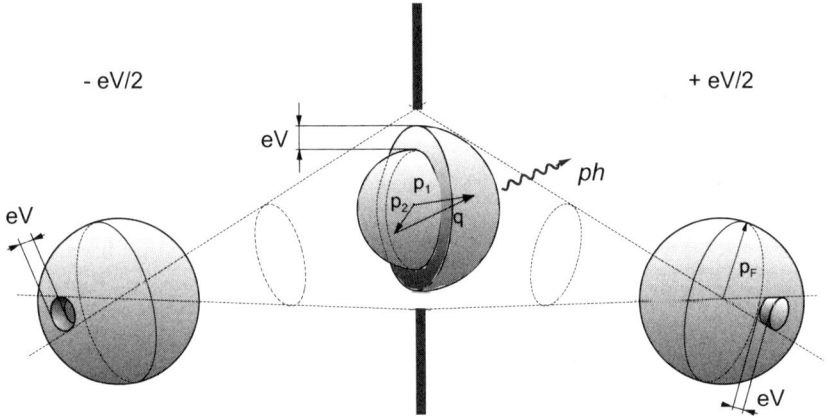

Fig. 3.2. The zeroth-order distribution function of electrons for a voltage biased by V ballistic point contact in its center and on the periphery (on the left and on the right side of the contact) at $T = 0$. Here, $\mathbf{p_1}$ is the initial momentum of electron, $\mathbf{p_2}$ is the final momentum of electron, $\mathbf{q}$ is the momentum of excited phonon (ph), and $\mathbf{p_F}$ is the Fermi momentum.

is the Fermi velocity. The summation over a $\mathbf{k}$-space in previous formula can be replaced by integration over the energy ϵ and over the surface of constant energy, which leads to

$$j_z = e \int_{\epsilon_F - eV/2}^{\epsilon_F + eV/2} d\epsilon \int \frac{d\Omega}{4\pi} f(\epsilon)\, v_z(\epsilon)\, N(\epsilon), \qquad (3.2)$$

where $d\Omega$ is the solid angle differential. Taking into account that electron velocity $v(\epsilon)$ is inversely proportional to the density of electronic states $N(\epsilon)$, then we do not expect any influence of the energy-dependent DOS on the net current and Ohm's law is valid for the point contact. The resistance of contact (the Sharvin resistance) as we already calculated for this case [see Section 2.1, Eq. (2.1)] can be written as:

$$R_{\mathrm{Sh}} = \frac{16\rho l}{3\pi d^2} = \frac{16\, R_{\mathrm{q}}}{(k_{\mathrm{F}} d)^2},\tag{3.3}$$

where $\rho l = p_{\mathrm{F}}/ne^2$, p_{F} is the Fermi momentum, n is the density of electrons, e is the electron charge, $R_{\mathrm{q}} = G_0^{-1} = h/2e^2 \simeq 12.9\,\mathrm{k\Omega}$, and G_0 is the conductance quantum. From (3.3), the contact diameter d can be determined in the ballistic regime by measuring only the normal-state resistance at zero bias R_0. In the case of copper, $d \approx 30/\sqrt{R_0\,[\Omega]}\,\mathrm{nm}$, what can be used for the estimations of the d value for most other simple metals and compounds.

By considering the energy-dependent scattering processes in the point contact, the absolute energy values of the initial and the final states are important. That means the scattering into the high DOS region will be more preferable and vise versa. This can lead, in the case of the strongly energy-dependent DOS, to the energy-dependent point-contact resistance so that at some circumstances, dV/dI can mimic the DOS image (see Fig. 14.7 and corresponding text for details). This means that PCS of the electronic spectrum of the metals might be realized in some specific cases.

3.2 Backflow current

Electrons passing through the constriction gain an excess energy eV and can scatter with creating of quasiparticle excitations. The electron–phonon interaction (EPI) in the ballistic point contact result in backflow scattering so that some of electrons are reflected by creating of phonons altering the net current. This leads to a nonlinear $I - V$ characteristic, whose second

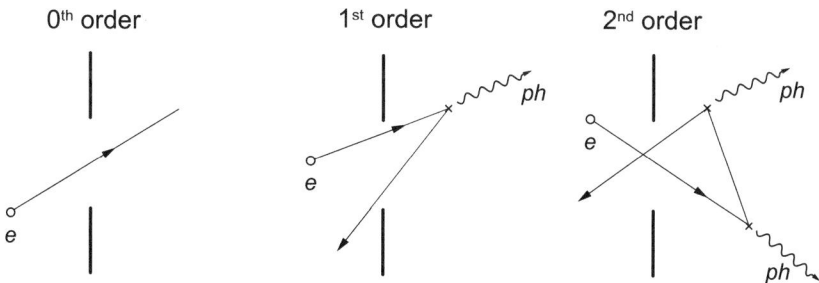

Fig. 3.3. Contributions to the current in a point contact. From the left to the right: zeroth-order Sharvin current, first-order single collision backflow current, and second-order double collision backflow current.

derivative[1] is expressed as [Kulik et al. (1977); Kulik (1992)]

$$\frac{1}{R_0}\frac{\mathrm{d}R(V,T)}{\mathrm{d}V} = \frac{8ed}{3\hbar v_F}\int_0^\infty \frac{\mathrm{d}\omega}{k_B T}g(\omega)\chi\left(\frac{\hbar\omega - eV}{k_B T}\right), \tag{3.4}$$

where $g(\omega)$ is the point-contact EPI function, and

$$\chi(x) = \frac{\mathrm{d}^2}{\mathrm{d}x^2}\left(\frac{x}{\exp x - 1}\right) \tag{3.5}$$

is the smearing or thermal resolution function. At finite temperature, $\chi(x)$ represents a bell-shaped peak with a width of $5.44\,k_B T$ at half-height [see Fig. 4.4(a)]. Thus, any features in $g(\omega)$ will be broadened by the temperature on this value, analogously as in the tunneling spectroscopy. At zero temperature, (3.4) reduces to

$$\frac{1}{R_0}\frac{\mathrm{d}R(V)}{\mathrm{d}V} = \frac{8ed}{3\hbar v_F}g(\omega)|_{\hbar\omega = eV}. \tag{3.6}$$

The point-contact EPI function $g(\omega)$ is given by

$$g(\omega) = \frac{\int \frac{\mathrm{d}^2\mathbf{p}}{(2\pi)^3 v_\mathbf{p}}\int \frac{\mathrm{d}^2\mathbf{p}'}{(2\pi)^3 v_{\mathbf{p}'}}W(\mathbf{p},\mathbf{p}')\,\delta(\omega - \omega_{\mathbf{p}-\mathbf{p}'})K(\mathbf{p},\mathbf{p}')}{\int \frac{\mathrm{d}^2\mathbf{p}}{(2\pi)^3 v_\mathbf{p}}}, \tag{3.7}$$

with the electron–phonon matrix element

$$W(\mathbf{p},\mathbf{p}') = \frac{2\pi}{\hbar}\left|(\mathbf{p}-\mathbf{p}')\mathbf{e}_{\mathbf{p}-\mathbf{p}'}\left(\frac{\hbar}{2NM\omega_{\mathbf{p}-\mathbf{p}'}}\right)^{1/2}V(\mathbf{p},\mathbf{p}')\right|^2.$$

Here $V(\mathbf{p},\mathbf{p}')$ is the electron–ion pseudopotential, $\mathbf{e}_{\mathbf{p}-\mathbf{p}'}$ is the polarization vector of phonons, M is the ion mass, and $K(\mathbf{p},\mathbf{p}')$ is the geometrical form-factor taking into account the kinematic restriction of the electron scattering processes in the point contact. For a circular orifice [see, e. g., Fig. 3.1(a)]

$$K(\mathbf{p},\mathbf{p}') = \frac{4\,|v_z v_z'|\,\Theta(-v_z v_z')}{|v_z \mathbf{v}_\perp' - v_z'\mathbf{v}_\perp|}, \quad \mathbf{v} = \partial\epsilon_\mathbf{p}/\partial\mathbf{p}, \tag{3.8}$$

---

[1] In an experimental situation, usually the $\frac{\mathrm{d}^2 V}{\mathrm{d}I^2}$ derivative is measured, whereas theoretical calculations are made mainly for $\frac{\mathrm{d}^2 I}{\mathrm{d}V^2}$ or $\frac{1}{R_d}\frac{\mathrm{d}R_d}{\mathrm{d}V}$ with $R_d = \mathrm{d}V/\mathrm{d}I$ being differential resistance. The relation for these derivatives is the following: $\frac{\mathrm{d}^2 V}{\mathrm{d}I^2} = R_d\frac{\mathrm{d}R_d}{\mathrm{d}V} = -R_d^3\frac{\mathrm{d}^2 I}{\mathrm{d}V^2}$. The theory usually deals with small nonlinearities where relative change of the differential resistance is small ($\Delta R_d/R_d \ll 1$ or $R_d \simeq \mathrm{const}$), and all of these derivatives have the same behavior. Nevertheless, one should keep in mind the possible difference between mentioned derivatives, especially in the case of large nonlinearity of a $I-V$ curve.

with $\Theta(x)$ being the Heaviside step function. K-factor depends on the scattering angle θ between the initial $\mathbf{p}$ and the final $\mathbf{p}'$ momenta of electrons and after averaging on the spherical Fermi surface reduces to [van Gelder (1978)]:

$$K(\theta) = 1/2(1 - \theta/\tan\theta). \tag{3.9}$$

It goes to infinity at $\theta = \pi$ (Fig. 3.4); however, this singularity is integrable and underlines the backscattering processes in PCS.

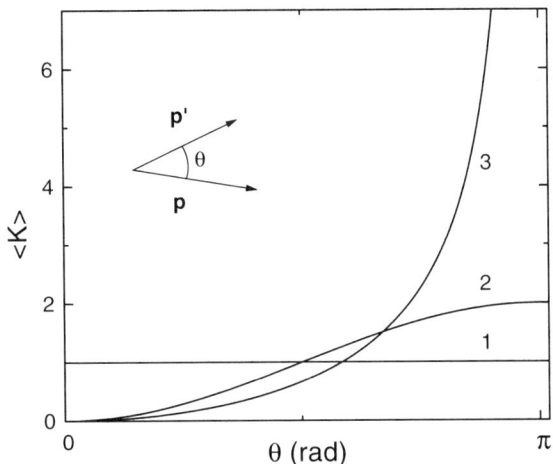

Fig. 3.4. Dependence of the averaged form-factor $\langle K \rangle$ on the scattering angle θ for the different EPI functions: thermodynamic (1), transport (2), and point contact (3).

Point-contact spectrum (usually $\mathrm{d}^2V/\mathrm{d}I^2(V)$ derivative is measured), as is seen from (3.6), is proportional to the EPI spectral function $g(eV)$, which is often introduced as $g(eV) = \alpha_{PC}^2 F(eV)$, with α_{PC}^2 being the averaged EPI matrix element with kinematic restrictions imposed by the contact geometry and $F(\omega)$ is the phonon density of states. The magnitude of nonlinearity in $I - V$ curve is estimated as $R_0^{-1}\,\mathrm{d}R(V)/\mathrm{d}V \sim d/l_{\mathrm{in}}$.

In this section, formulas, as we mentioned, are noncritical with respect to the constriction geometry. For example, as shown by MacDonald and Leavens (1984), representation of the orifice like a square changes averaged geometrical K-factor only about 1% compared with the value for a circle. If orifice has a rectangular shape with the length-to-width ratio w, then at $w = 2$, K-factor decreases about 5%, whereas underestimation by a factor 2 is only possible for a nonrealistic situation with $w \simeq 50$. The geometrical shape factor $C(w)$ for a rectangular orifice is shown in Fig. 3.5.

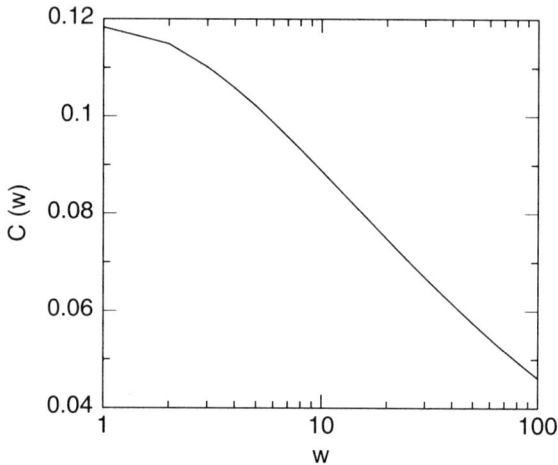

Fig. 3.5. Behavior of the geometrical factor $C(w)$ for a rectangular-shaped orifice with the length-to-width ratio w and constant surface. Data taken from McDonald and Leavens (1985).

The point-contact EPI function $g(eV)$ is related to the Eliashberg thermodynamical EPI function $g_{\mathrm{th}}(\omega)$ given by the same equation as (3.7) with $K = 1$. The Eliashberg EPI function is used in canonical theory of superconductivity, and it enters in so-called Eliashberg gap equations. The Eliashberg EPI function determines the EPI constant λ:

$$\lambda = 2 \int_0^\infty \frac{g(\omega)}{\omega} d\omega, \tag{3.10}$$

which allows us to estimate the superconducting transition temperature using the well-known McMillan (1968) equation

$$T_{\mathrm{c}} \simeq \frac{\hbar\omega_{\mathrm{D}}}{1.45 k_{\mathrm{B}}} \exp\left(-\frac{1.04(1 + \lambda)}{\lambda - \mu^*(1 + 0.62\lambda)}\right), \tag{3.11}$$

where μ^* is the constant of repulsive Coulomb interaction with the typical value of about 0.1 and variation between 0 and 0.25.

There is also a known transport EPI function $g_{\mathrm{tr}}(\omega)$ with the form-factor $K = (1 - \cos\theta)$ entered in the temperature-dependent resistivity [Grimvall (1981)]

$$\rho = \frac{4\pi m_{\mathrm{b}}}{ne^2} \int_0^{\omega_{\max}} \frac{\beta\omega g_{\mathrm{tr}}(\omega)}{(\exp^{\beta\omega} - 1)(1 - \exp^{-\beta\omega})} d\omega, \tag{3.12}$$

Table 3.1. Behavior of the form-factor K and low-energy part for point-contact $g_{PC}(\omega)$, transport $g_{tr}(\omega)$, and Eliashberg (thermodynamic) $g_{th}(\omega)$ EPI functions for a spherical Fermi surface depending on the prevalence of normal (N) and umklapp (U) processes [Jansen et al. (1980)].

	$g_{th}(\omega)$	$g_{tr}(\omega)$	$g_{PC}(\omega)$
form-factor K	1	1-$\cos\theta$	$\frac{1}{2}(1 - \frac{\theta}{\tan\theta})$
N - processes	ω^2	ω^4	ω^4
U - processes	ω^3	ω^3	$\omega^{5/2}$

where m_b is the electron band mass, and $\hbar\omega$ is the phonon energy, $\beta = 1/k_B T$.

Table 3.1 shows the behavior of the above-mentioned EPI functions at low energy. The contribution of the large angle scattering increases for the transport and especially for the point-contact EPI function as seen from Fig. 3.4.

The electron distribution function in the point contact differs radically from the related one in the case if current flows through a bulk sample (Fig. 3.6). In the latter case, it is impossible to impart an excess energy to the electrons that is comparable with the Debye energy. It turns out that the current density should exceed $10^9 \text{A}/\text{cm}^2$ so that the electron energy acquired in the electric field could reach the characteristic phonon energies. Such current densities are unattainable in the bulk metal owing to the Joule heating and melting. It is not the case with the point contacts. The contact area remains cold as a result of a large mean free path of electrons compared with the contact size: The electrons transfer their energy into the surrounding medium. There is another important difference between the distributions in Fig. 3.6(a) and (c). In the latter case, the distribution is smeared with a width of the order of the excess energy even in spite of the fact that there is no heating. This smearing is caused by the statistical nature of the electron mean free path.

3.3 Background

The point-contact EPI function $g(\omega)$ defined by (3.7) vanishes above the maximum phonon energy $\hbar\omega_{max}$ close to the Debye energy $k_B T_D$ because of the lack of phonons with the larger energy. Therefore, according to (3.6), the point-contact spectrum should go to zero at $eV \geq \hbar\omega_{max}$. In fact,

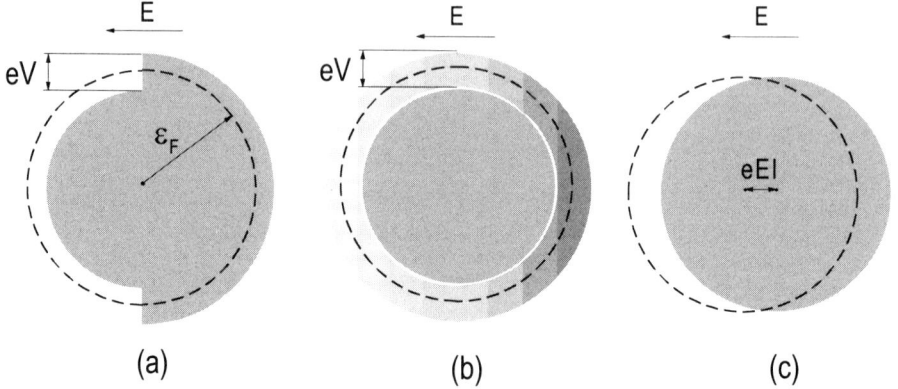

Fig. 3.6. (a) Electron distribution function (dark area) for a voltage-biased contact in the ballistic regime, (b) the same function for the diffusive, and (c) for the thermal regime or in the bulk material at an electric field E and a mean free path l. The equilibrium electron distribution function is shown by dashed circles in each case. All for $T = 0$.

point-contact spectra measured always have a nonzero almost constant value above the Debye energy, so-called background. A first explanation for the background concerned a successive generation of phonons by energized electrons was given by van Gelder (1978). Accounting double-collision processes, showed in Fig. 3.3, van Gelder found that they produce a nonzero signal at twice the energy of the Debye one.

$$\frac{\mathrm{d}^2 I}{\mathrm{d}V^2}(eV) = \mathrm{const}\frac{\int_0^{eV} g(\omega)g(eV - \omega)\,\mathrm{d}\omega}{\int_0^{eV} g(\omega)\,\mathrm{d}\omega}. \tag{3.13}$$

In the ballistic regime, this results in correction of the order of d/l_{in} to (3.6). The higher order n-collision processes extend a contribution to energy scale up to neV, but their relative amplitude behaves as $(d/l_{\mathrm{in}})^n$ and they are too small to explain a background signal that often is of the same order as $g(\omega)$.

The general nature of the background was understood by taking into account nonequilibrium phonons in the point-contact region created by energized electrons [van Gelder (1980)]. These phonons can be accumulated in the constriction by low group velocity and scattering from imperfection and grain boundaries. The phonon distribution function N_ω was calculated by Kulik (1985):

$$N_\omega \simeq \frac{eV - \omega}{4(\omega + \omega_0)}\theta(eV - \omega), \tag{3.14}$$

where ω_0 is the phonon escape frequency

$$\omega_0 \simeq \omega_{\mathrm{max}}\frac{l_{\mathrm{ph}}l_{\mathrm{r}}}{d^2}, \tag{3.15}$$

with l_{ph} and l_r being the inelastic phonon–electron and elastic phonon–defect scattering lengths. In the case of a strong phonon reabsorption, ω_0 is of the order or less than the maximum phonon frequency: $\omega_{\mathrm{D}} \sim \omega_{\mathrm{max}}$. Considering reabsorption of phonons, the point-contact spectrum is given by

$$\frac{1}{R_0}\frac{\mathrm{d}R(V)}{\mathrm{d}V} = \frac{8ed}{3\hbar v_{\mathrm{F}}}\left(g(eV) + \gamma \int_0^\infty \frac{g(\omega)\mathrm{d}\omega}{\omega + \omega_0} + \frac{\gamma}{2}\frac{eV}{eV + \omega_0}g(eV)\right), \quad (3.16)$$

where $\gamma = 0.58$ for the orifice [Kulik (1985, 1992)]. The last two terms in (3.16) correspond to the background signal. In the case of the weak reabsorption ($\omega_0 \gg \omega_{\mathrm{max}}$), (3.16) coincides with an empirically proposed by Yanson (1983) behavior of the background:

$$B(eV) = \mathrm{const}\int_0^{eV}\frac{g(\omega)}{\omega}\mathrm{d}\omega. \quad (3.17)$$

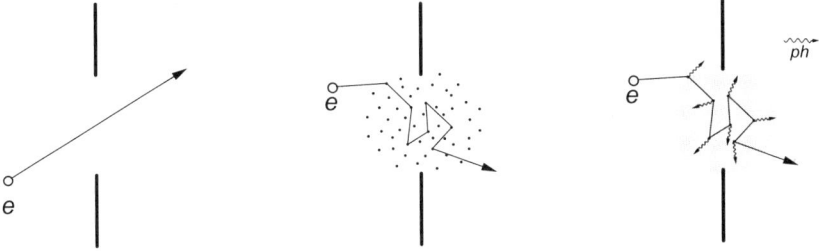

Fig. 3.7. Electronic trajectories in a point contact in the ballistic, diffusive, and thermal regime (from the left to the right).

3.4 Diffusive regime

Depending on the relationship between the elastic l_{el}, the inelastic l_{in} mean free path of electrons and the point-contact dimension d various regimes have been established. They are ballistic ($l_{\mathrm{el}}, l_{\mathrm{in}} \gg d$), diffusive ($l_{\mathrm{el}} \ll d \ll \sqrt{l_{\mathrm{in}}l_{\mathrm{el}}}$), and thermal ($l_{\mathrm{el}}, l_{\mathrm{in}} \ll d$) (Fig. 3.7).

It has been previously noted that the spectral regime realizes only in the very small shorts, such that the electron mean free path l_{el} and l_{in} should exceed the contact size. A question develops of whether PCS is possible if the elastic mean free path becomes less than the short size. This question is not idle because many alloys and compounds have a short l_{el}. The positive answer permits us to apply the PCS technique not only to the pure metals whose number is, although large, but limited, but also to practically unlimited collection of more or less complex alloys or compounds of different metals. It

Table 3.2. Basic point-contact relations for different geometries and regimes. Here, $\mathbf{n} = \mathbf{v}/v_F$, KOS = Kulik, Omelyanchouk, and Shekhter (1977) and KSS = Kulik, Shekhter, and Shkorbatov (1981). Note: The EPI function $g(\omega)$ determined by Kulik (1992) has $\langle K\rangle$=1 for clean orifice, and KOS, KSS, and other theoretical papers of Kulik et al. before 1992 used $G(\omega)$ with $\langle K\rangle$=1/4. Correspondingly, equations like (3.4), (3.6), (3.16), and so on have coefficient 8/3 for $g(\omega)$ and 32/3 for $G(\omega)$. In this table, K-factor and $\langle K\rangle$ are presented for $G(\omega)$. The last column shows expression for a second derivative for different geometries and regimes [compare with (3.6)].

	R_0	$K(\mathbf{n}, \mathbf{n}')/\langle K\rangle$	$\langle K\rangle$	Reference	$R^{-1}dR/dV$
clean orifice	$\frac{16\rho l}{3\pi d^2}$	$\frac{4\left|n_z n'_z\right|\Theta(-n_z n'_z)}{\left|n_z \mathbf{n}'_\perp - n'_z \mathbf{n}_\perp\right|}$	$\frac{1}{4}$	KOS	$\frac{8ed}{3\hbar v_F}g(\omega)$
dirty orifice	$\frac{\rho}{d}$	$\frac{3}{8}((n_z - n'_z)^2 + (\mathbf{n} - \mathbf{n}')^2)$	$\frac{3\pi}{16}\frac{l_{el}}{d}$	KSS	$\frac{2\pi e l_{el}}{\hbar v_F}g(\omega)$
clean channel	$\frac{16\rho l}{3\pi d^2}$	$2\Theta(-n_z n'_z)$	$\frac{3\pi}{16}\frac{L}{d}$	KSS	$\frac{2\pi eL}{3\hbar v_F}g(\omega)$
dirty channel	$\frac{4L\rho}{\pi d^2}$	$\frac{3}{2}(n_z - n'_z)^2$	$\frac{3\pi}{16}\frac{l_{el}}{d}$	KSS	$\frac{2\pi e l_{el}}{\hbar v_F}g(\omega)$

turned out that the technique works under less strict conditions, namely, if the inelastic relaxation length during the diffusion motion of electrons in the contact $\Lambda = \sqrt{l_{el}l_{in}/3}$ is greater than the size of the contact. The condition $d \ll \Lambda$ is possible to satisfy for point contacts made of many complex substances, and the spectral information about electron scattering mechanisms can be obtained. In the diffusion regime, as was shown by Kulik and Yanson (1978), Kulik et al. (1985), the proportionality between d^2I/dV^2 and the spectral function $g(eV)$ is preserved. In this case, K-factor in the right-hand part of (3.7) starts to depend on l_{el} (see Table 3.2) so that K is proportional to the l_{el}/d ratio at $l_{el} \ll d$. This leads to a decrease of the point-contact spectrum intensity that is proportional to $\langle K\rangle$. The behavior of the average $\langle K\rangle$ versus l_{el}/d is shown in Fig. 3.8.

By arbitrary relation between the mean free path and the point-contact diameter, a simple interpolation formula was derived by Wexler (1966) for the contact resistance:

$$R_{PC}(T) = \frac{16\rho l}{3\pi d^2} + \beta\frac{\rho(T)}{d}, \qquad (3.18)$$

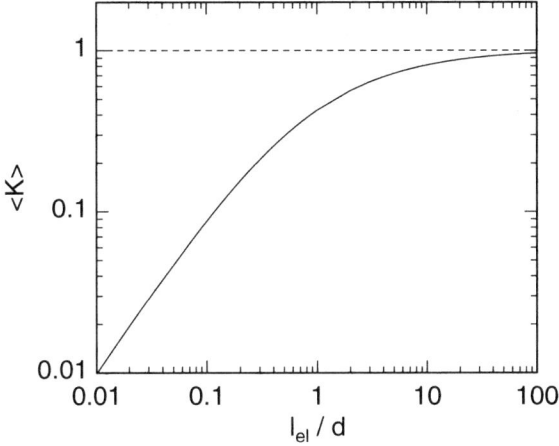

Fig. 3.8. Averaged K-factor versus the normalized to the contact diameter elastic mean free path calculated by expressions (8), (17), and (19) from Yanson and Shklyarevskii (1986).

with $\beta \simeq 1$ at $l_{\mathrm{el}} \ll d$. It turned out that the Wexler formula represents simply a sum of the ballistic Sharvin ($l \gg d$) and the diffusion Maxwell ($l \ll d$) resistances. The first term in (3.18) corresponds to the ballistic transport of electrons in the constriction and the resistance depends only on the size of the contact. The prevalence of the one or another term in (3.18) depends on the relation between d and electron mean free path. For high-resistivity materials and large contacts, the Maxwell contribution plays a predominant role and vice versa for clean metals and small constrictions the Sharvin resistance dominates (see Fig. 3.9).

Differentiation of (3.18) yields an alternative formula to determine the size (diameter) d of the contact:

$$d = \frac{\mathrm{d}\rho/\mathrm{d}T}{\mathrm{d}R_{\mathrm{PC}}/\mathrm{d}T} \,, \qquad (3.19)$$

experimentally verified for PCs with simple metals by Akimenko et al. (1982). It gives more reliable values for d compared with (3.18), whereas residual resistance in point contact can differ from the bulk ρ and decreased transmission of the interface adds some undetermined coefficient to the Sharvin resistance.

The second term in (3.18) meets diffusive electron flow inside the point contact, and its resistance is governed by ordinary electron transport as in the bulk. We can rewrite (3.18) as before [see (2.4)]. It is seen that all processes resulting in the energy-dependent scattering time $\tau(eV = \epsilon)$ lead to the energy-dependent point-contact resistance. Thus, the second derivative will have maxima at those energies at which the mean free path of electrons varies rapidly. In the case of an electron–phonon scattering, the electron mean

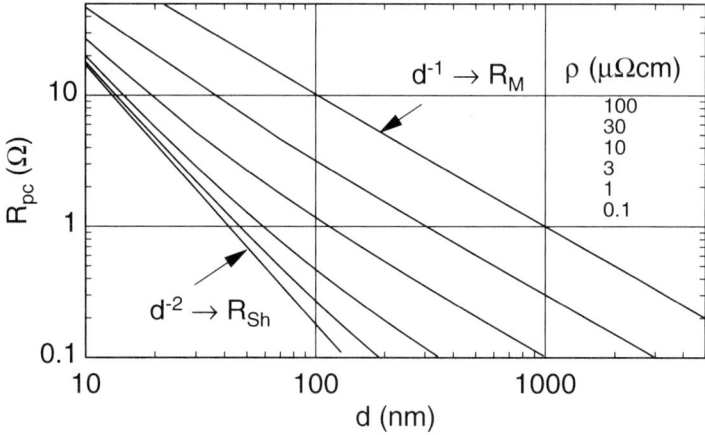

Fig. 3.9. Point-contact resistance versus diameter calculated according to (3.18) at different ρ values and a standard for metals $\rho l \simeq 10^{-11}\,\Omega\,\mathrm{cm}^2$. For typical resistances between 1 and 10 Ω, the point-contact diameter for a clean metal (ballistic regime) is below 40 nm, whereas that for the high-resistivity ($\rho \geq 10\,\mu\Omega\,\mathrm{cm}$) materials (diffusive regime) is between 100 and 1000 nm.

free path considerably decreases at those energies where a large number of phonons exists, i. e., at the phonon spectrum maxima. That is why a peak in $\mathrm{d}^2V/\mathrm{d}I^2$ can be expected at this energy. In principle, any energy-dependent mechanism of electron scattering in metals can be studied by means of point contacts. To put it another way, the processes influenced the inelastic mean free path of electrons in the lattice should be affected in the point-contact spectra (see, e. g., Chapter 6).

3.5 Thermal regime

The opposite case of the ballistic regime is the thermal regime (Fig. 3.7). It takes place if $l_{\mathrm{in}} \ll d$ or $\Lambda = \sqrt{l_{\mathrm{el}}l_{\mathrm{in}}/3} \ll d$. The resistance is given here by the Maxwell formula as in the diffusive regime [see (2.2)]. The heat removal from such large contacts with the small mean free path lengths is not so efficient as in the ballistic regime; therefore, whole energy dissipates in the constriction, what results in the Joule heating and increase of the temperature inside the contact with the voltage rise. As in the metals, the heat transfer from the hot region to the cold one is realized by the conducting electrons; then, assuming the fulfillment of the Wiedemann–Franz condition, the temperature of the point contact is determined solely by the applied voltage by virtue of the Kohlrausch (1900) equation:

$$V^2 = 8 \int_{T_{\text{bath}}}^{T_{\text{PC}}} \rho\lambda \, \mathrm{dT} \,. \tag{3.20}$$

Considering $\rho\lambda = LT$, where L is the Lorenz number, and λ is the thermal conductivity, it leads to the simple relation between the temperature in a center of the contact T_{PC} and an applied voltage V:

$$T_{\text{PC}}^2 = T_{\text{bath}}^2 + \frac{V^2}{4L} \,. \tag{3.21}$$

At $T_{\text{bath}} \to 0$ or $T_{\text{bath}} \ll T_{\text{PC}}$, this yields a linear connection between T and V: $T = V/2\sqrt{L}$, which is often represented as

$$eV = (2\pi/\sqrt{3}) \, k_{\text{B}} T_{\text{PC}} = 3.63 \, k_{\text{B}} T_{\text{PC}} \tag{3.22}$$

so that $T_{\text{PC}}[\text{K}] \simeq 3.2 \, [\text{K/mV}] \, V[\text{mV}]$ using the standard Lorenz number $L = L_0 = 2.45 \times 10^{-8} \, \text{V}^2/\text{K}^2$, i.e., a bias voltage of $1 \, \text{mV}$ raises a temperature at the contact center by $3.2 \, \text{K}$. The situation seems to be paradoxical: A sample is at liquid helium temperature, but if we apply to the contact a voltage of only $100 \, \text{mV}$, the temperature of the constriction becomes higher than the room temperature. The experimental verification of these conclusions will be given in Chapter 7. This leads to establishing of the so-called modulating temperature spectroscopy [Verkin et al. (1980)] because by measuring of derivatives of the current-voltage characteristics, the contact is supplied with a small alternating voltage and the contact temperature in the thermal limit changes (modulates) in step with the ac voltage. Note that in this case, the contact size has no upper bound and the result does not depend on the constriction geometry. There is only one necessary condition that the heat should be removed from the contact mainly by electrons.

In the thermal regime nonlinearity of the $I - V$ characteristic of a contact is caused by the temperature-dependent resistivity $\rho(T)$ [Verkin et al. (1979), Kulik (1992)]:

$$I(V) = Vd \int_0^1 \frac{\mathrm{d}x}{\rho(T\sqrt{1-x^2})|_{T=eV/3.63k_{\text{B}}}}. \tag{3.23}$$

This allows both calculation of the $I - V$ characteristic using $\rho(T)$ and vise versa reconstruction of the $\rho(T)$ dependence in the constriction from the measured $I - V$ curve.

The point-contact spectrum in the thermal regime was calculated by Kulik (1984) supposing that phonon part of the resistivity is small compared with the residual resistivity. As a result,

$$\frac{1}{R_0}\frac{\mathrm{d}R}{\mathrm{d}V}(V) = C \int_0^\infty \frac{\mathrm{d}\omega}{\omega} g_{\text{tr}}(\omega) \, S(eV/\omega) \,, \tag{3.24}$$

where $C = \pi\sqrt{3}m/ne\hbar\rho$ and $S(x)$ is given by

$$S(x) = \frac{2\pi}{3} \frac{d^2}{dx^2} \int_0^{\pi/2} \frac{dy}{\sinh^2(\pi/\sqrt{3}\, x \sin y)} \tag{3.25}$$

and shown in Fig. 3.10. Actually, in the thermal regime, the point-contact

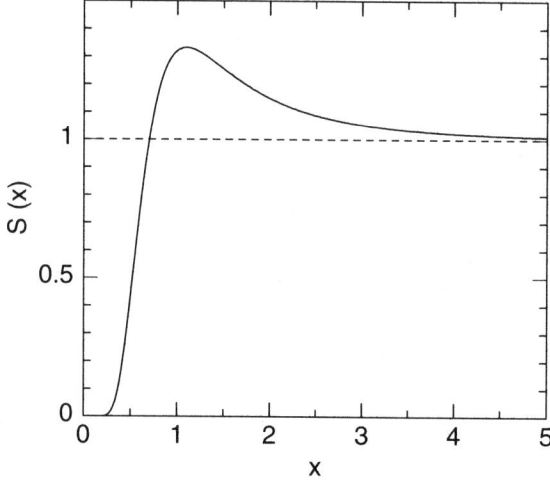

Fig. 3.10. Calculated according to (3.25) $S(x)$ function, which represents the shape of one-mode (Einstein) phonon spectrum in the thermal regime.

spectrum of an Einstein oscillator with the energy $\hbar\omega_0$ looks like $S(x)$ function exhibiting a smeared step with a gentle maximum at $eV = 1.09\,\hbar\omega_0$ (Fig. 3.10). Hence, in the thermal limit, it is still possible to resolve degraded phonon features. Moreover, as shown by Kulik (1984), there is a possibility to determine $g_{tr}(\omega)$ from (3.24) by measuring d^2I/dV^2 of the contact in the thermal limit followed by inverting of the linear equation relating these quantities. By 0.1% accuracy in d^2I/dV^2, it gives about 5% accuracy in the reconstructed $g_{tr}(\omega)$. This is important for metals and alloys with a short inelastic mean free path, where it is not possible to realize ballistic or diffusive regime. Figure 3.11 shows an example of the model spectra reconstruction from the second or third derivative.

3.6 Heterocontacts

In the case of a heterocontact between two different metals (1 and 2), its resistance is naturally defined by a sum of the contributions from both electrodes. Supposing a geometrically symmetric heterocontact and almost equal Fermi parameters of the metals, one can write

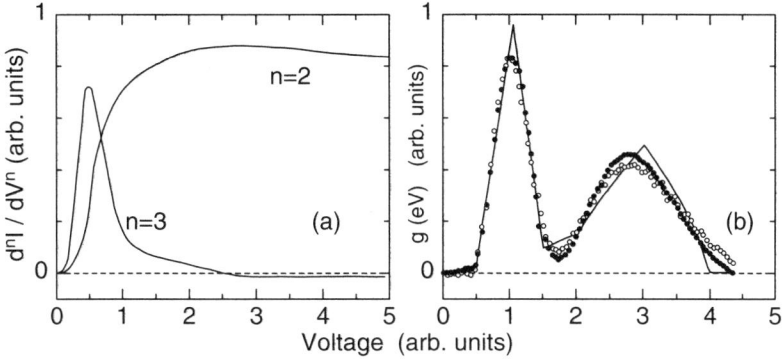

Fig. 3.11. (a) Second (n = 2) and third (n = 3) derivatives calculated from (3.24), using $g(eV)$ as shown by solid line in (b). (b) Reconstructed $g(eV)$ spectrum from the second (open circles) and third (closed circles) derivatives from (a). Data taken from Kulik (1984).

$$R_{PC}(T) \simeq \frac{16\rho l}{3\pi d^2} + \frac{\rho_1(T) + \rho_2(T)}{2d}. \qquad (3.26)$$

It turns out [see Shekhter and Kulik (1983)] that point-contact spectrum of the heterocontact also represents a sum of the contributions from both metals 1 and 2:

$$\frac{1}{R_0}\frac{dR}{dV} \propto v_1 \frac{g_1(\omega)}{v_{F1}} + v_2 \frac{g_2(\omega)}{v_{F2}}. \qquad (3.27)$$

The partial contribution of each metal in the spectrum is proportional to the "strength" of EPI and to the relative volume $v_{1,2}$ ($v_1 + v_2 = 1$) occupied by this metal. Thus, it is possible to carry out an estimation of the quantity of EPI, for example, in the unknown metals as compared with some standard one. It is seen from (3.27) that for different Fermi velocities, the spectrum of the material with the smaller Fermi velocity v_F is more intensive in the ballistic contact. In other words, electrons moving slower through the contact have a higher probability to scatter by creating phonons. Moreover, study of heterocontacts gives some additional information regarding more delicate features of the transport of nonequilibrium electrons and phonons in the constriction, which connected with the presence of the boundary between metals (see below and Section 5.5).

The reflection and refraction of the electron trajectories at the boundary in the heterocontact become important because of the difference in the Fermi momenta or the width of the conduction bands. This leads both to the suppression of the spectrum for the metal with larger v_F and to the modification of its shape [Shekhter and Kulik (1983), Baranger et al. (1985)]. The presence of the potential barrier caused by difference between the bandwidths restricts the minimal momentum projection on the contact axis for electrons passing through the boundary (Fig. 3.12). Consequently, only the

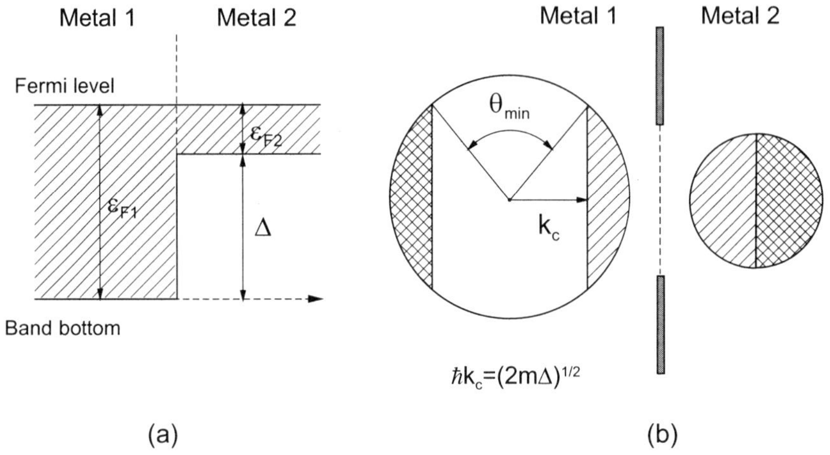

Fig. 3.12. (a) Conduction bands of metals in heterocontact in equilibrium with the intrinsic barrier Δ for electrons coming from the left. (b) Equilibrium distribution of electrons in the contact center for both metals with a critical wave vector $\mathbf{k}_c$. Electrons in the shaded regions can cross the interface, those in the crosshatched regions have crossed the interface, those in the blank region stay in metal 1. Adapted from Baranger et al. (1985).

electron–phonon scattering processes with the scattering angle more than $\Theta_{\min} = 2\arcsin(k_c/k_F) = 2\arcsin((v_{F1}^2 - v_{F2}^2)^{1/2}/v_{F1})$ are possible in the constriction. Thus, some of the phonons will be cut off from EPI. Therefore, it lets extraction from the point-contact spectra an information about both correlation between the Fermi parameters of metals and electron passing conditions through the boundary. In the event of a large difference between the Fermi momenta of electrons, focusing of electron flux occurs (Fig. 3.13). It results in transformation of the point-contact spectrum of the metal with the larger $\mathbf{p}$ into narrow peaks [Shekhter and Kulik (1983)]. It is pertinent to note that strong elastic scattering hinders such a transformation; however, it does not affect the partial contribution of the metals in the spectrum.

Phonons generated in the contact by energized electrons move into the bulk metal realizing efficient heat transfer from the "hot" contact center to the "cold" bath so long as $d \ll l_{\text{ph}-\text{el}}$ ($l_{\text{ph}-\text{el}}$ is the phonon inelastic relaxation length). But in the case of a small elastic phonon mean free path l_r ($l_r \ll d$), nonequilibrium phonons are accumulated in the contact. This leads to a reabsorption of electrons by nonequilibrium phonons and results in so-called background in the point-contact spectra (Section 3.3). The reabsorption is more effective at the small phonon elastic mean free path l_r, so that, at $l_r < d$, the background component becomes comparable with the spectral part of dV^2/dI^2 [Kulik (1985)]. The next order effect is that phonons accumulated in the orifice drift from the point-contact center isotropically into the bulk metal and drag electrons. In a homocontact, this effect is compen-

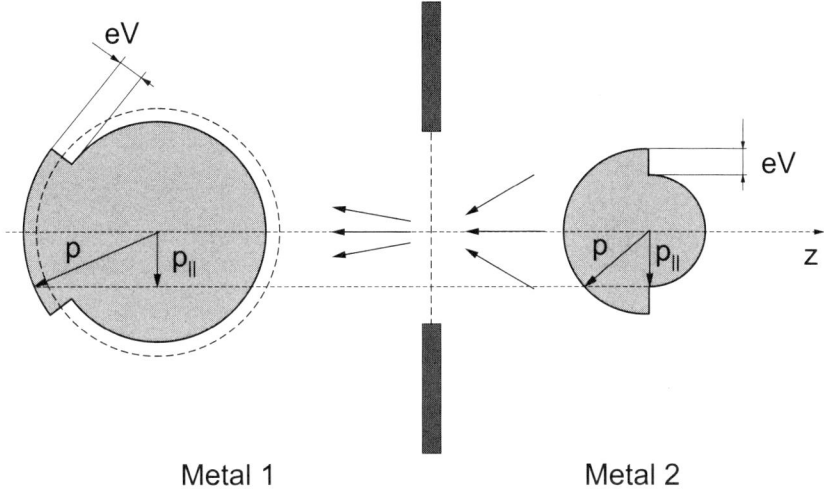

Fig. 3.13. Focusing of the electron trajectories in heterocontact for metals with sufficiently different Fermi momenta **p**. Note: The electron distribution functions are shown for both metals in the center of the contact.

sated by the identical phonon characteristics of the metals, but in the case of heterocontacts, it results in proportional l_r asymmetry of the point-contact spectra with respect to the polarity of the applied voltage (see Itskovich and Shekhter (1985) and Fig. 5.29).

For heterocontact in the thermal regime, while temperature of the contact increases compared with the bath one, the thermoelectric voltage caused by the difference of the Seebeck coefficients of the contacted metals results in an asymmetry of the $I - V$ or $dV/dI(V)$ curves versus a bias voltage V. As was shown by Itskovich et al. (1985, 1987),

$$\frac{1}{R_0}\left(\frac{dV}{dI}(V)\right)^{as} \equiv \frac{1}{2R_0}\left(\frac{dV}{dI}(V > 0) - \frac{dV}{dI}(V < 0)\right) \simeq$$

$$\simeq \sqrt{\frac{\rho_1\rho_2}{(L_1\rho_2 + L_2\rho_1)(\rho_1 + \rho_2)}} \ (S_1(T) - S_2(T)), (3.28)$$

where $S_{1,2}$ are Seebeck coefficients and the temperature in the contact center is

$$T_{PC}^2 = T_{bath}^2 + V^2 \frac{\rho_1\rho_2}{(L_1\rho_2 + L_2\rho_1)(\rho_1 + \rho_2)}. \tag{3.29}$$

It is interesting to note that the maximal temperature in the contact is still determined by (3.21). However, it is reached not in the center of the constriction, but in the metal with larger resistivity. In this fashion, a voltage-dependent asymmetry of the differential resistance dV/dI is determined by the temperature dependence of the difference between thermopower of two metals. In the case of metal (1) with the high resistivity and thermopower

$(\rho_1(T) \gg \rho_2(T), S_1(T) \gg S_2(T))$, we have

$$dV/dI(V) \propto \rho_1(T) , \quad (dV/dI)^{as}(V) \propto S_1(T) .$$

3.7 Superconducting contacts

3.7.1 S-c-N contacts

In an S-c-N contact, a contribution from the Maxwell term in the superconducting (SC) electrode vanishes, because $\rho_S = 0$, whereas a ballistic resistance decreases up to two times as a result of an Andreev reflection [Andreev
(1964)] at the N-S boundary (see Fig. 3.14). The reason is that provided the
excess energy eV of an electron is smaller than the SC gap Δ, electron should
form a Cooper pair with another electron to enter in the superconductor. To
conserve momentum and charge, a hole with an opposite velocity appears in
the normal metal. This retroreflection doubles the net current in the contact
or, in other words, decreases two times its resistance. Blonder, Tinkham,

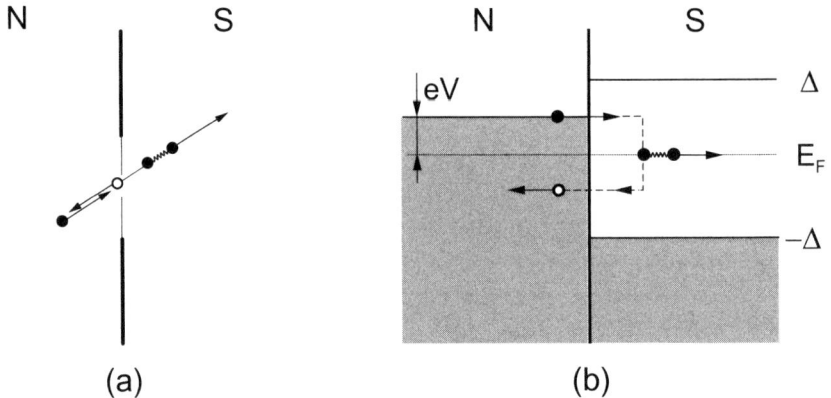

(a)   (b)

Fig. 3.14. (a) Andreev reflection of an electron (closed circle left) as a hole (open
circle) at the normal metal (N) – superconductor (S) interface by creating a Cooper
pair in the superconductor (connected closed circles right). (b) Energetic schema
represents Andreev reflection processes for semiconductor-like model.

and Klapwijk (BTK) (1982), exploiting the concept of Andreev reflection,
have calculated the reflection and transmission probabilities of the charged
particles passing through the N–S interface using the Bogoliubov–de Gennes
equations. Supposing a one-dimensional model for the S-c-N contact with a

dimension smaller than the SC coherence length ξ, they derived the $I - V$ characteristic[2] :

$$I(V) \sim \int_{-\infty}^{\infty} T(\epsilon)[f(\epsilon - eV) - f(\epsilon)]d\epsilon, \qquad (3.30)$$

$$T(\epsilon) = \frac{2\Delta^2}{\epsilon^2 + (\Delta^2 - \epsilon^2)(2Z^2 + 1)^2}, \quad |\epsilon| < \Delta,$$

$$T(\epsilon) = \frac{2|\epsilon|}{|\epsilon| + \sqrt{\epsilon^2 - \Delta^2}(2Z^2 + 1)}, \quad |\epsilon| > \Delta,$$

where $f(\epsilon)$ is the Fermi distribution function, Δ is the SC order parameter, and Z is the barrier strength at the N–S interface. Z can be related not only to some oxide barrier, but also it is a measure of elastic scattering in the constriction, whether it originates from lattice distortion or from impurities at the interface. Normal reflection of quasiparticles at the interface caused by mismatch of the Fermi velocities v_{F} yields $Z \neq 0$, namely,

$$Z = (1 - r)/2r^{1/2}, \qquad (3.31)$$

where $r = v_{\mathrm{F1}}/v_{\mathrm{F2}}$, even if there is no natural barrier [Blonder and Tinkham (1983)].

Equation (3.30) results in a flat zero-bias minimum in $dV/dI\,(V)$ with the width of 2Δ at $T \to 0$ and $Z = 0$ [Fig. 3.15(a)]. Additionally, a maximum at $V=0$ appears for $Z \neq 0$ so that double-minimum structure develops in $dV/dI\,(V)$. This feature is considered as the hallmark of Andreev–reflection processes at the N–S interface. The minima smear out with increasing of temperature and vanish above T_c [Fig. 3.15(b)].

The BTK theory was modified by Srikanth and Raychaudhuri (1992) and Plecenik et al. (1994) by including the broadening of the quasiparticle DOS $N(\epsilon)$ (see Fig. 3.16) in the superconductor. Following Dynes et al. (1978),

$$N(\epsilon, \Gamma) = \mathrm{Re}\left\{ \frac{\epsilon - i\Gamma}{\sqrt{(\epsilon - i\Gamma)^2 - \Delta^2}} \right\}. \qquad (3.32)$$

Here Γ is the broadening parameter,[3] which accounts for the finite quasiparticle lifetime τ ($\Gamma = \hbar/\tau$). $\Gamma \neq 0$ results in a broadening of the minimum in $dV/dI\,(V)$ and a decreasing of its depth [Fig. 3.16(c)].

The $I - V$ characteristic of the S-c-N contact at $eV \gg \Delta$ can be written as

$$I(V) = \frac{V}{R_{\mathrm{N}}} + I_{\mathrm{exc}}, \qquad (3.33)$$

---

[2] The microscopic justification of the BTK theory was provided by Zaitsev (1984).

[3] The same expression as (3.32) appears if one assumes an imaginary part in the energy-gap parameter Δ. Therefore, any depairing interaction in a superconductor may be formally accounted by including Γ.

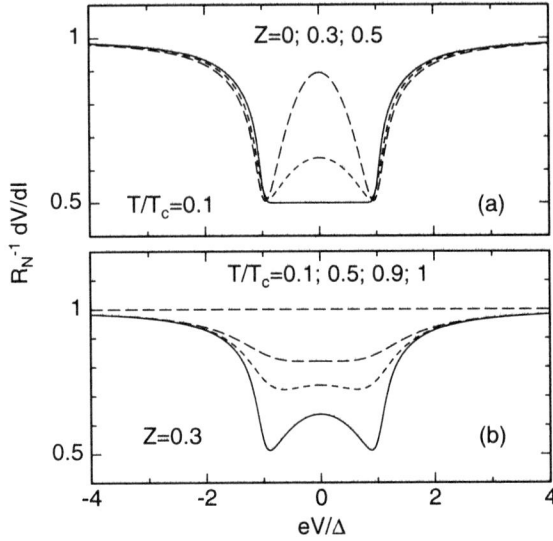

Fig. 3.15. $dV/dI\,(V)$ curves calculated according to (3.30): (a) for different values of the barrier parameter Z at $T/T_c = 0.1$, and (b) for $Z = 0.3$ and different reduced temperatures.

where I_{exc} is the so-called excess current, which is expressed in the ballistic regime [Zaitsev (1980, 1984)] as

$$I_{exc} = \frac{4\Delta}{3eR_N} \qquad (3.34)$$

and R_N is the resistance in the normal state [see Fig. 3.17(a)]. Artemenko et al. (1979) were the first to link excess current in superconducting contacts with Andreev reflection. For the dirty contacts, I_{exc} is approximately two times smaller than in the ballistic regime:

$$I_{exc} = \frac{1}{2}\left(\frac{\pi^2}{4} - 1\right)\frac{\Delta}{eR_N}. \qquad (3.35)$$

The electron–phonon interaction in the S-c-N contact leads to a backscattering current in the $I - V$ characteristics at the energy well above Δ [Khlus (1983)]. As a consequence, the point-contact spectrum represents a sum of the electron–phonon interaction functions according to (3.27). The only difference is the modification of the smearing function (3.5), so that a δ-shaped phonon peak smears out in d^2V/dI^2 by about Δ and shifts on the value of Δ to the lower energy.

3.7.2 S-c-S contacts

In the ballistic S-c-S contact at $eV \gg \Delta$, the excess current [Zaitsev (1984)]

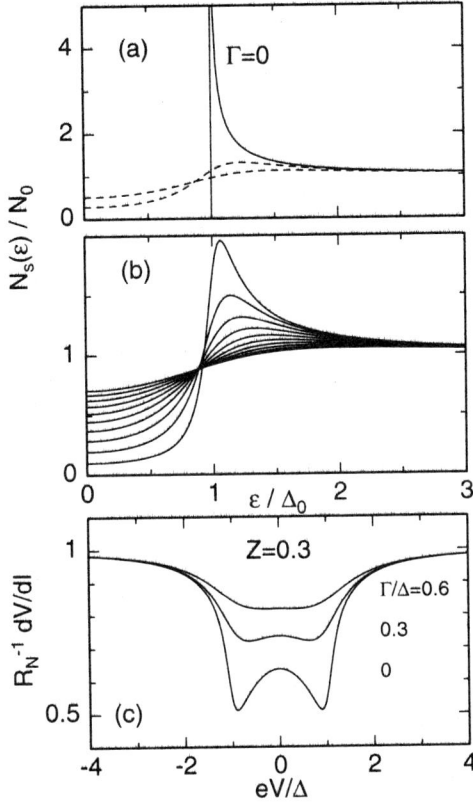

Fig. 3.16. (a) Quasiparticle density of states for superconductor according to (3.32) at $\Gamma = 0$ (solid curve), and $\Gamma = 0.3\Delta$ and 0.6Δ (dashed curves). (b) The same for $\Gamma = (n/10)\Delta, n = 1\ldots10$. (c) $dV/dI\ (V)$ curves calculated by (3.30) and (3.32) for three values of Γ with quasiparticle density of states shown in (a) for the barrier parameter $Z = 0.3$ at $T/T_c = 0.1$.

$$I_{\text{exc}} = \frac{4(\Delta_1 + \Delta_2)}{3eR_N} \tag{3.36}$$

is presented in $I - V$ curves [Fig. 3.17(b)].

A characteristic feature of the S-c-S contact is the occurrence of a Josephson supercurrent I_c [see Fig. 3.17(b)] with a value [Kulik and Omelyanchouk (1977, 1978)]:

$$I_c = \frac{\pi\Delta}{eR_N} \tag{3.37}$$

for a ballistic contact and smaller by a factor of 3/2 for a dirty constriction [Kulik and Omelyanchouk (1975)]. An unambiguous test for the presence of a Josephson current is the observation of the Fraunhofer diffraction pattern

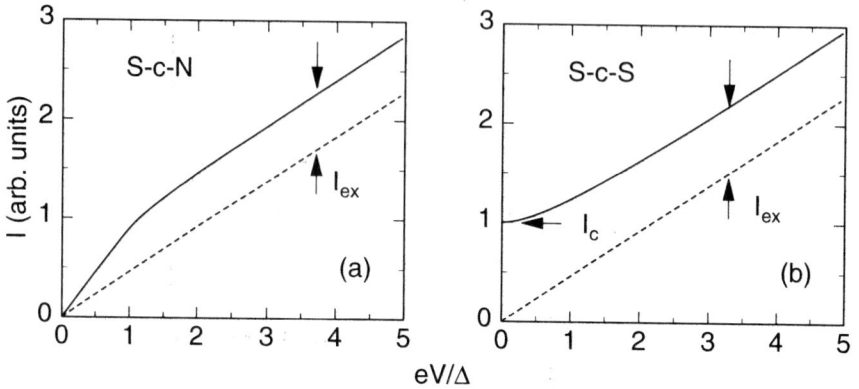

Fig. 3.17. (a) $I - V$ curves calculated by (3.30) for $Z = 0$, $T/T_c = 0.1$ (solid line) and $T/T_c = 1$ (dashed line). (b) Sketch of the $I - V$ curve of a Josephson contact in the RSJ model according to Likharev (1985) (solid line) and in the normal state (dashed line). The arrows indicate definition of the excess I_{exc} and the critical I_c current.

of the critical current in a magnetic field[4]:

$$I_c(H) = I_c(0) \left| \frac{\sin(\pi \Phi / \Phi_0)}{(\pi \Phi / \Phi_0)} \right|, \tag{3.38}$$

where $\Phi_0 = 2.07 \times 10^{-15}\,\mathrm{T\,m^2}$ and Φ is the magnetic flux penetrated the contact.

The $\mathrm{d}^2 V / \mathrm{d} I^2$ characteristic of the S-c-S contact at the energy well above Δ should also reflect the electron–phonon interaction [Khlus and Omelyanchouk (1983)]. The theory also predicts that in the case of smooth behavior of $g(eV)$ compared with Δ, the difference of $\mathrm{d} I / \mathrm{d} V$ in the normal and superconducting states is proportional to the EPI function:

$$\left(\frac{\mathrm{d} I}{\mathrm{d} V} \right)_S - \left(\frac{\mathrm{d} I}{\mathrm{d} V} \right)_N = -\frac{32 \Delta_0 d}{3 \hbar R_0} \left(\frac{g_1(eV)}{v_{F1}} + \frac{g_2(eV)}{v_{F2}} \right). \tag{3.39}$$

If the interface of the S-c-S contact represents a normal region with relative large mean free path, the processes of multiple Andreev reflection (MAR) occur for such a sandwich. As shown by Klapwijk et al. (1982) and Octavio et al. (1983), this leads to peculiarities in the $I - V$ curves at energies $2\Delta/n$ ($n = 1, 2, 3 \ldots$) developing by increasing of Z parameter or, in other words, by decreasing transparency of the N-S interface. Figure 3.18 shows calculated $\mathrm{d} V / \mathrm{d} I$ curves by virtue of the following theory of MAR by Flensberg et al. (1988). The distinct minimum determines the position of 2Δ, although

---

[4] In (3.38), it is supposed that the contact has a well-defined cross section with a uniform distribution of the Josephson current density.

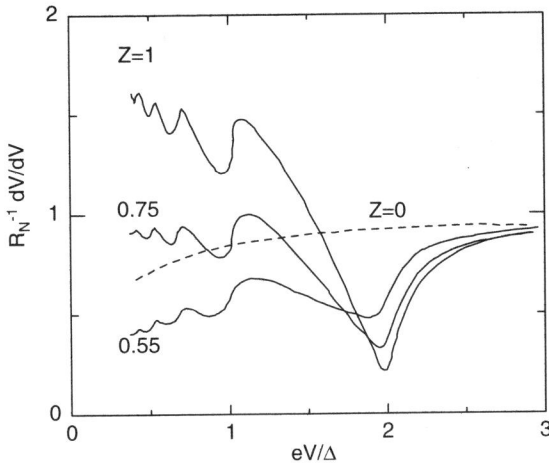

Fig. 3.18. Normalized $\mathrm{d}V/\mathrm{d}I$ of S-c-S contacts versus reduced voltage for $T = 0$ at a different barrier strength parameter Z. Data taken from Flensberg et al. (1988).

for $Z < 1$, the minimum occurs at a voltage somewhat below 2Δ. The subharmonic gap structure at about $2\Delta/n$ becomes also progressively sharper with increasing of Z or reflectivity of the interface and finally exponentially disappears for large Z.

References

Akimenko A. I., Verkin A. B., Ponomarenko N. M. and Yanson I. K. (1982) Sov. J. Low Temp. Phys. **8** 130.

Andreev A. F. (1964) Sov. Phys. - JETP **19** 1228.

Artemenko S. N., Volkov A. F. and Zaitsev A. V. (1979) Sol. State Communs. **30** 771; JETP Lett. **49** 924.

Baranger H. U., MacDonald A. H. and Leavens C. R. (1985) Phys. Rev. B **31** 6197.

Blonder G. E. and Tinkham M. (1983) Phys. Rev. **27** 112.

Blonder G. E., Tinkham M. and Klapwijk T. M. (1982) Phys. Rev. **25** 4515.

Dynes R. C., Narayanamurti V. and Garno J. P. (1978) Phys. Rev. Lett. **41** 1509.

Flensberg K., Hansen J. B. and Octavio M. (1988) Phys. Rev. B **38** 8707.

Grimvall G. (1981) *The Electron–phonon Interaction in Metals* North-Holland Publ. Co., Amst., N.-Y., Oxf.

Itskovich I. F., Kulik I. O. and Shekhter R. I. (1985) Sov. J. Low Temp. Phys. **11** 588.

Itskovich I. F., Moskalets M. V., Shekhter R. I. and Kulik I. O. (1987) Sov. J. Low Temp. Phys. **13** 588.

Itskovich I. F. and Shekhter R. I. (1985) Sov. J. Low Temp. Phys. **11** 649.

Jansen A. G. M., van Gelder A. P. and Wyder P. (1980) J. Phys. C **13** 6073.

Khlus V. A. (1983) Sov. J. Low Temp. Phys. **9** 510.

Khlus V. A. and Omelyanchouk A. N. (1983) Sov. J. Low Temp. Phys. **9** 189.

Klapwijk T.M., Blonder G.E., and Tinkham M. (1982) Physica B, **109 & 110** 1657.

Kohlrausch N. (1900) Ann. Physik Leipzig **1** 132.

Kulik I. O. (1984) Phys. Lett. **106 A** 187.

Kulik I. O. (1985) Sov. J. Low Temp. Phys. **11** 516.

Kulik I. O. (1992) Sov. J. Low Temp. Phys. **18** 302.

Kulik I. O. and Omelyanchouk A. N. (1975) JETP Lett. **21** 96.

Kulik I. O. and Omelyanchouk A. N. (1977) Sov. J. Low Temp. Phys. **3** 459.

Kulik I. O. and Omelyanchouk A. N. (1978) Sov. J. Low Temp. Phys. **4** 142.

Kulik I. O., Omelyanchouk A. N. and Shekhter R. I. (1977) Sov. J. Low Temp. Phys. **3** 840.

Kulik I. O., Shekhter R. I. and Shkorbatov A. G. (1981) Sov. Phys.- JETP **54** 1130.

Kulik I. O. and Yanson I. K. (1978) Sov. J. Low Temp. Phys. **4** 596.

Likharev K. K. (1985) *Introduction to the Dymamics of Josephson Junctions* Moscow: Nauka (in Russian).

MacDonald A. H. and Leavens C. R. (1984) Solid State Commun. **50** 467.

McMillan W. L. (1968) Phys. Rev. **167** 331.

Octavio M., Tinkham M., Blonder G. E. and Klapwijk T. M. (1983) Phys. Rev. **27** 6739.

Plecenik A., Grajcar M., Beňačka Š., Seidel P. and Pfuch A. (1994) Phys. Rev. **B 49** 10016.

Shekhter R. I. and Kulik I. O. (1983) Sov. J. Low Temp. Phys. **9** 22.

Srikanth H. and Raychaudhuri A. K. (1992) Physica **C 190** 229.

van Gelder A. P. (1978) Solid State Commun. **25** 1097.

van Gelder A. P. (1980) Solid State Commun. **35** 19.

Verkin B. I., Yanson I. K., Kulik I. O., Shklyarevskii O. I., Lysykh A. A. and Naidyuk Yu. G. (1979) Solid State Commun., **30** 215.

Verkin B. I., Yanson I. K., Kulik I. O., Shklyarevskii O. I., Lysykh A. A. and Naidyuk Yu. G. (1980) Izv. Akad. Nauk SSSR, Ser. Fiz. **44** 1330.

Wexler A., (1966) Proc. Phys. Soc. (London) **89** 927.

Yanson I. K. (1983) Sov. J. Low Temp. Phys. **9** 343.

Yanson I. K. and Shklyarevskii O. I. (1986) Sov. J. Low Temp. Phys. **12** 509.

Zaitsev A. V. (1980) Sov. Phys. - JETP **51** 111.

Zaitsev A. V. (1984) Sov. Phys. - JETP **59** 1015.

4 Experimental techniques

4.1 Fabrication of point contacts

4.1.1 Thin films

Different types of the point contacts have been developed during elaboration of PCS method. The pioneering measurements were carried out by Yanson (see also Section 2.2) at the beginning of the 1970s exploiting thin film junctions. Here two evaporated in-sequence metallic films (the first one being oxidized before evaporating of the next one) were electrically isolated by oxides. The oxide layer on a metal, as a rule, is a few nanometers thick and serves as a good and strong dielectric. This structure is called a sandwich-like tunnel junction, and it is widely used as a tool by physical researches (see Section 2.2). If the oxidation is strong, the dielectric layer will be thick and the tunneling current negligibly small. By reducing of oxidation time, the appearance of "holes" in the oxide becomes possible; i. e., the conductivity between metallic films is not stipulated by the tunneling but by the current flowing through the metallic shorts. Such contacts have naturally to be rejected as far as they are not suitable for the tunneling investigations.

However, such contacts with metallic shorts might work successfully for PCS. Schematic representation of a contact with a short in the dielectric layer is shown in Fig. 4.1(a). A conducting bridge between films can often appear spontaneously as a result of metallic dendrites intergrowth through the weak spots or defects of the dielectric film. The bridge can be created artificially by means of mechanical loads causing the oxide film damage or by means of electric breakdown. In the former case, a sharp needle creating high local stress can be used, and in the latter case, electric voltage is raised up to the level exceeding the breakdown value at the weakest spot of the dielectric. This case resembles electric microwelding when metal instantaneously melts in the short and then solidifies. During this process, metal in the short can be purified from foreign inclusions or impurities. Certainly, depending on the film material, dielectric interlayer properties, magnitude, and duration of the current pulse, the degree of structure perfection and the purity of metal in the constriction can vary in a wide range. Using the film technology, small point contacts only of a few nanometers in size with sufficiently high stability versus thermocycling and mechanical vibrations may be obtained.

Those are the advantages of thin film technology. However, fabrication of point contacts by film method involves an awkward technique of vacuum deposition of metals. Besides, the structure of metal in the films is usually worse than that of the bulk samples.

4.1.2 Needle-anvil

The creation of contacts of so-called "needle-anvil" type between bulk electrodes used by Jansen et al. (1977) gave an impetus to the further development of the research in this field. A clamping contact of this type is shown in Fig. 4.1(b). It is obtained by gradually approaching two electrodes directly in the liquid helium by means of a precision mechanism. Generally, the radius of curvature of the needle tip equals tens of micrometers. Before measurement, the electrodes are processed mechanically, then chemically or electrochemically, to obtain clean unperturbed surfaces of required shape. At this time, the surface is coated by a thin film of oxide or another nonconducting layer. Because of the dielectric layer, at the first light touch of electrodes, the contact conductivity is often of the tunneling character; i.e., the contact resistance decreases with the voltage increase. A metallic bridge appears after increase of the pressing force either as a consequence of dielectric cracking or because of the breakdown by the current pulse. The dielectric film covering the electrodes is necessary to stiffen the whole structure as far as in this case the mechanical-contact area is much larger than the size of the metallic short. In any event, for a specific metal, its own methods of contact preparation are desirable. For example, in the case of clamping Ta point contacts, Bobrov et al. (1987) developed a so-called "electro-formation" of contact by dc current and showed that using Cu as counter-electrode result in better Ta spectrum (see, e.g., Fig. 5.14).

4.1.3 Shear method

In the needle-anvil configuration, a metal in the contact area is obviously deformed and polluted with pressed-in oxide residues, as well as with various impurities absorbed on the surface. Better results can be obtained if the contacts are created in the way shown in Fig. 4.1(c), first described by Chubov et al. (1982). In this instance, the electrodes are brought together until they are touching; then they are shifted one relatively to another in the plane of their crossing. As a result of this, the deformation of a metal is much less here, and the oxide and surface impurities can be removed out of the contact area. Thus, shorts are formed in those places where the cleaned areas of electrodes come in the contact. With this technique, the samples can have a different form, e.g., of small bulk cylinders or metallic slabs as well as films deposited on dielectric rods. By shifting the electrodes independently in the mutually perpendicular directions or rotating them around their axes, it is possible to change the places of the contact and obtain a number of shorts with different resistances within the same measurement cycle.

4.1.4 Lithography

The basic idea here, realized by Ralls et al. (1989), is to fabricate a tiny hole in the membrane using nanolithography and reactive ion etching. A single hole is patterned in the 30-nm-thick silicon nitride membrane by e-beam lithography and dry etching. The structure is formed by evaporating under UHV conditions a 200-nm metal layer onto both sides of the membrane. The hole is filled with a metal so that a nanobridge is formed between two metallic layers as shown in Fig. 4.1(d). The method has the advantage of described thin films techniques. Additionally, it is very important that both the geometry and the size of the contact are well fixed. Of course, the disadvantages are the same: (1) a metal structure in the film is not perfect, (2) only a restricted number of metals can be used, and (3) this is a still complicated and time-consuming technology.

4.1.5 Break junctions

The exploiting of Moreland and Ekin (1985) approach, Muller et al. (1992) developed a mechanically controllable break-junction (MCBJ) technique that is suited for investigation of metallic bridges up to an atomic scale. Its advantage is perfect vibration stability together with the possibility of tuning of the resistance. Moreover, freshly broken clean surfaces form the contact. Additionally, there is the opportunity to variate contact resistance from a few milliohm up to about $10\,\mathrm{k}\Omega$, what corresponds to the contact dimension from a tenth of a millimeter down to the atomic size. Also important the fact that a vacuum tunneling regime between two broken pieces can be realized. A schematic representation of MCBJ is shown in Fig. 4.1(e). The sample made into the shape of a long (from a few to 10 mm) bar or rod is glued on a flexible substrate (bending beam), which is usually made of phosphor bronze covered with insulating capton foil. A deep knife or spark-cut notch defines the break position. By bending the substrate with the screw at low temperatures and high vacuum conditions, the filament will break. Thus, the freshly exposed electrodes can be brought back into the contact. A piezotube serves for fine adjustment with about 10^{-11}-m resolution at low temperatures. The vertical motion of the piezotube transfers to a horizontal displacement of the two parts of the sample against each other by tuning of a piezovoltage, reduced by a factor of order 100. This results in an excellent mechanical stability of MCBJ and allows a fine control of its resistance in many orders of magnitude. The disadvantage here is the deformed metal structure caused by the low-temperature mechanical stress (break) getting worse after repeated regulation of MCBJ from the close to open case and vice versa.

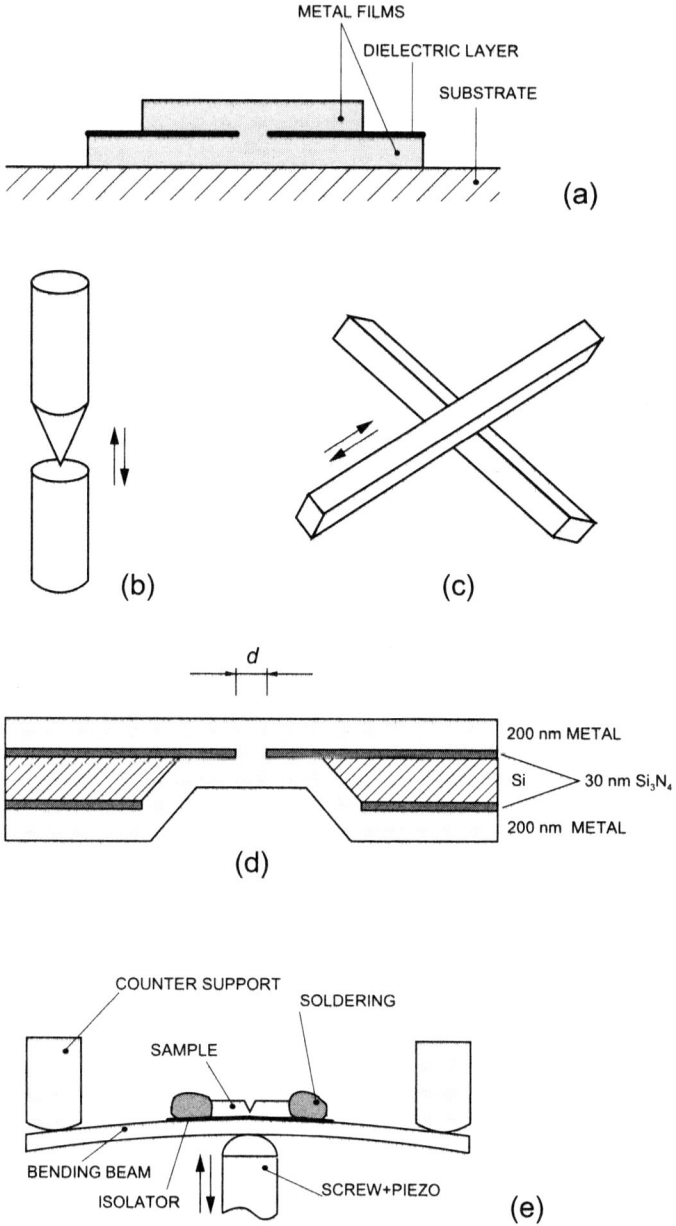

Fig. 4.1. Schematic view of different methods of point-contact formation: (a) thin films, (b) needle-anvil, (c) shear, (d) lithography, and (e) break-junction (see text for details).

4.2 Sampling of point contacts

PCS would be impossible if the measured characteristics were critically dependent on the preparation method and the constriction geometry, which,
in general, cannot be well controlled. However, the experience showed that
if the contacts were chosen properly by means of established experimental
criteria, the reproducibility of point-contact spectra is none the worse than
in the tunneling spectroscopy (see Section 2.4), where such selection rules are
generally accepted. Let us formulate the quality criteria for clean metal point
contacts, which are suitable for investigation of the electron–phonon interaction; that is, they meet the requirements of the theoretical model (Chapter
3). Point-contact spectrum of a simple metal sodium (Fig. 4.2) may be used
as an example.

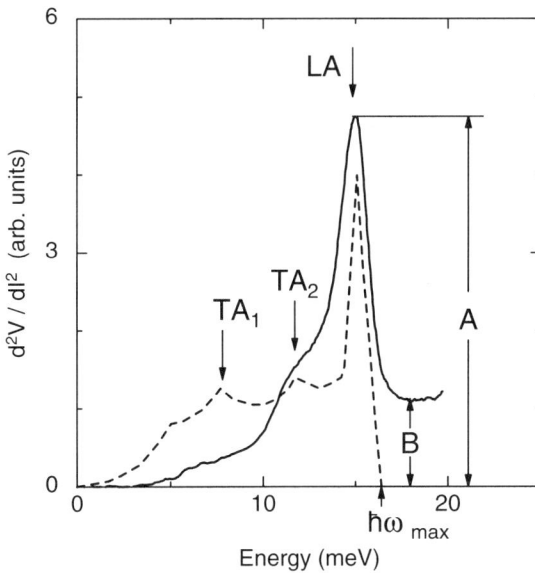

Fig. 4.2. Point-contact spectrum of Na (solid line) at temperature 1.6 K along
with the phonon density of states (dashed line). The position of transverse (TA)
and longitudinal (LA) acoustic peaks is marked by arrows.

1. For pure metals, the contact resistance is within 1 to 100 Ω and the
   short has the metallic conductivity; i. e., its resistance increases with the
   voltage rise so that $d^2V/dI^2 > 0$ over the whole energy range.
2. Distinct maxima are observed in the PC spectrum at energy between zero
   and maximal one $\hbar\omega_{max}$. They are reproducible for different contacts of
   the particular metal. At energy $eV > \hbar\omega_{max}$, the spectrum contains a

certain constant background value. The background parameter $\gamma = B/A$ (Fig. 4.2) is different for each metal, but as a rule, it should not be greater than 0.5.

3. Relative variation of the differential resistance within the spectrum has to be maximal among the spectra with the same other characteristics. Around zero-bias at $eV \ll \hbar\omega_{max}$, the $\mathrm{d}^2V/\mathrm{d}I^2$ curve behaves monotonously as V^n (see Table 3.1) or zero-bias anomaly should be as weak as possible.

The above-mentioned attributes provide a possibility, judging by the shape of the point-contact spectrum, to select the contacts for the further investigation of the electron–phonon interaction function. Applying these criteria, e. g., to the spectra shown in Fig. 2.3, one can see that they are fulfilled only for two bottom curves.

When studying the other, non-electron–phonon mechanisms of electron scattering in solids, in other words, different quasiparticle excitations, of course, some additional criteria should be involved based mainly on statistical analysis and reproducibility of the point-contact spectra. As far as the electron–phonon interaction exists in all conductors, the above-mentioned criteria should be taken into account in each case. Finally, the quality of the contact, whether it is suitable for further study can be determined only after scrupulous analysis and classification of the complete collection of the experimental results.

4.3 Electronic methods and schema

One of the widely used methods to measure derivatives of the $I - V$ characteristic is the so-called modulation technique. It consists of the following: Besides the dc current I, the sample is supplied with the small ac current i at the definite frequency ω. In this case, the voltage can be presented as a Taylor series expansion of the $I - V$ curve:

$$V(I + i\cos(\omega t)) = V(I) + \frac{\mathrm{d}V}{\mathrm{d}I} i\,\cos(\omega t) + \frac{1}{2}\frac{\mathrm{d}^2V}{\mathrm{d}I^2} i^2\,\cos^2(\omega t) + \cdots \quad (4.1)$$

$$= V(I) + \frac{\mathrm{d}V}{\mathrm{d}I} i\,\cos(\omega t) + \frac{1}{4}\frac{\mathrm{d}^2V}{\mathrm{d}I^2} i^2\,(1 + \cos(2\omega t)) + \cdots.$$

If the current i is sufficiently small, the higher order in i terms can be neglected. It is seen that a signal measured at the frequency ω will be proportional to the first derivative $\mathrm{d}V/\mathrm{d}I$ of the $I-V$ characteristic, whereas a signal at the frequency 2ω will be proportional to the second derivative $\mathrm{d}^2V/\mathrm{d}I^2$. The next problem is to measure a weak voltage signal at the frequency ω or especially at 2ω. This is achieved by phase-sensitive detection at the frequency ω or 2ω, respectively, by spectroscope, whose block diagram is shown in Fig. 4.3. The sample is held in a cryostat with liquid helium. Connected to the sample are four wires, a pair of which serves as an electric circuit, through which the contact is supplied with a current from the dc-source together with ac modulation current from a low-distortion ac-generator of the harmonic signal. The dc-voltage V created on the sample by current I is measured by multimeter by means of the other pair of wires. This pair is also used for measuring small ac voltage preamplified by a low-noise amplifier. Then the ac signal is recovered by phase-sensitive detection at the fundamental frequency ω or at twofold of the fundamental frequency 2ω, which is proportional to the first dV/dI or to the second d^2V/dI^2 derivative of the $I - V$ curve, correspondingly. Often a resonance (or notch) filter tuning to ω or 2ω is used instead of the preamplifier, also necessitating matching of the low-resistance sample with the high-resistance input of the amplifier. The ac voltage is converted into the dc one and transmitted to the Y coordinate of the recorder, whereas the dc bias voltage is transmitted to the X coordinate. Thus, when a linearly increasing current I from sweep generator passes through the sample, the recorder will automatically plot the first or second derivative of the $I - V$ characteristics as a function of the applied voltage V. Certainly, instead of the recorder, the measured signals can be transferred via IEEE interface to a computer.

So far as a variation of the contact resistance, in other words, a nonlinearity of the $I - V$ curve can be small, a bridge circuit is often used for measuring dV/dI at the fundamental frequency ω to compensate the $dV/dI(V = 0)$ signal in such a way that only changes of the difference resistance will be measured with the look-in amplifier. Figure 4.3 shows a typical electronic setup in the direct and in the bridge mode.

The choice of the fundamental frequency ω is also important. A high frequency about $100\,\mathrm{kHz}$ decreases sufficiently the $1/f$ noise, but this leads to the capacitive currents, which cause increasing difficulties by measuring the low ohmic contacts with the resistance lower than the resistance of the connecting wires. Empirically, the frequency range between 1 and $10\,\mathrm{kHz}$ is mostly used in PCS. In all cases, a care must be taken not to choose a multiple to 50-Hz frequency. For more details of harmonic detection and bridge circuit for derivative measurements, see Wolf (1985).

From the contact resistance R_0 and the rms modulating signal $V_1 = (i/\sqrt{2})R_0$, both taken at zero-bias voltage, the absolute value of the second derivative can be obtained via (4.1) by measuring of the rms signal V_2 at 2ω:

$$\frac{d^2V}{dI^2} = \frac{4R_0^2}{\sqrt{2}V_1^2}V_2. \tag{4.2}$$

As seen from (4.1), the signal at the frequency 2ω is proportional to the square of the modulating current or voltage amplitude. Therefore, it seems there are no special problems in measuring this signal by merely increasing the ac component V_1. However, it cannot be too large inasmuch as the ac voltage

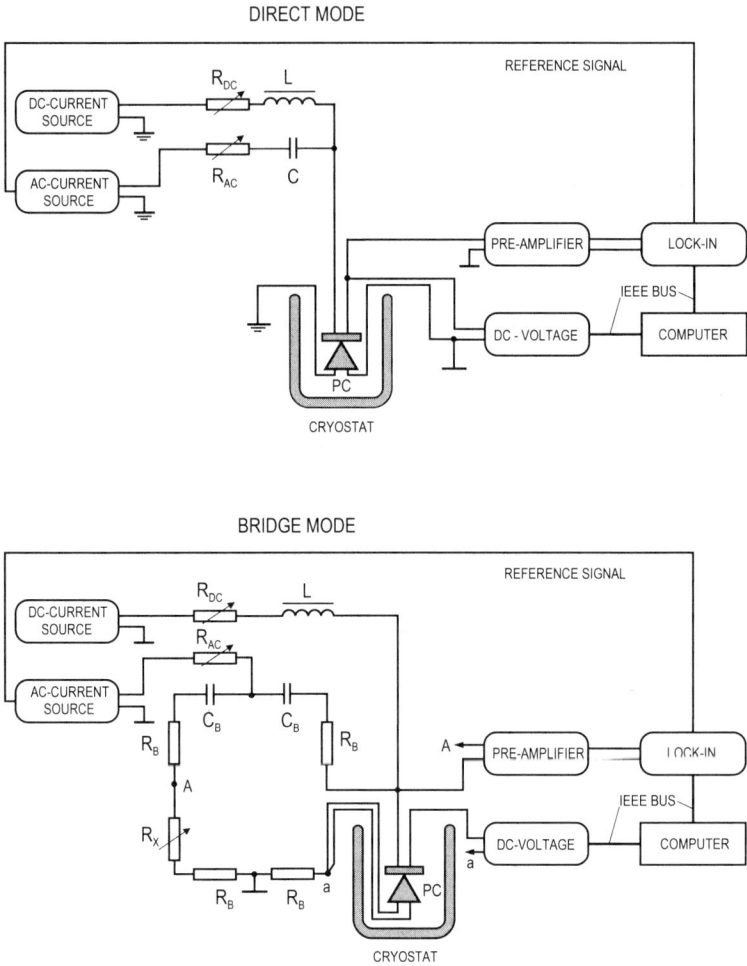

Fig. 4.3. Electronic circuits for measuring the point-contact $I - V$ characteristics and their derivatives in the direct and bridge mode.

amplitude determines the resolution of this technique. Exact calculations [Klein et al. (1973)] indicate that an infinitely narrow spectral feature when measuring d^2V/dI^2 using the modulation technique will be smeared into a bell-shaped $1.22\,V_\omega$ wide maximum ($V_\omega = \sqrt{2}V_1$ is the amplitude of the modulating voltage) [Fig. 4.4(b)]. In most cases, the value of V_1 is as a rule not higher than a few millivolts, and in some cases, it can be even several tenths of a millivolt. In this case, the value of measured signal that characterizes the nonlinearity of the $I - V$ characteristics associated with the electron–phonon interaction is of the order of 1μV. At the same time, the circuit sensitivity

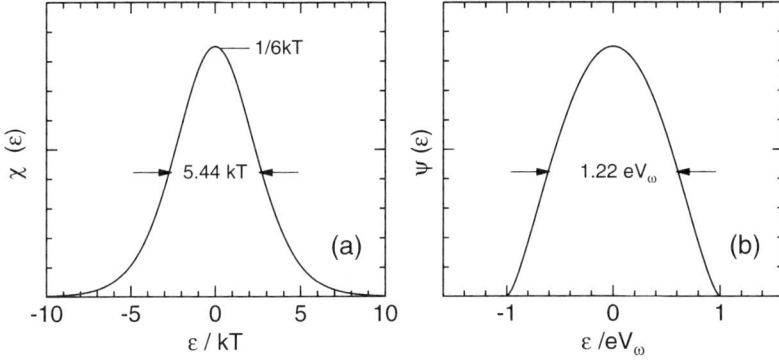

Fig. 4.4. (a) Thermal broadening function for d^2V/dI^2 calculated according to (3.5). (b) Broadening function for d^2V/dI^2 caused by finite modulation voltage $V_1 = V_\omega/\sqrt{2}$. Data taken from Klein et al. (1973).

is two orders of magnitude higher. That makes it possible to register reliably the point-contact spectra.

It should be also noted that the resolution in PCS depends on the temperature T caused by the Fermi level smearing of the order of $k_B T$. In this case, the infinitely narrow spectral feature smears into a bell-shaped maximum [Fig. 4.4(a)]. Therefore, to obtain clearcut spectral features, the condition $k_B T \ll \hbar\omega_{max}$ must be satisfied; i.e., measurements should be carried out at a temperature considerably lower than the characteristic temperature (or energy $\hbar\omega_{max}$) of excitations being under investigation. Under the action of the above factors, alternating voltage and temperature, the resulting resolution in the case of tunneling spectroscopy is expressed by Lambe and Jaklevic (1968) as

$$\delta_2 = [(5.44 \, k_B T/e)^2 + (1.22\sqrt{2} \, V_1)^2]^{1/2}, \qquad (4.3)$$

which is also valid for PCS. Generally, measurements are carried out in liquid helium at the temperature between $1.5 \, K$ and $4.2 \, K$ and modulating voltage between $0.3 \, mV$ and $1.5 \, mV$. As a result, the resolution calculated from (4.3) is between $0.9 \, mV$ and $2.9 \, mV$ or, expressed in degrees, between $10 \, K$ and $34 \, K$. This is at least one order of magnitude below the Debye temperature of typical metals, what enables us to obtain a sufficiently detailed spectrum of the electron–phonon interaction. Of course, to resolve the fine feature of the electron–quasiparticles interaction, both the lower temperature and modulating voltage are desirable. It is also worth noticing that the resolution by measuring the first derivative dV/dI signal is different from that mentioned above and according to Duif (1983) is

$$\delta_1 = [(3.53 k_B T/e)^2 + (1.73\sqrt{2}V_1)^2]^{1/2}. \qquad (4.4)$$

The shape of the mentioned broadening functions is shown in Fig. 4.5.

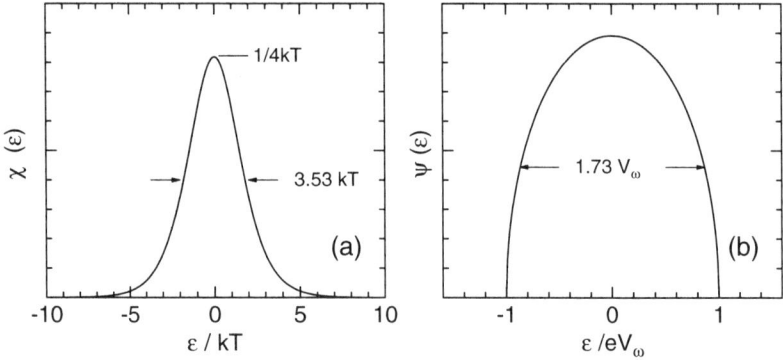

Fig. 4.5. (a) Thermal and (b) modulating voltage broadening functions for the first derivative dV/dI of the $I - V$ characteristic. $V_1 = V_\omega/\sqrt{2}$. Data taken from Duif (1983).

4.4 Point-contact drivers

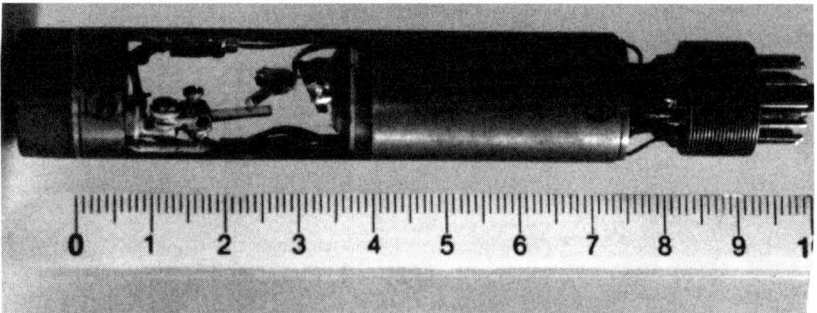

Fig. 4.6. Experimental setup for point contact establishing at low temperatures based on the differential screw mechanism arranged inside of a cylinder in the middle of the image. The samples in the shape of small bars are situated in the window center between two and three marks. Electrical connectors are situated on the right-hand part of the insert between eight and ten marks. Setup diameter is below 20 mm, what allows us to use this system with most of standard solenoids.

To regulate the resistance of clamping point contacts or break junctions at low temperatures, some mechanisms are needed. The most exploited mechanical equipment to produce a controlled movement of one electrode in

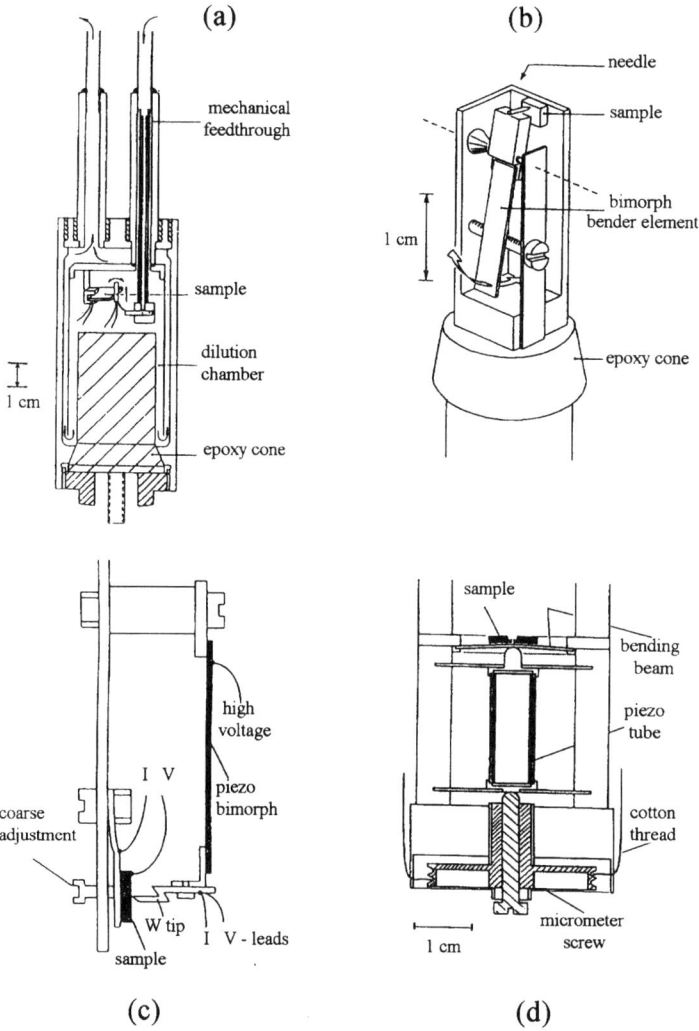

Fig. 4.7. Schematic drawing of the different setups using mechanical as well as piezodrivers for the point-contact formation in a dilution refrigerator directly in the mixing chamber (a), (b) and outside in the vacuum chamber (c), (d). See text and corresponding references for the further details.

liquid helium has various technical solutions, but mainly they are built on the base of the differential screw mechanism. The typical example of the low-temperature part of the suitable inset is shown in Fig. 4.6.

Additionally, a piezodriver is employed, what is especially desirable for the fine control of the resistance of mechanically controlled high-ohmic break junctions. In all cases, there is no an essential technical problem to build an inset for the contact establishing at a liquid He^4 temperature. The difficulties increase sufficiently by carrying out measurements at very low temperatures, positioning the whole mechanism in a dilution refrigerator. Up to now, there have been a few examples showing the establishing of contacts in the mK range, which are shown in Fig. 4.7. Goll (1993) developed the construction to produce point contacts mechanically directly in the mixing chamber of the dilution refrigerator [Fig. 4.7(a)], and Heil et al. (1993) presented a piezoelectric mechanism for micromovements in the dilution chamber [Fig. 4.7(b)]. Gloos et al. (1996) used piezo setup [Fig. 4.7(c)] as well as the latter one combined with a screw driven by two cotton threads both sitting in the vacuum part of the fridge [Gloos et al. (1999)]. The latter construction [Fig. 4.7(d)] was used for producing break junctions.

References

Bobrov N. L., Rybaltchenko L. F., Yanson I. K. and Fisun V. V. (1987) Sov. J. Low Temp. Phys. **13** 344.

Chubov P. N., Yanson I. K. and Akimenko A. I. (1982) Sov. J. Low Temp. Phys. **8** 32.

Duif A. M. (1983) Doctoraalscriptie, Nijmegen. (unpublished)

Goll G. (1993) Ph. D. Thesis, Karlsruhe. (unpublished)

Gloos K. and Anders F. B. (1999) J. Low Temp. Phys. **116** 21.

Gloos K., Anders F. B., Buschinger B., Geibel C., Heuser K., Järling F., Kim J. S., Klemens R., Müller-Reisener., Schank C. and Stewart G. R. (1996) J. Low Temp. Phys. **105** 37.

Heil J., de Wilde Y., Jansen A. G. M. and Wyder P. (1993) Rev. Sci. Instrum. **64** 1347.

Jansen A. G. M., Mueller F. M. and Wyder P. (1977) Phys. Rev. **16** 1325.

Lambe J. and Jacklevic R. C. (1968) Phys. Rev. **165** 821.

Klein J., Leger A., Belin M. and Defourneau D. (1973) Phys. Rev. B **7** 2336.

Moreland J. and Ekin J. W. (1985) J. Appl. Phys. **58** 3888.

Muller C. J., van Ruitenbeek J. M. and de Jongh L. J. (1992) Physica C **191** 485.

Ralls K. S., Buhrman R. A. and Tiberio R. C. (1989) Appl. Phys. Lett **55** 2459.

Wolf E. L. (1985) *Principles of Electron Tunneling Spectroscopy* Oxford University Press, Inc. New York.

5 PCS of quasiparticle excitations

Recovering of the electron–phonon interaction (EPI) function for metals is the most attractive and impressive side of PCS. Direct proportionality between the second derivative of the $I - V$ characteristic and the EPI function admits to using point contact as a spectroscopic instrument that develops point-contact spectroscopy in a new kind of solid-state spectroscopy. In this chapter, we are going to survey the study of EPI in metals[1] with different crystal structure, topology of the Fermi surface, exhibiting of ferro- or antiferromagnetic order as well as related topics concerning electron–magnon interaction, localized phonon vibrations, influence of boundary between metals on the transport processes in contact, study of EPI in compounds and so on.

5.1 Electron–phonon interaction

5.1.1 Alkali metals

For a long time, alkali metals played a role of touchstone in metal physics. Their thermodynamic and transport properties can be well understood in the free electron model. The Fermi surface of the alkali metals (Na and K) is practically very nearly spherical, what simplifies calculations in many cases. In our case, it permits the direct comparison between measured and calculated point-contact spectra. Only from the experimental point of view, a high chemical activity of the alkali metals makes it difficult to find a relevant method to produce high-quality contacts. Nevertheless, some experimental tricks allow us to produce clean contacts in the case of Na and K with perfect spectra (Fig. 5.1). The main peculiarities in the spectra correspond to the maxima in the phonon density of states, although their relative intensities differ strongly in different metals. For Na and K, peaks caused by the transverse acoustic (TA) phonons are almost absent or strongly depressed.

---

[1] The detailed spectra of EPI, calculated point-contact EPI functions with tabulation, dependencies of the phonon density of states, related thermodynamic and transport EPI functions, constants of EPI, and so on for 32 studied by PCS clean metals are given in the handbook written by Khotkevich and Yanson (1995) (see Chapter 1 and the references therein).

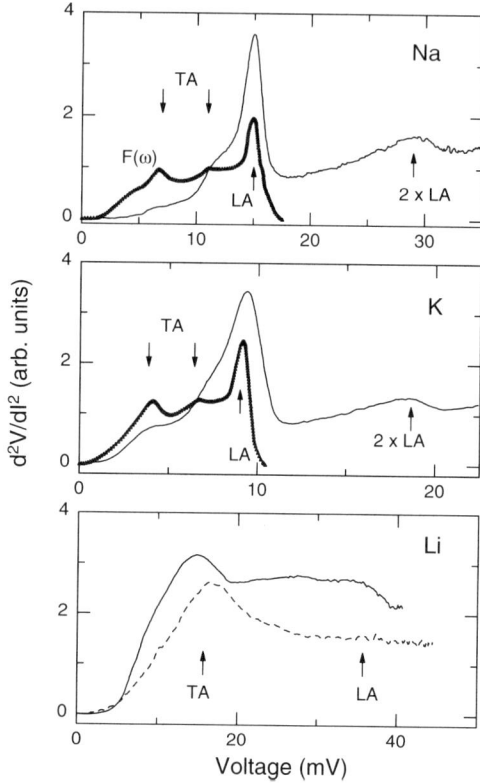

Fig. 5.1. Point-contact spectra of Na, K, and Li (thin solid lines) measured at helium temperatures by Naidyuk et al. (1980). Two spectra for Li correspond, apparently, to FCC (solid) and HCP (dashed) lattice structure. The arrows show the position of the transverse acoustic (TA) and the longitudinal acoustic (LA) phonon peaks. The gentle maxima at higher energy correspond to two phonon processes ($2 \times$LA; see Section 5.6). For Na and K, the phonon density of states $F(\omega)$ calculated by Zhernov et al. (1982) is shown.

We recall that for the normal electron–phonon scattering processes typical for a metal with a closed Fermi surface, the interaction between TA phonons and electrons is negligible. At deviation of the Fermi surface from the sphere, it approaches to the boundary of the Brillouin zone and then the intensity of the TA peaks increases (compare the amplitude of TA phonon features on PC spectra for Na and K in Fig. 5.1). The tendency of the Fermi sphere distortion increase goes in direction Na, K, and Li, and for the latter metal, TA peaks prevail in the spectra. Moreover, Li is apparently a fine crystalline mixture of two phases FCC (face-centered cubic) and HCP (hexagonal close-packed) caused by martensitic transformation at nitrogen temperatures. This leads, on the one hand, to the additional scattering of electrons and phonons, what results in the higher background in the spectra and smearing of the fine structure. On the other hand, the prevalence of the FCC or HCP phase causes the variation of the phonon peaks position and their relative intensity as shown in Fig. 5.1. The spectrum with a lower position of the TA peak and some additional structure (see also Fig. 5.3) at the higher energy, apparently, corresponds to the FCC phase, whereas the spectra with only one pronounced TA peak correspond to the HCP one. The calculated point-contact EPI func-

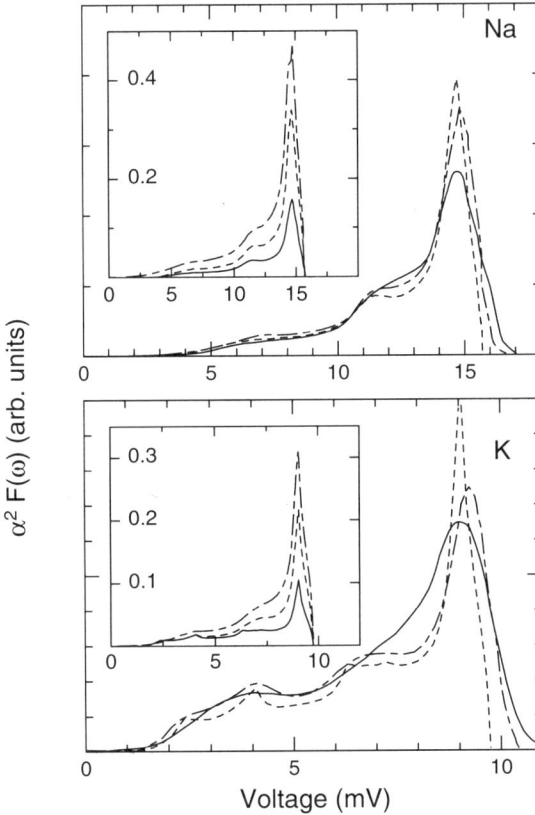

Fig. 5.2. Experimental point-contact EPI functions for Na and K (solid curves) from Naidyuk et al. (1980) in comparison with calculated ones by MacDonald and Leavens (1982) (dashed curves) and Ashraf and Swihart (1982) (dash-dotted curves). Insets: Point-contact (solid curves), transport (dashed curves) and thermodynamic (dash-dotted curves) EPI functions in absolute units for sodium and potassium calculated by MacDonald and Leavens (1982).

tions for Na and K (Fig. 5.2) correspond satisfactorily to the experimental ones. As was demonstrated by Zhernov et al. (1982), the shape of the calculated spectra critically depends on the pseudopotential form factor $v(\mathbf{q})$

behavior, especially on their value in the region of $q \leq 2k_F$. It can be used for correction of the pseudopotential behavior in the mentioned region by comparison of the calculated and the experimental spectra.

Although calculated point-contact g_{PC}, transport g_{tr}, and thermodynamic EPI functions g_{th} (see definition of these functions (3.7) and Table 3.1) for Na and K are similar regardless of their intensity (Fig. 5.2, insets), calculations for Li demonstrate remarkable difference between the mentioned dependencies (Fig. 5.3). It is seen that g_{PC} underlines the low-energy transverse peaks

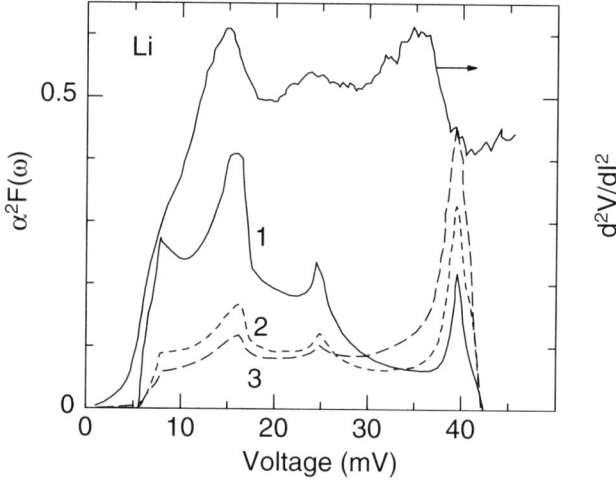

Fig. 5.3. Point-contact (curve 1), transport (curve 2), and thermodynamic (curve 3) EPI functions for lithium calculated by MacDonald and Leavens (1982). The upper curve is point-contact spectrum in Li measured by Jansen et al. (1980).

compared with the other EPI functions. It is caused by the difference in the transport form-factor for these dependencies (see Table 3.1). The Fermi surface of Li deviates more from the sphere compared with Na and K, what increases the contribution by virtue of Umklapp scattering processes, which are allowed for transverse phonons. This situation reflects the pseudopotential form-factor value at $2k_F$, which increases from 0.025 for Na and 0.057 for K to 0.273 for Li [MacDonald and Leavens (1982)]. Correspondingly, the electron–phonon interaction constant λ determined by analogy with (3.10) is larger for a point-contact EPI function in Li. The experimental measurements for Li displayed in Fig. 5.3 in general reproduce the theoretical calculation underlying the low-energy part of the spectra.

The calculated and measured EPI constants λ_{PC} for investigated metals are presented in Table 5.1.1. It is seen that calculations yield reasonable value λ or intensity of g_{PC} for sodium, exceed experimental data for lithium, and underestimate λ for potassium. It is not excluded that a more precise choice or correction of the pseudopotential form-factor can improve the results for the mentioned alkali metals.

Table 5.1. EPI constant λ for different simple metals. Here λ_{PC} is the point-contact experimental value in the free electron model, λ_{PC}^{cal} is calculated for point-contact EPI function, λ_{th} is a recommended value for the thermodynamic EPI function from Grimvall (1976, 1981). λ_{PC} is according to Khotkevich and Yanson (1995), with exception of the cases indicated in Ref. column for Li.

Metal	λ_{PC}	λ_{PC}^{cal}	Reference	λ_{th}
Li	0.25±0.1		Naidyuk et al. (1980)	0.4±0.1
	0.45±0.2		Jansen et al. (1980)	
		0.665	MacDonald and Leavens (1982)	
		0.6	Zhernov and Kulagina (1992)	
Na	0.068			0.16±0.03
		0.051	Zhernov et al. (1982)	
		0.166	Ashraf and Swichart (1982)	
		0.06	MacDonald and Leavens (1982)	
K	0.11			0.13±0.03
		0.051	Zhernov et al. (1982)	
		0.079	Ashraf and Swichart (1982)	
		0.058	MacDonald and Leavens (1982)	
Cu	0.24			0.15±0.03
		0.39	Lee et al. (1985)	
Ag	0.1			0.13±0.04
Au	0.095			0.17±0.05
Be	0.24			0.24±0.05
		0.42	Zhernov and Kulagina (1992)	
Mg	0.28			0.35±0.05
		0.13	Zhernov and Kulagina (1992)	
Zn	0.57			0.37±0.05
		0.4	Zhernov and Kulagina (1992)	
Cd	0.44			0.40±0.05
Al	0.45			0.43±0.05
		0.42	Zhernov and Kulagina (1992)	
Ga	0.31			0.97±0.05
In	0.85			0.80±0.15
Tl	0.78			0.80±0.05
Sn	0.71			0.72±0.05
Pb	1.7			1.55±0.05

5.1.2 Noble metals

Copper is the most comprehensively investigated by PCS metal. It plays a role like aluminum in the conventional tunnel spectroscopy. This is connected with the fact that Cu point contacts independently of the preparation method every time have spectra with more or less distinct phonon features and a relatively low background level. Even if one takes simple copper electrical wires, removes the isolation, and touches them to each other at helium temperature, the main TA phonon maximum, more or less distinct, will always be present in d^2V/dI^2. The spectra of Cu and other noble metals Ag and Au are shown in Fig. 5.4. The spectra look similarly and contain a broad maximum

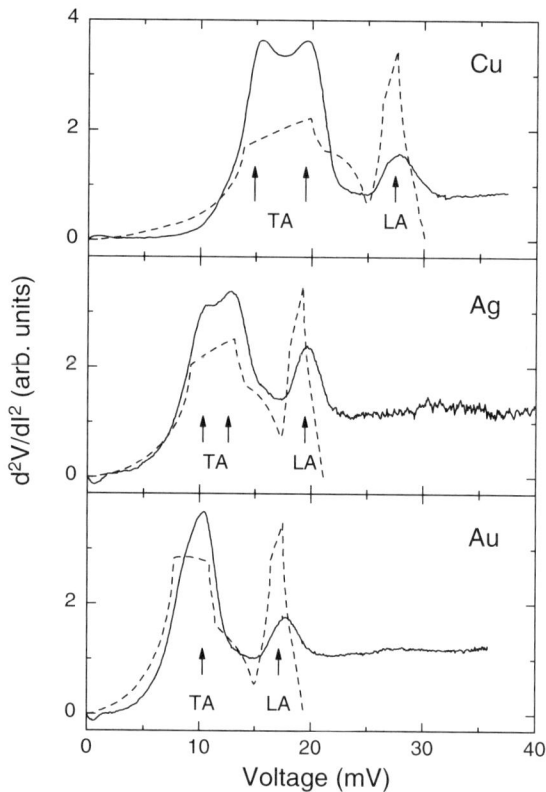

Fig. 5.4. Point-contact spectra of Cu, Ag, and Au (solid curves) measured at 1.6 K. The arrows show the position of the transverse (TA) and longitudinal acoustic (LA) phonon peaks. The phonon density of states $F(\omega)$ (dashed curves) measured by Lynn et al. (1973) are shown.

(which often looks for Cu and Ag like a double maximum) at the energies of TA phonons and less intensive maximum at higher energies corresponding to the LA phonons. The position of the maxima come out right with the maxima in the calculated phonon density of states (Fig. 5.4), but their relative intensities are different. As was shown by Akimenko et al. (1982a), the LA peak amplitude distinctly diminishes with decreasing of the mean free path of electrons in the contacts. The LA peak intensity is maximal for clean contacts; therefore the relative intensity of LA versus TA peak (LA/TA) can be used as a direct test to distinguish between regimes of the current flow in the constriction. This ratio between LA and TA peaks (provided that background is subtracted) in the d^2V/dI^2 spectra for clean contacts is maximal for Ag^2 (LA/TA$\simeq$0.47). Exactly for Ag deviation of the Fermi surface from the sphere and correspondingly the size of the neck in the [111] direction is minimal compared with other noble metals Cu and Au [Cracknell and Wong (1973)]. Thus, the trend in the decrease of the ratio LA/TA as the Fermi surface approaches the boundary of the Brillouin zone, mentioned earlier for the alkali metals, is even more distinct for noble metals, the Fermi surface of which intersects the boundary of the Brillouin zone. Just as mentioned above, the normal scattering processes for TA phonons in metal with a closed Fermi surface are depressed, and the opening of the Fermi surface for some directions invokes the Umklapp scattering, which significantly increases interaction between electrons and TA phonons. We should note that Umklapp scattering is mainly a large-angle scattering process emphasized in the case of point contact because of the effect of the geometrical form-factor (see Fig. 3.4). As follows from Table 3.1, Umklapp processes result in $V^{5/2}$ dependence of the initial part of the point-contact spectra. This is observed for point-contact spectra of Cu [Akimenko et al. (1982a)] and Ag, Au [Naidyuk et al. (1982b)].

Calculation of the point-contact EPI function of Cu by Lee et al. (1985) is in good agreement with the experimental data (see Fig. 5.5). The authors have shown that the predominance of the TA phonon maximum relative to the LA one can be explained by the energy and polarization dependence of the large-angle scattering. They argued that the sensitivity of the point-contact spectrum to the magnitude of the matrix element involved in large-angle scattering provides the possibility of determining the form factor of the electron–ion interaction at large momentum transfer. Lee et al. (1985) elicited a fact that Umklapp large-angle scattering processes prevail in the PC spectrum of Cu in the whole energy range. At the same time, it is interesting to note that for the Eliashberg EPI function, calculated by the same authors, LA peak for Cu is more intensive compared with the TA one, whereas the calculated EPI constant $\lambda = 0.15$ is much lower than $\lambda_{PC} = 0.39$.

---

[2] Just this spectrum is presented in Fig. 5.4.

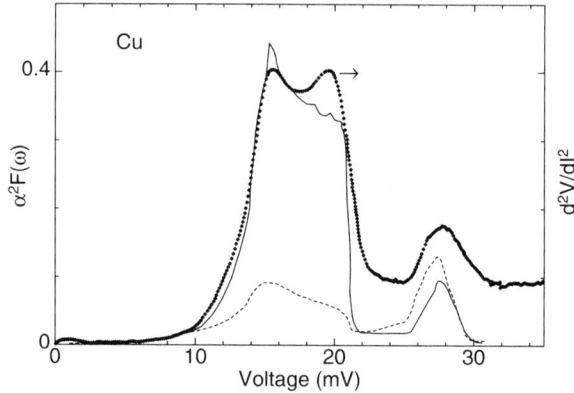

Fig. 5.5. Point-contact (solid) and thermodynamic (dashed) EPI functions for copper calculated by Lee et al. (1985). The symbols depict point-contact spectrum of Cu from Fig. 5.4.

5.1.3 Bivalent metals

Studied by PCS, bivalent metals Be, Mg, Zn, and Cd crystallize in the HCP structure. Although the c/a ratio for Mg is close to the ideal for HCP structure value 8/3, the lattice of other elements becomes distorted in such a way that the c/a ratio for Be is about 10% smaller and it is 15% higher for Zn and Cd than the ideal. Figures 5.6 and 5.7 show the point-contact spectra of discussed metals. It is pertinent to note that Be has the highest Debye temperature among the metals, so that the phonon structure in d^2V/dI^2 vanishes only close to the energy of 100 meV. The comparison of the point-contact spectra of Be and Mg with the phonon density of states, obtained by neutron experiments (Fig. 5.6), shows that both dependencies are more similar compared with that in the alkali or noble metals. Namely, along with the coincidence of the main peak position, also the relative intensity of the low-energy part of point-contact spectra compared with the high-energy part is nearly the same as in the phonon density of states. The Eliashberg EPI function for Mg obtained by proximity tunneling effect reveals virtually the same peculiarities as the point-contact spectrum. However, the high-energy peak intensity is larger and its position is lower by 2 mV compared with the point-contact and the neutron data (Fig. 5.6). This discrepancy in the peak position is, apparently, because in the tunnel experiments, Mg was in the form of thin film.

In the case of the mentioned HCP metals, as it is seen in Figs. 5.6 and 5.7, the neutron scattering data yield insufficient resolution of the phonon features (maxima) in the spectra. In such a manner, point-contact data permit refining the position of singularities in $F(\omega)$ and therefore can be useful in selection of

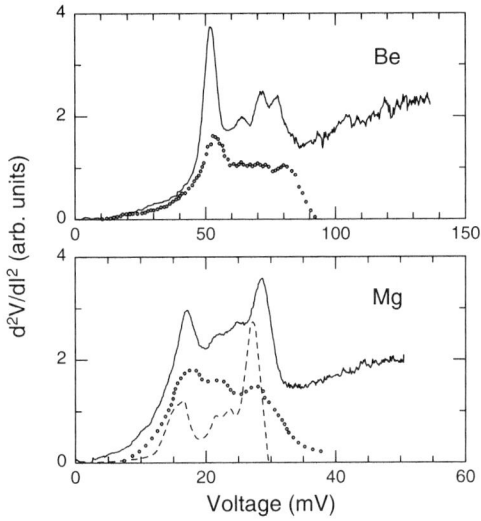

Fig. 5.6. Point-contact spectra (solid curves) of Be at 4.2 K [data taken from Naidyuk and Shklyarevskii (1982)] and Mg at 1.7 K [data taken from Naidyuk et al. (1981)]. The phonon density of states $F(\omega)$ determined by inelastic coherent neutron scattering on a polycrystal by Bulat (1979) for Be and by Eremeev et al. (1976) for Mg are shown by symbols. Dashed curve represents Eliashberg EPI function for Mg obtained from an Ag-MgO-MgNb proximity junction with 880-Å-thick Mg layer by Burnell and Wolf (1982).

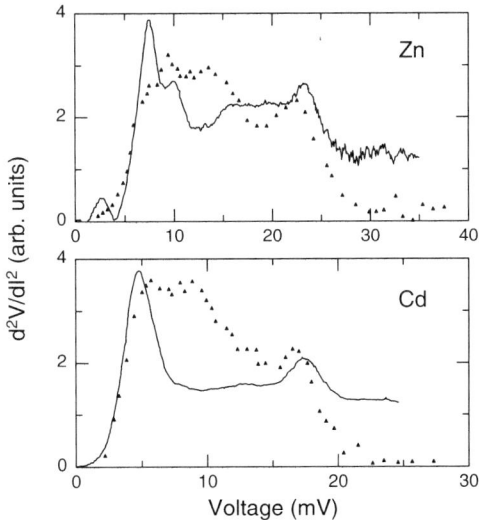

Fig. 5.7. Point-contact spectra of Zn and Cd (solid curves) at 1.5 K after Khotkevich and Yanson (1995). The phonon density of states $F(\omega)$ determined by inelastic coherent neutron scattering on polycrystals by Eremeev et al. (1976) are shown by symbols.

appropriate model of the phonon density of states calculation, which describes a lattice dynamic of the real metal more accurately.

Deviation of the c/a ratio from the ideal one for the HCP structure leads to a strong anisotropy of the metal properties as measured along the hexagonal c axis and for the perpendicular direction. Indeed, the measurements on single crystalline Zn samples along the hexagonal axis and in the base plane reveal remarkable different d^2V/dI^2 curves (see Fig. 5.8). This is also

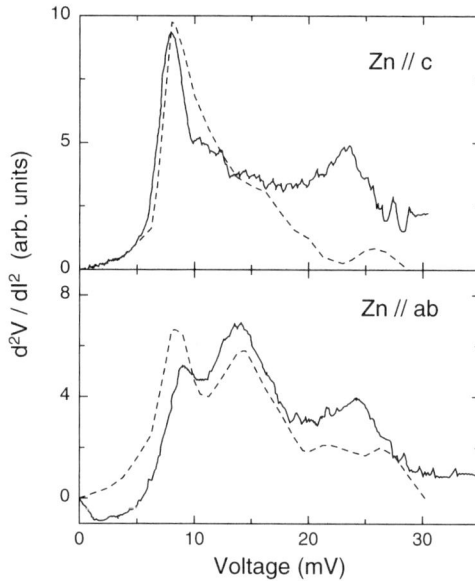

Fig. 5.8. Point-contact spectra of Zn single crystal measured along the hexagonal c-axis and in the perpendicular direction at 1.5 K (data taken from Yanson and Batrak (1979)). The dashed curves represent $g_{PC}(\omega)$ calculated by Zhernov and Kulagina (1992). The intensity and x-axis for calculated curves were scaled to better fit the experimental curve.

in agreement with the theoretical calculation of the point-contact EPI function by Zhernov and Kulagina (1992). The point-contact EPI parameter λ for Zn and Cd agrees well with the literature data (see Table 5.1.1). As mentioned by Zhernov and Kulagina (1992) because of anisotropy, the λ value for Zn can differ by a factor of two, whereas calculated anisotropy in λ is 10–20% for the other HCP metals like Mg and Be and is much lower (5–7%) in the case of alkali metals and Al.

5.1.4 Magnetically ordered metals

As a consequence of the magnetic order, new kind of quasiparticles, in addition to the phonons, appear in the solid connected with a spin wave excitation

– magnons. The latter will be discussed in Section 5.2. Here we talk about point-contact study of EPI for the well-known ferromagnetic metals Fe, Co, and Ni and antiferromagnetically ordered Cr and Mn. As the Curie temperature of the ferromagnetic metals is much higher than the Debye one, characteristic phonon and magnon features are believed to be well separated on the energy scale. Hence, for these metals, the phonon structure in $\mathrm{d}^2V/\mathrm{d}I^2$ curves is expected to be observed in the typical range of a few tens of millivolt without admixture of magnons. Figure 5.9 shows the point-contact spectra

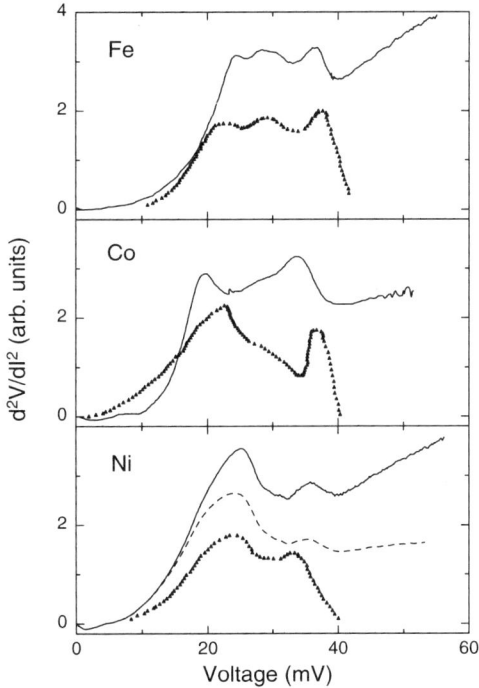

Fig. 5.9. Point-contact spectra (solid curves) of Fe, Ni (data taken from Lysykh et al. (1980a)) and Co (data taken from Gribov (1984)) measured at helium temperatures. The phonon density of states $F(\omega)$ (symbols) is determined by inelastic coherent neutron scattering on a polycrystal by Gorbachev et al. (1976) for Fe, by Gorbachev et al. (1973) for Ni and calculated by Rao and Ramanand (1979) for Co. Dashed curve is - $\mathrm{d}^2I/\mathrm{d}V^2$ for Ni recovered from $\mathrm{d}^2V/\mathrm{d}I^2$ dependence.

of Fe, Co, and Ni. In spite of the difference in a crystal structure, BCC, HCP, and FCC for Fe, Co and Ni, respectively, these metals have almost the same Debye energy [Kittel (1986)] and their phonon features in the spectra vanish above 40 mV. Additionally, the point-contact spectra of Fe and Ni are similar to the density of the phonon states measured by neutron scattering.

The remarkable difference from calculated $F(\omega)$ is seen for Co, where phonon spectrum is shifted to the higher energy. The common feature in the spectra of ferromagnetic metals is a high value of the background ($\gamma > 0.5$) in d^2V/dI^2 and its nearly linear high-slope increase above 40 mV. This increase diminishing for the -d^2I/dV^2 dependence (see Fig. 5.9, dashed curve), corresponds to the high nonlinearity of the $I - V$ characteristic so that differential resistance of contact in the range of the Debye energy rises sufficiently ($>10\%$). The high value of the background was found to be independent on the purity of the electrodes used, and it is apparently because of the scattering of electrons by spin waves. It will be shown in Chapter 7 that the transition into the thermal regime at higher energy for the contacts with Fe, Co, and Ni occurs with a sharp anomaly at the voltage corresponding to the Curie temperature. As likely as not, that temperature in the point contact may increase already in the region of the phonon structure. This can be a reason for additional smearing and intensity decreasing of the high-energy peculiarities (e. g. LA peak) in the spectra. Considering the fact that dV/dI of these metals increases more than 10%, measured d^2V/dI^2 curves remarkably deviate from the theoretically exploited d^2I/dV^2 dependencies (see curves for Ni in Fig. 5.9). This difference between d^2V/dI^2 and d^2I/dV^2 always should be kept in mind, especially for the spectra with large nonlinearity.

Cr has about two times lower Néel temperature compared with the Curie temperature of the mentioned ferromagnetic metals. The thermal effects for Cr contacts were difficult to overcome; nevertheless, Khotkevich and Krainyukov (1991) succeeded in reproducing the basic features of the Cr EPI function [Fig. 5.10(a)]. They found the main maximum at 25 mV and a gentle maximum at 38–40 mV, which correlate with the phonon density of states. The typical feature of d^2V/dI^2 for Cr contacts is a negative zero-bias anomaly corresponding to a zero-bias maximum in the differential resistance dV/dI [Fig. 5.10(b)]. The latter was interpreted by Meekes (1988), exploiting the presence of the spin-density wave gap in the electron spectra of Cr below antiferromagnetic transition.

Mn has the lower Néel temperature compared with Cr and the highest resistivity both among the metals and even amid the semimetals [Kittel (1986)]. Up to now, there has been no success in obtaining of Mn contacts close to the ballistic or diffusive regime to establish EPI function. Nonlinear behavior of the conductivity of the voltage-biased Mn contacts is mainly because of the heating effects, analogously like for the ferromagnetic metals Fe, Co, and Ni at higher biases.

5.1.5 Polyvalent metals

This section outlines the measurements of well-known metals Al, Ga, In, Tl, Sn, and Pb, which undergo transition to the superconducting state at liquid helium temperature. Therefore, for all of these metals (except Ga), the thermodynamic EPI functions have been established from the tunnel

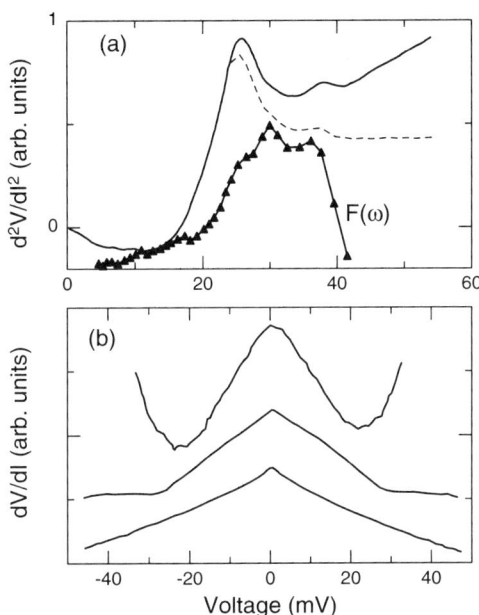

Fig. 5.10. (a) Point-contact spectrum of Cr ($R_0 = 3.1\,\Omega$, $T = 4.2\,\mathrm{K}$). The dashed curve is recalculated $-\mathrm{d}^2 I/\mathrm{d}V^2$. The symbols show neutron data for the phonon density of states $F(\omega)$. Data taken from Khotkevich and Krainyukov (1991). (b) $\mathrm{d}V/\mathrm{d}I$ of a few Cr point contacts with resistance 1.3, 3, and 9 Ω from the top curve to the bottom one. The zero-bias maximum intensity is about a few percents of the zero-bias resistance. Data taken from Meekes (1988).

measurements by the method described in Section 2.4. Figure 5.11 shows point-contact EPI spectra for Al and In. Both spectra have similar behavior because of similarities in their Fermi surface shape and crystal structure, although their Debye energy differs more than by a factor of two. For Al, there is a number of thermodynamic and transport EPI function calculations. It is seen from Fig. 5.11 that the transport EPI function is similar in details to the point-contact one showing the same three well-outlined maxima. Moreover, the shape of the peaks and other singularities is qualitatively similar, and their position in the energy is the same. However, the relative intensity of high-energy peak in point-contact spectrum is appreciably lower (after subtracting of background), which is a common feature observed for many other metals.

Qualitatively, the same behavior is typical for point-contact spectrum of In (Fig. 5.11). Here, the thermodynamic EPI function, experimentally derived from the tunneling measurements, is presented for comparison. Except for the smaller intensity of the high-energy longitudinal peak (after subtracting of the background), for the point-contact spectrum, both curves are equal.

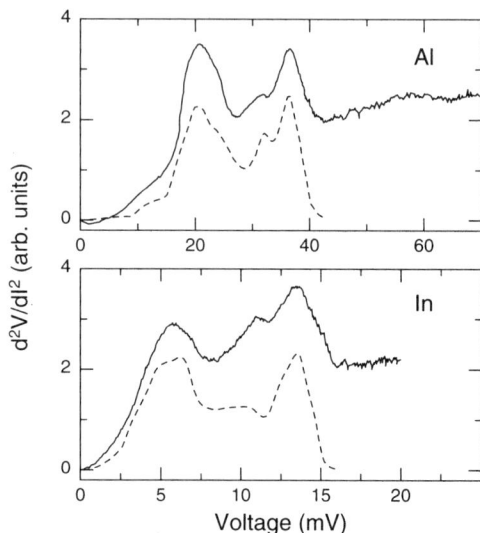

Fig. 5.11. Point-contact spectrum (solid curve) at 1.8 K along with transport EPI function (dashed curve) for Al from Chubov et al. (1982). Bottom panel shows the point-contact spectrum for In (solid curve) at 1.55 K from Khotkevich and Yanson (1995). In was driven in the normal state by a small magnetic field. Dashed curve is the Eliashberg EPI function in In after Dynes (1970).

Additionally, the calculated point-contact EPI constant λ agrees well with the textbook data (see Table 5.1.1).

With decreasing of the lattice symmetry, more complicated spectra are observed for gallium having orthorhombic structure. Typical spectrum of Ga shows four well-defined maxima (Fig. 5.12). Additionally, Sato et al. (1985) observed that for the better quality spectra, asymmetric first and third maxima split into two separate peaks. All of these peaks correspond to the phonon as can be seen by comparison of the spectra with the experimental phonon density of states for Ga. However, the latter dependency has a low resolution with an energy rise. After subtracting of background even a gap at about 25 mV appears between the last two maxima in PC spectrum, which corresponds to the gap between optical and acoustic phonons presented also in the dispersion curves.

Figure 5.12 shows the point-contact spectrum of Tl. As Tl crystallizes in the HCP structure, its spectrum is similar to the spectra of HCP metals, viz. Zn, only with about two times smaller Debye energy. On the other hand, the thermodynamic EPI function for Tl determined from the tunneling measurements is also similar to the point-contact ones.

Lead, tin, and alloys on the base of both metals were among the most intensively studied by tunnel effect metals with a goal to receive their EPI function. These metals are related to superconductors with strong ($\lambda \geq 1$) or

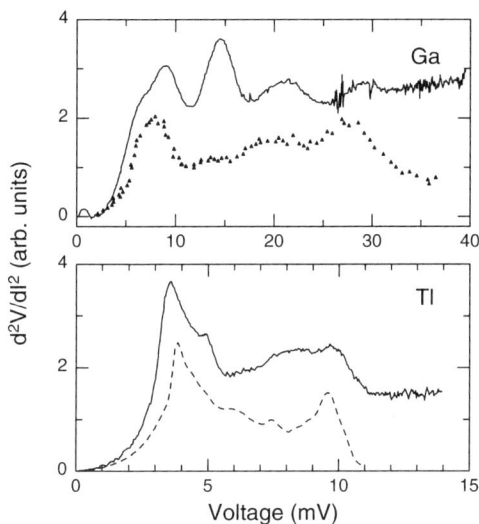

Fig. 5.12. Point-contact spectrum of Ga at 4.2 K from Shklyarevskii et al. (1983) and Tl at 1.5 K from Khotkevich and Yanson (1995). Magnetic field was used in both cases to suppress superconductivity. The phonon density of states $F(\omega)$ (symbols) determined by inelastic coherent neutron scattering on a polycrystal by Bogomolov et al. (1971) is shown for Ga. Dashed curve depicts Eliashberg EPI function for Tl reconstructed from tunneling data after Dynes (1970).

moderate coupling; therefore, the nonlinearity in the tunneling conductance is pronounced to extract the Eliashberg (or in other words thermodynamic) EPI function $\alpha^2 F(\omega)$ reliably. These metals were the first studied by PCS using short-circuit tunnel junctions with metallic holes. This method allows for preparing high ohmic (above a few hundreds Ω) stable contacts showing $d^2V/dI^2(V)$ spectra with a very low background (see Fig. 2.3). However, phonon peaks are appreciably broadened for high ohmic contacts. This broadening, apparently, is caused by the quantum diffraction effects considering small size, of about 1 nm diameter, of such contacts. With the contact resistance decrease, the phonon structure becomes sharper (see Fig. 2.3); however, enlargement of the background manifests increasing of the nonequilibrium phonons in the point-contact region. At the further lowering of the contact resistance, progressive broadening of the spectra along with increasing of the background signal occurs, what testifies most likely to a temperature increase in the constriction.

It should be noted that the relative intensity of the peaks in the lead point-contact spectrum is virtually independent of the degree of the contact quality, whereas for many other metals, the intensity of the high-energy peaks is usually more sensitive as to the method of the contact preparation compared with the low-energy peaks. This correlates with the stability of the superconducting critical temperature for Pb in regard to the crystal lattice perturbation. In general, the EPI functions for lead received by carrying out point-contact and tunnel measurements practically coincide (Fig. 5.13) if one includes a small additional broadening for the high-energy LA peak, e. g., because of the temperature increase on 1.5-2 degrees [Khotkevich and Yanson 1981)].

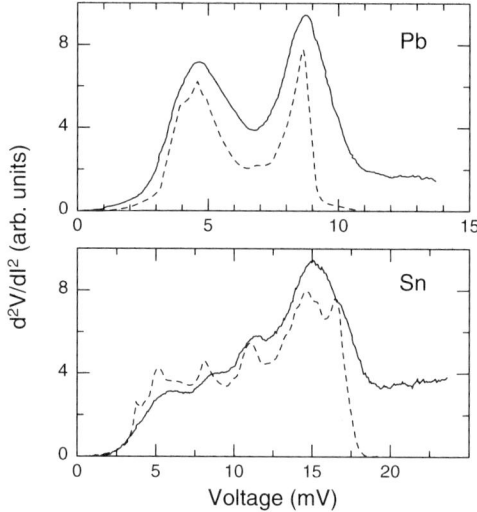

Fig. 5.13. Point-contact spectra in Pb and Sn at about 2 K from Khotkevich and Yanson (1995). Both metals were driven in the normal state by a small magnetic field. The Eliashberg EPI functions (dashed curves) are derived from the tunnel data by Rowell et al. (1969) for Pb and by Rowell et al. (1971) for Sn.

Tin has a tetragonal lattice with two atoms in the elementary cell and correspondingly a number of peaks in the spectrum along with a sufficient anisotropy (Fig. 5.13). The observed "softening" of the spectrum along the [001] direction correlates with the higher value of the reduced gap for this direction compared with the [110] one. The point-contact and tunnel results for the EPI function are similar. Only the peculiarities in the point-contact spectrum are more broad because of the temperature and alternating signal smearing. The small deviation in the peak position between the mentioned dependencies is connected, likely, with anisotropy effects. The calculated EPI constants λ for tin and lead well correspond to the textbook data (see Table 5.1.1).

Khotkevich et al. (1990) studied spectra of tin with twinning planes produced by the sample extension at liquid nitrogen temperatures. The separation between the twin planes was estimated to be 1–10 μm, what is much larger than the contact diameter. Along with typical point-contact spectra of tin, a nontypical spectrum with a considerable softening in the low-energy part in some cases was observed. After subtracting of the background, the intensity of the first low-energy peak in this spectrum was higher than the last one at higher energy; that is, the relation between intensity of these peaks is opposite to the typical spectra. The authors speculated that the phonons localized at the twin boundary are responsible for the softening of the spectra. Such explanation is in line with the observed increase of the critical temperature for twinned tin samples (see Khlyustikov and Buzdin (1988) for review of superconductivity at twinning plane).

5.1.6 Transition metals

Transition metals are the most interesting from the point of view of the metal physics. Their specific properties determined by partially filled d shell vary widely from superconducting to magnetic. In the simplified model of the transition metals band structure, the valence electrons are often divided into two groups, so-called s- and d-electrons. The former constitutes an s band with nearly free electron properties, and the latter one forms a d band with complicated Fermi surface and the lower Fermi velocity. The s-band electrons are believed to play a dominate role in the basic transport properties of the transition metals, whereas the electron scattering both inside s band and s-d band exchange scattering contribute to the fine features in the electron transport. It is different from the free electron model, and therefore, the theoretical description of the transition metals is more complicated. In this case, the experimental information plays an important role for the better understanding of the d-metals properties, especially if this information is spectral that shows the characteristic energies of quasiparticles like phonons, magnons and so on or peculiarities of the electron spectra. Thereby, PCS is the most suitable method to receive such information and approve theoretical approaches for these metals, having regard that experimental EPI functions for the transition metals are limited only by a few superconductors V, Nb, and Ta with a relative large electron–phonon coupling or T_c. The point-contact spectra of these metals along with the EPI Eliashberg function extracted from the tunnel measurements are shown in Fig. 5.14. All mentioned metals crystallized in the BCC lattice; therefore, their phonon densities of states $F(\omega)$ have to be similar. As shown by neutron scattering measurements and calculations cited by Khotkevich and Yanson (1995), $F(\omega)$ consists mainly of two expressed low- and high-energy peaks. This is characteristic also for the Eliashberg EPI function measured by the tunnel technique (Fig. 5.14). Additionally, for V and Nb, a low-energy shoulder in the EPI function is seen and for Ta, a small maximum is resolved between the low- and high-energy peaks. Unfortunately, the point-contact spectra for V and Nb do not exhibit much detail. The phonon maxima are very broad. Supposedly, the oxide layers on the surface of V and Nb do not possess good dielectric and mechanical characteristics required for obtaining high-quality metallic shorts.

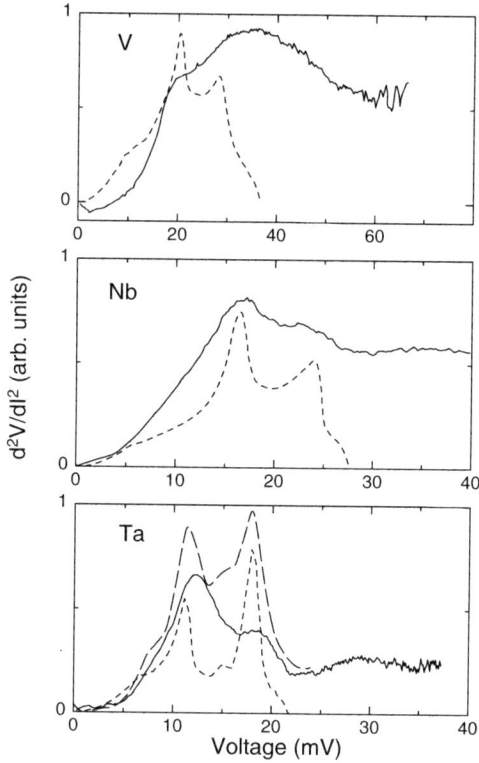

Fig. 5.14. Point-contact spectrum (solid curves) at 4.2 K of V [data taken from Rybaltchenko et al. (1980)], Nb [data taken from Yanson et al. (1983)] and Ta [data taken from Rybaltchenko et al. (1981)]. The weak broad maximum around 30 mV for Ta corresponds to two phonon processes (see Section 5.6). Long-dash curve in the Ta panel is spectrum of Ta-Cu contact at 1.88 K and 0.3 T from Bobrov et al. (1987) with negligible Cu contribution (see text for explanation). Magnetic field was used to drive metals in the normal state. Dashed curves represents EPI Eliashberg functions extracted from the tunnel measurements by Zasadzinski et al. (1982) for V, by Wolf et al. (1980) for Nb, and by Shen (1970) for Ta.

For example, niobium oxide NbO has metallic properties that make it difficult to separate the contribution of a clean metal from the conducting oxide in the measured d^2V/dI^2 curves. It may be assumed that the wide maximum at about 40 mV in the spectra for V is partially from the mentioned oxide layer as well.

Contrary to V and Nb, d^2V/dI^2 curves for Ta contain distinct maxima. It is interesting that the more detailed structure for the Ta spectrum was received using Ta-Cu heterocontacts. In the latter case, it turned out that the point-contact EPI function has the shape more closer to the Eliashberg EPI function (see Fig. 5.14). In this instance as shown by Bobrov et al. (1987), contribution of Cu for a geometrically symmetric contact is not more as 10%. It should be mentioned that the calculations of the point-contact EPI function for Ta by Al-Lehaibi et al. (1987) give a high-energy longitudinal peak about two times higher and broader than the low-energy transverse one contrary to the point-contact data, pointing to the difficulties in the theoretical describing of the transition metals. However, more recent calculations by Savrasov and Savrasov (1996) are already in good agreement with the EPI Eliashberg function for Ta given by Shen (1970).

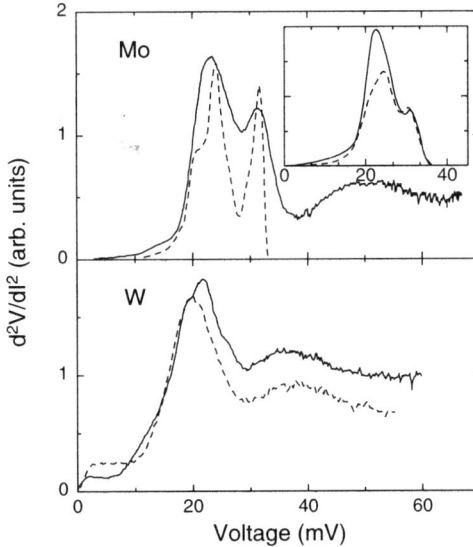

Fig. 5.15. Point-contact spectra of Mo and W (solid curves) at 4.2 K from Rybaltchenko et al. (1981). The broad maximum above 40 mV for Mo and 30 mV for W corresponds to two phonon processes (see Section 5.6). Dashed curve in the upper panel shows the calculated by Savrasov and Savrasov (1996) EPI function for Mo. Dashed curve in the bottom panel represents spectrum of W measured by Gribov et al. (1996) using lithography method. Inset: comparison of the point-contact EPI function for Mo (solid curve) with the calculated Eliashberg function (dashed curve) smeared by value (4.3) accounting for the temperature increase according to (5.1). Data taken from Caro et al. (1981).

Mo and W belong to the VI group elements with relative low value of the superconducting transition temperature (T_c <1 K). Both metals have a BCC crystal structure. The phonon density of states for these metals displays, like in the previous case of BCC metals, two main peaks corresponding to the transverse and longitudinal phonons. For Mo, these two peaks are resolved in the d^2V/dI^2 curves, and for W, the transverse peak looks like one broad peak with a small splitting. The longitudinal phonons in W manifest themselves as a gentle shoulder at about 26 mV (Fig. 5.15). The W spectrum received by nanofabrication technique (see Section 4.1) exhibits the similar structure [Gribov et al. (1996)]. For both spectra, a broad maximum is seen above the Debye energy at 50 mV for Mo and 35 mV for W, which are caused by the double-phonon processes (see Section 5.6).

Caro et al. (1981) compared the measured point-contact EPI function $g_{\mathrm{PC}}(\omega)$ for Mo and W with the calculated thermodynamic EPI functions. To receive a better agreement, they included an additional smearing according to (4.3) for the calculated EPI function supposing a temperature increase in the contact with a voltage bias similar to the thermal regime:

$$T_{\mathrm{PC}} = (T_{\mathrm{bath}}^2 + AV^2)^{1/2}. \tag{5.1}$$

The best consent between the broadened EPI function and the point-contact experimental curve was obtained for the A value being in the range 0.04–0.08 (here voltage is indicated in millivolts and temperature in Kelvin). For these A values, the temperature at the high-energy LA phonon peak could be as high as 8 K, although the bath temperature was only 1.5 K. Apparently analogous phenomenon is the reason of broadening of the high-energy peaks for many other metals. We have already mentioned this for Pb and ferromagnetic metals.

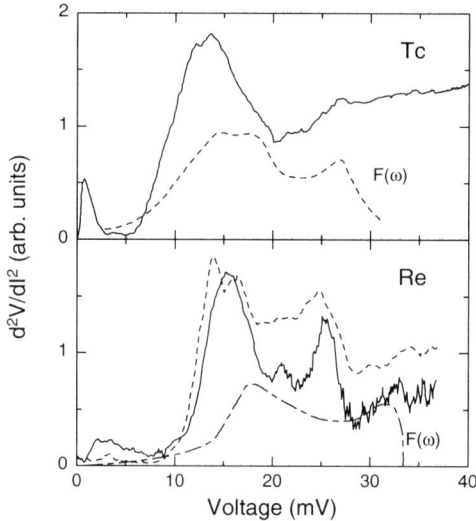

Fig. 5.16. Point-contact spectrum (solid curve) and phonon density of states (dashed curve) of Tc at $T = 4.2$ K and $B = 4.5$ T from Zakharov et al. (1984). For Re, the point-contact spectra at 2.2 K for two main crystallographic directions along (solid curve) and perpendicular (dashed curve) to the hexagonal axis are shown [data taken from Tulina (1982)]. Bottom dash-dotted curve represents phonon density of states in Re calculated by Rao and Ramanand (1977).

The EPI function for the above-mentioned five transition metals was also calculated by Zhernov and Kulagina (1992). They used the simple model of EPI in these metals supposing that a matrix element of EPI varies slightly over the Fermi surface and that tightly bound electrons in the one-band approximation interact with the phonons. For definition of the phonon modes, the neutron scattering data were used. In general, the results of calculations are similar in shape for all metals. Spectra possess transverse and longitudinal maxima more or less equal in the amplitude (the transverse maximum is broader than the longitudinal one in the case of V and Nb). Surprisingly, calculated by Zhernov and Kulagina (1992), the Eliashberg, transport, and point-contact EPI functions are almost the same in shape. The authors also mentioned that the theoretical description of the experimental data for the point-contact EPI function is unsatisfactory except in the case of W.

The next two superconducting metals, namely, Tc and Re from the VII group, both have HCP crystal lattice. Although for Re a rich structure in d^2V/dI^2 is measured with anisotropy typical for a metal with hexagonal lattice, the Tc spectra contain one broad pronounced peak along with a gentle maxima at higher energy (Fig. 5.16). The position of these peculiarities correspond more or less to the phonon maxima in Tc measured by neutron scattering. The calculated $F(\omega)$ for Re poorly agree with the point-contact spectrum, especially in the high-energy region.

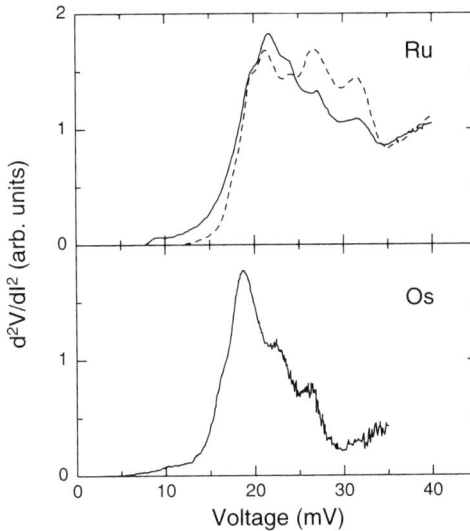

Fig. 5.17. Point-contact spectra of Ru [data taken from Khotkevich (1996)] and Os [data taken from Khotkevich et al. (1984)] at helium temperatures. The spectra for two main crystallographic directions along (solid curve) and perpendicular (dashed curve) to the hexagonal axis are shown for Ru.

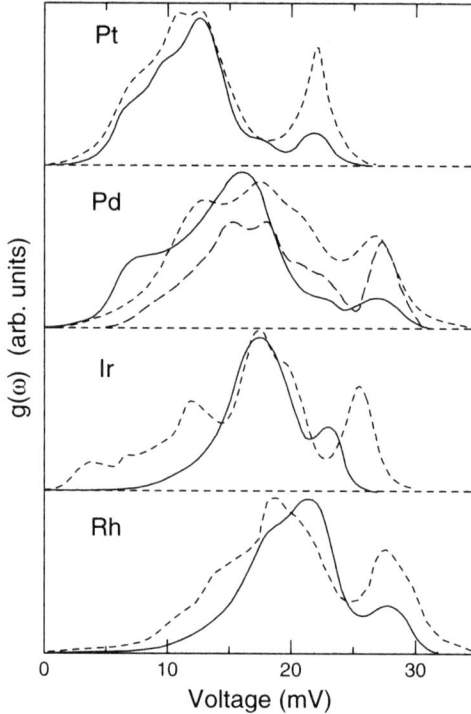

Fig. 5.18. Point-contact EPI functions for Pt, Pd, Ir, and Rh [solid curves, data taken from Khotkevich (1996)]. Dashed curves represent calculated point-contact EPI functions for Pt [data taken from Krainyukov et al. (1988)], Pd [data taken from Antonov et al. (1991)], Ir [data taken from Krainyukov et al. (1989)] and Rh [data taken from Zhalko-Titarenko et al. (1989)]. For Pd, the calculation of the Eliashberg EPI function by Savrasov and Savrasov (1996) is shown by a long-dash curve for comparison.

Now let us consider the six platinum metals from the VIII group of the periodic table of elements. Four of them, namely, Rh, Pd, Ir, and Pt, crystallize in the FCC lattice, whereas Ru and Os possess the HCP one. The latter metals reveal d^2V/dI^2 curves with a few maxima showing an anisotropy between the axial and basal plane orientation (Fig. 5.17). The spectra of these two metals are similar. The common feature of the point-contact spectra for FCC metals (Rh, Pd, Ir, and Pt) is the presence of two main maxima. The intensity of the high-energy peak is suppressed compared with the low-energy one (Fig. 5.18). At the same time, there is a difference between the spectra. Thus, the spectrum of Pd has a pronounced shoulder at lower energy with respect to the main peak. Pt has two, Rh has one, and Ir has no shoulders. Additionally, a gentle maximum is resolved in Pt and Pd between two main peaks. Peculiarities of the phonon dispersion curves of these metals is responsible for the discussed fine structure on the spectra as mentioned by Khotkevich (1996). The calculations of the point-contact EPI function for Pt (see Fig. 5.18) are in good agreement with experimental data, whereas for the other platinum metals, accordance is satisfactory.

Figure 5.19 shows the main steps in the transformation of the experimental point-contact spectrum to the point-contact EPI function. The first step is the transformation of the d^2V/dI^2 curve into $-d^2I/dV^2$. These two derivatives differ remarkably for contacts with a large ($>10\%$) differential resistance (dV/dI) increase as was shown above for the ferromagnetic metals. The next step is the background subtracting. A widely exploited empirical formula for the background signal is presented by (3.17). The theoretical calculation of the background caused by nonequilibrium phonons yields a different expression (3.16), which produces spectral features in the background signal (compare curves 3 and 4 in Fig. 5.19). However, the subtracting of the background by the different methods does not change remarkably the shape of the EPI function mainly modifying the absolute value of λ_{PC}.

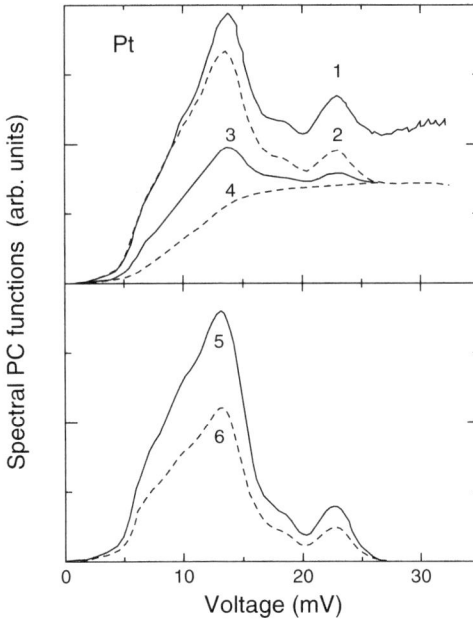

Fig. 5.19. Point-contact spectrum d^2V/dI^2 for Pt (curve 1) at $1.5\,\mathrm{K}$. Curve 2 is the $-d^2I/dV^2 \equiv \frac{d^2V}{dI^2}(\frac{dV}{dI})^{-3}$ for the same contact. Curves 3 and 4 show calculated background according to (3.17) and (3.16) correspondingly. Curves 5 and 6 display EPI spectra after subtracting of background represented by curves 3 and 4. Data taken from Krainyukov et al. (1988).

5.2 Electron–magnon interaction

Another type of quasiparticle excitations in magnetically ordered systems is magnons or spin waves. Phonons or lattice vibrations represent oscillations of the lattice ions relative position, and magnons or spin waves are oscillations of the relative orientation of the ion spins. By analogy with the phonons, a magnon density of states is used for characterization of the magnetic state. Therefore, it is almost obvious that point-contact spectra should also reflect a magnon density of states for magnetically ordered metals, shown theoretically by Kulik and Shekhter (1980). In the case of ferromagnetic and antiferromagnetic d-metals discussed above, the Curie or Néel temperature exceeds the Debye one, what results in characteristic magnon energy much higher than the phonon one. Thereby with the voltage increase, EPI leads at first to the shortening of the mean free path in the constriction and then the electron–magnon scattering, additionally, forces transition to the thermal regime. For the latter case, no spectral features are seen except singularities of the thermal origin, which correspond to transition from the magnetically ordered state to the nonmagnetic one (see Fig. 7.7). Hence, to observe electron–magnon interaction spectra, the characteristic energy of magnon or Curie (Néel) temperature is desirable to be lower than the Debye one. This takes place mainly for the magnetically ordered rare-earth metals from the lanthanum group.

The first measurements of Gd, Ho, and Tb by Akimenko and Yanson (1980) yielded point-contact spectra with singularities interpreted as a manifestation of the energy-dependent electron–magnon interaction. Figure 5.20(a) depicts the d^2V/dI^2 curve for Gd. As the phonon spectrum of Gd vanishes around 13–15 mV, a part of the curve at higher voltage between 15 and 25 mV was attributed to the interaction of electrons only with the magnons. Indeed, as derived from neutron-diffraction data, the dispersion curves of magnons show [Fig. 5.20(b)] that the maximal energy of the spin waves is about 25 meV. The features in the point-contact spectrum at 14 and 20 mV correspond well to the Van-Hove singularities $\partial\omega/\partial q = 0$ in the dispersion curves for the magnons. Appreciable phonon features in the spectrum were not observed even below 15 mV. Additionally, an initial part of the d^2V/dI^2 curve is linear in accordance with the theoretical prediction by Kulik and Shekhter (1980) in the case of the electron–magnon interaction. All of this indicates that the electron–magnon interaction in Gd is much stronger than EPI and the point-contact spectra reflect the density of magnon states for Gd.

The inset in Fig. 5.20 displays that d^2V/dI^2 first rises and then saturates and finally falls above 100 mV for the further increase of the voltage. The latter is connected with the heating of contact to the Curie temperature (T_C= 292 K) like in the case of Fe, Co, and Ni (see Fig. 7.7), indicating that the contact region is indeed in the magnetically ordered state at biases lower than 100 mV.

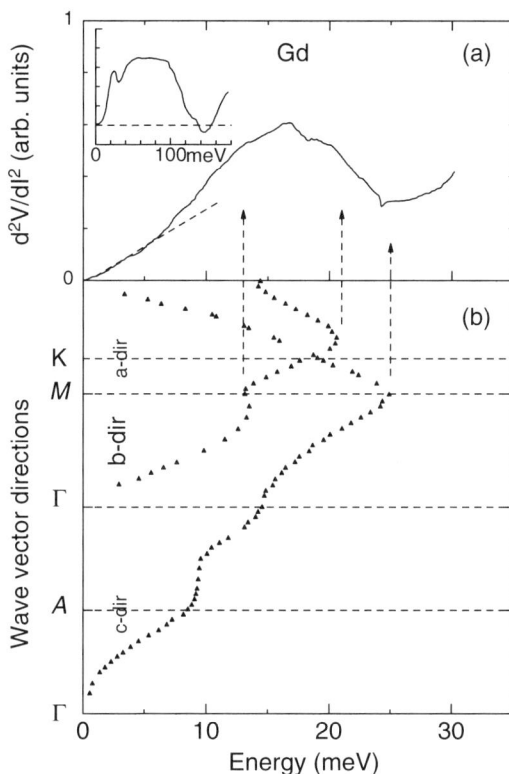

Fig. 5.20. (a) Point-contact spectrum of Gd at $4.2\,\mathrm{K}$ ($R_0 = 35\,\Omega$). Inset: the same spectrum at higher voltages with thermal transition feature around $140\,\mathrm{meV}$. Data taken from Akimenko and Yanson (1980). (b) Dispersion curves of magnons received by neutron scattering experiment by Koehler et al. (1970). Arrows indicate position of magnons with zero group velocity $\partial\omega/\partial q = 0$ resulting in peculiarities in the spectra.

Figure 5.21 illustrates point-contact measurements for another rare-earth metal Tb. Functions of the phonon and magnon density of states for this metal are also shown. The intensive maximum in the point-contact spectrum at about $11\,\mathrm{mV}$ coincides with the main peak in the magnon density of states. The gentle shoulders at $6.5\,\mathrm{mV}$ and $13\,\mathrm{mV}$, apparently, correspond to the phonons. Above $13\,\mathrm{mV}$, only the background without any spectral features is seen. Thereby, the $\mathrm{d}^2V/\mathrm{d}I^2$ curve reflects both phonon and magnon spectra for Tb.

Akimenko et al. (1982b) measured $\mathrm{d}^2V/\mathrm{d}I^2$ for contacts with Ho and Dy as well. The Ho spectra demonstrate pronounced negative zero-bias anomaly at low voltages with a small maximum at about $4\,\mathrm{mV}$. The position of this maximum corresponds to the singularities $\partial\omega/\partial q = 0$ in the dispersion curves

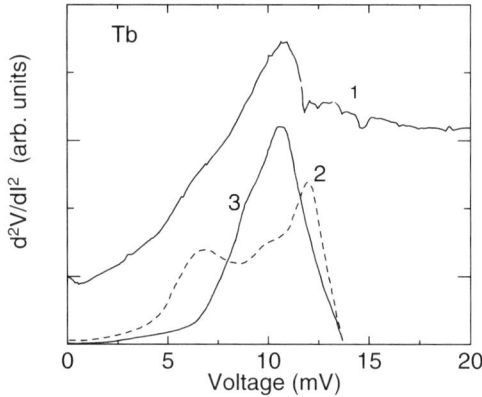

Fig. 5.21. Point-contact spectrum of Tb at 1.9 K (curve 1, $R_0 = 40\ \Omega$) along with the density of states of phonons (curve 2) and magnons (curve 3) both broadened by the experimental resolution. Data taken from Akimenko et al. (1982b).

for magnons. Inasmuch as the first phonon peak in the density of phonon states is located at 6–7 mV, the mentioned maximum may be related to the electron–magnon interaction.

For other studied rare-earth metals, a characteristic feature in the spectra is the negative minimum close to zero bias as mentioned by Akimenko et al. (1982b). A similar type of anomalies is found for point-contact spectra of the noble metals with a small amount of magnetic impurities like Fe or Mn (see Section 6.1). Therefore, for contacts with magnetic rare-earth metals destruction of the magnetic order, e. g., at the surface, can produce a local magnetic moments and consequently the negative minimum in $\mathrm{d}^2V/\mathrm{d}I^2$ corresponding to the maximum in $\mathrm{d}V/\mathrm{d}I$ at $V = 0$. Another important problem of the rare-earth metals is their high chemical activity. This leads to the thick and tough oxide layer, which hinders creation of pure metallic contacts and contaminates the contact region.

5.3 Alloys and compounds

As was shown in Section 3.4, an energy-resolved spectroscopy is possible even in the case of short elastic mean free path of electrons, however, providing that the inelastic relaxation length is still larger than the contact dimension. The experimental confirmation of the energy-resolved spectroscopy in the diffusive regime was obtained for the first time by Lysykh et al. (1980b) during investigation of CuNi alloys. Figure 5.22 shows point-contact spectra of Cu with a gradually increasing concentration of Ni. The following modification of the spectra takes place while alloying Cu with Ni: (1) the transverse phonon peak of Cu broadens while the longitudinal phonon peak intensity decreases,

(2) both peaks shift to the higher energies, (3) the background rises, (4) the absolute intensity of spectra diminishes (see inset in Fig. 5.22), and (5) the zero-bias anomaly appears. The broadening and shifting of the phonon struc-

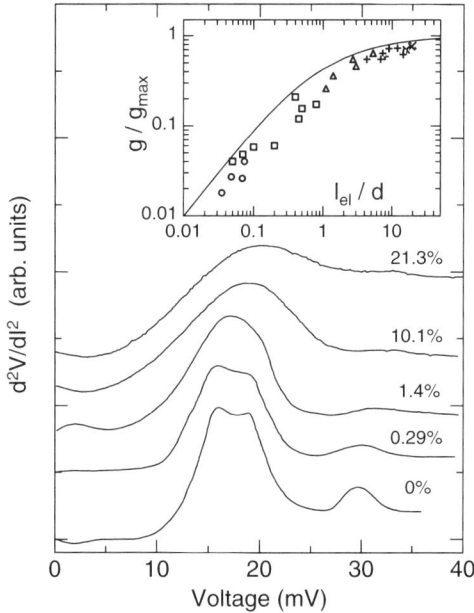

Fig. 5.22. Point-contact spectra of CuNi alloys at T = 4.2 K. The Ni concentration is shown by each curve. Inset: dependence of the absolute intensity of the spectra versus l_{el}/d ratio (symbols are experimental data, curve is from Fig. 3.8). Data taken from Lysykh et al. (1980b).

ture may be caused by inhomogeneity of alloys on the scale roughly equal to the contact diameter and/or by increasing of the Ni phonons contribution having larger characteristic vibration energy. The background rise is apparently caused by the decreasing of the phonon mean free path for a more disordered crystal structure, which can also cause zero-bias anomalies. Diminishing of the absolute intensity of the spectra is expected from the theory of diffusive regime because of K-factor dependence on the elastic mean free path (see Table 3.2 and Fig. 3.8). The experimental data qualitatively behave as theoretical dependence of the K factor for the diffusive regime (see inset Fig. 5.22). The data deviation from the calculated curve and its scattering is likely because of overestimation of the mean free path in the contact as a consequence of using a bulk value of the resistivity, whereas in the constriction owing to defects and additional contamination from the surface dirties, l_{el} is usually smaller than in the bulk.

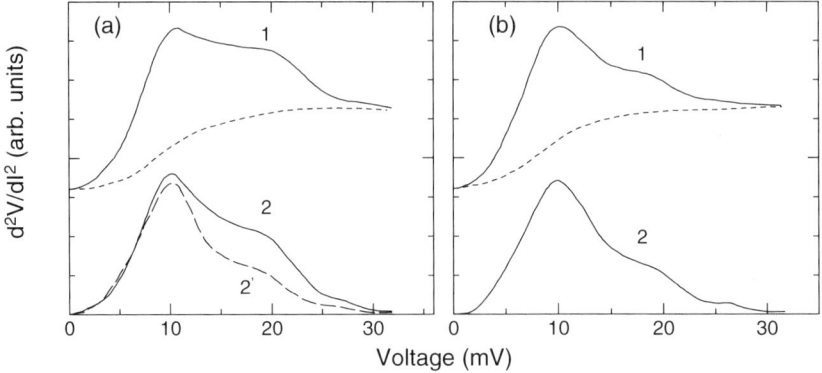

Fig. 5.23. Point-contact spectra (curves 1) of Cu_3Au alloy in the ordered (a) and disordered (b) states at 4.2 K. Curves 2 represent spectra after subtracting of background shown by dashed line for curves 1. Curve 2′ is curve 2 from (b) for comparison. Data taken from Shklyarevskii et al. (1982).

Point-contact spectra of Cu_3Au intermetallic alloy for the ordered and disordered state were investigated by Shklyarevskii et al. (1982). The room temperature resistance to the helium temperature resistance ratio was 1.22 for the disordered (quenched) alloy and about 3 for the ordered (annealed) one. Characteristic features of the spectra are the presence of a broad maximum at 10 mV, a shoulder in the region of 20 mV, and a inexpressive hump slightly above 25 mV (Fig. 5.23). In general, the spectra of ordered and disordered alloys are similar. The main difference is the relative increase of the high-energy part of the spectrum for the ordered alloy (compare curves 2 and 2′). Additionally, the absolute spectrum intensity for the ordered sample is in average two times higher. The latter is caused by the increase of the mean free path and the ratio l_{el}/d for the ordered lattice, correspondingly. The enlargement of the relative intensity of the spectra in the region of 12–25 mV is most probably related to the appearance of a number of additional optic phonon modes for the ordered lattice with more than one atom in the unit cell. The EPI constant λ calculated from the spectra is 0.13 for the ordered alloy, which is between its value for pure Au and Cu (see Table 5.1.1).

The ordered intermetallic compound Cd_3Mg with HCP crystal lattice was investigated by Khotkevich and Khotkevich (1985). The residual resistivity ratio for this compound was about 16 testifying a relatively high ordering. The point-contact spectra are very similar to the spectra of pure Cd (Fig. 5.7). Only the main low-energy peak is slightly shifted upward to 5 mV instead of 4.5 mV for pure Cd and the high-energy peak around 17 mV is broader compared with the same peak for pure Cd. No significant change of the spectra was observed depending on the mutual orientation of electrodes. The absolute value of the EPI function for Cd_3Mg was found between those in pure Cd and Mg with the EPI constant λ about 0.33.

Table 5.2. Studied by point-contact spectroscopy intermetallic compounds and alloys, for which EPI function was obtained.

Compound	Reference
Cd_3Mg	Khotkevich and Khotkevich (1985)
Cu_3Au	Shklyarevskii et al. (1982)
CuNi	Lysykh et al. (1980b)
$LaAl_3$	Ponomarenko et al. (1992)
LaB_6	Frankowski and Wachter (1982)
	Kunii (1987)
	Samuely et al. (1988)
$LaBe_{13}$	Nowack et al. (1987)
$LaCu_6$	Sato et al. (1985)
	Brück et al. (1986)
$LaCu_2Si_2$	Naidyuk et al. (1985)
$LaNi_5$	Akimenko et al. (1990)
LaSe	Frankowski and Wachter (1982)
TB_2 (T=Zr,Ta, Nb)	Naidyuk et al. (2002)
TSi_2 (T=Ta, Nb, V)	Balkashin et al. (1996)
RS (R=La, Gd)	Frankowski and Wachter (1981)
$ThPd_3$	Moser et al. (1985)
YNi_5	Szabo et al. (1995)

Very impressive spectra were measured for the lanthanum hexaboride LaB_6 by Samuely et al. (1988). The phonon density of states in this compound with light boron atoms displays a rich structure up to the extremely high as for typical metals frequency (energy). Yet, point-contact characteristics reveal this structure with a number of well-separated sharp peaks up to 100 mV (Fig. 5.24).

Table 5.3 contains the list of intermetallic compounds for which EPI was investigated by point contacts.

5.4 Local vibrations

As follows from the previous section, the continuous shift of the phonon structure to the higher energy for the CuNi alloys with the Ni concentration increasing can be qualitatively explained by an increased rigidity of the lattice

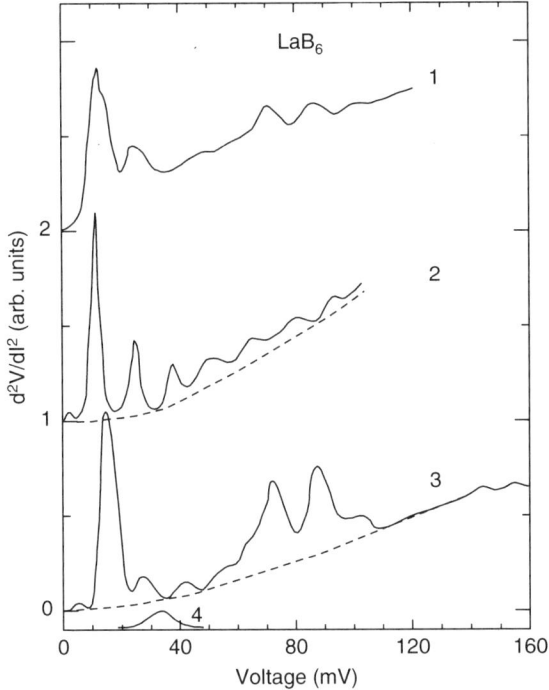

Fig. 5.24. Point-contact spectra of LaB$_6$ compound at 4.2 K for [100] (curve 2) and [111] (curve 3) directions. Curve 1 represents spectrum for two electrodes with different crystallographic orientation. Curve 4 is the calculated two-phonon contribution to the spectrum for [111] direction caused by the first peak. Dashed curves show a guide by eye background. Data taken from Samuely et al. (1988).

by adding of Ni atoms. A particular interest, however, provokes study of systems in which new vibration modes can appear because of the impurities. Intuitively, it is clear that the impurity with the mass m being different compared with the mass of the host atoms M will result in new phonon modes with sufficiently different frequency than the host lattice vibrations. First of all, it takes place because the vibrations frequency is inversely proportional to the square root of the atomic mass. Next, the atomic forces between the strange and the host atoms can be modified, what also causes change in the vibration frequency. If the impurity concentration is low, the interaction between strange atoms can be neglected and vibrations will be localized at the defect place; that is why the name of local vibrations is used. The latter were introduced theoretically by Lifshitz (1965). In the case of a light impurity atom, the vibration energy is higher than for the host atoms. This leads to the isolated single impurity vibrations producing δ-like peak in the phonon density of states at energy [see, e. g., textbook of Kittel (1986)]

$$\hbar\omega_{LV} \simeq \hbar\omega_{max}(M/2m)^{1/2}, \tag{5.2}$$

that is above the maximal phonon energy $\hbar\omega_{max}$. For systems with weakly interacting impurities, this peak broadens into a band, which resembles the phonon spectrum of the host metal.

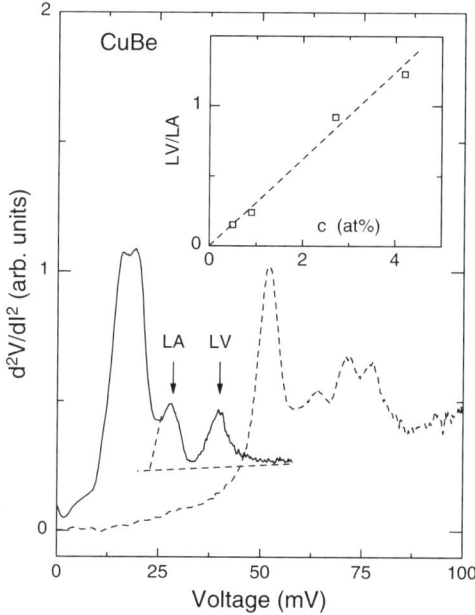

Fig. 5.25. Point-contact spectrum of CuBe (2.7 at% Be) alloy (solid curve). Dashed curve represents spectrum of pure Be. Arrows show the LA phonons in Cu and local vibration (LV) maximum. Region restricted by spectrum, and straight dashed lines was used to determine relative integral intensity of LA and LV peaks. Inset: dependence of the relative integral intensity of the LV peak with respect to the LA maximum for a few Be concentration for spectra shown in Fig. 5.26. The line is guide to the eye. Data taken from Naidyuk et al. (1982a).

The typical system with the local vibrations is the solution of Be atoms in the Cu lattice. The difference in the atomic mass here is high: about 7. The point-contact spectrum of CuBe alloy with 2.7 at% Be (Fig. 5.25) shows a new maximum at 40 mV appeared just between outermost LA peak of Cu at 30 mV and the first phonon peak in the Be spectrum at 50 mV. The energy position of this singularity agrees well with that of the local mode determined by the inelastic neutron scattering [Shitikov et al. (1980)].

Figure 5.26 depicts spectra for a number of samples with different Be contents. Both the local vibration peak intensity (see Fig. 5.25, inset) and the width increases (see Fig. 5.26, inset) with the adding of the Be atoms.

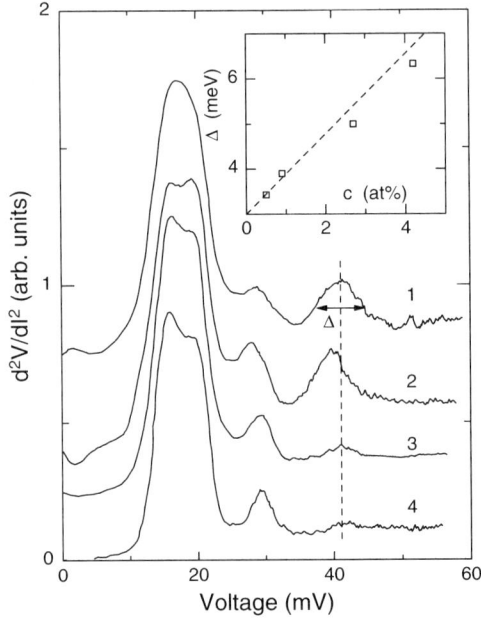

Fig. 5.26. Point-contact spectra of CuBe alloys with different Be concentration: 1 – 4.2 at%, 2 – 2.7 at%, 3 – 0.9 at%, and 4 – 0.5 at%. The vertical dashed line marks the position of local vibration (LV) peak. Inset: dependence of the width Δ of the LV peak versus Be contents. The line is guide to the eye. Data taken from Naidyuk et al. (1982a).

Naidyuk et al. (1982a) compared the integral intensity of the local vibration peak per Be impurity atom with the integral intensity of Cu spectrum with the subtracted background per Cu atom using relation:

$$\frac{1}{c}\frac{S_{LV}}{S_{Cu}} = \frac{1}{c}\frac{\int_{LV}\alpha^2 F(\omega)d\omega}{\int_{Cu}\alpha^2 F(\omega)d\omega} = \frac{1}{c}\frac{\langle\alpha^2\rangle_{LV}N_{LV}}{\langle\alpha^2\rangle_{Cu}N_{Cu}}. \qquad (5.3)$$

Here N_{LV} and N_{Cu} are the number of the phonon states corresponding to the local vibrations and to the lattice phonons. For the point-contact spectra with Be concentration 0.5, 0.9, 2.7, and 4.2 at% (5.3) gives, respectively, 3.87±0.2, 3.46±0.15, 3.24±0.4, and 2.74±0.15. The corresponding value of the relative intensity of the local vibration band by virtue of the neutron data [Shitikov et al. (1980)] is equal to 3.6. The latter calculation was made similar to the (5.3) formula but without α. Thus, the ratio $\langle\alpha^2\rangle_{LV}/\langle\alpha^2\rangle_{Cu}$ is close to unity for the two lower Be concentrations, testifying that effectiveness of the electron scattering by the vibrations of the isolated impurity and by vibrations of the host lattice is approximately the same. The decreasing of this ratio with the increasing of impurity contents is not clear. A similar intensity decrease of high-energy peaks in point-contact spectra occurs, e.g., for the

CuNi alloys (see previous section) or is observed for many pure metals with decreasing of the quality (i. e., the ratio l_{el}/d) of point contact. Nonetheless, if one takes the integral intensity ratio of the two neighbor LA and local vibration peaks, it will be linear in the whole concentration range. Hence, electron scattering effectiveness by the LA phonons and the local vibrations does not depend on both the concrete structure and the contact geometry.

Let us turn to the width of the local vibration peak [after subtracting of the temperature and modulation signal smearing, see (4.3)], which also increases linearly with the Be contents (Fig. 5.26, inset). The value of the width of about 3.1 meV obtained by extrapolation to $c \to 0$ is larger than according to neutron data. Moreover, the slope also exceeds the neutron data by 2.3 times. The additional broadening at $c \to 0$ can be explained by the influence of inhomogeneous mechanical stress in the contact, whereas a more quicker increase of the broadening with an impurity concentration is possible because of the temperature increase with the voltage bias by shortening of mean free path, as was considered for the ferromagnetic or transition metals. The latter is apparently the reason of lacking of the local vibrations peak in the point-contact spectra of NiBe alloys [Naidyuk et al. (1984)].

The impurity concentration increase leads to the interaction between impurities. Such collectivization of vibrations under certain conditions results in optical phonon modes. The systems PdH and PdD are the case. Point-contact spectra for these systems exhibit a broad band near 50 mV additionally to the maximum of Pd transverse phonons around 16 mV [Caro (1983)].

For the opposite case of heavy impurity with a mass a few times higher than the host atom mass, according to Kagan and Iosilevsky (1963), so-called quasilocal vibrations with the frequency

$$\omega_{QLV} = \omega_{max}(K|1 - m/M|)^{-1/2} \tag{5.4}$$

appear, where K is the numerical coefficient of about 3 for the Debye model and it depends on the specific form of the phonon spectrum. This energy is sufficiently below the Debye energy that is in the region of a smooth increase of the phonon density of states. Therefore, quasilocal vibrations should also be seen directly in the spectra well separated from the host EPI maxima.

The well-known system with quasilocal vibrations is Mg with Pb admixture where the difference in the mass exceeds 8. Point-contact spectra of this compound are presented by Yanson and Shklyarevskii (1986). Unfortunately, introducing of impurity here gives a rise to the practically complete smearing of the Mg spectrum, which looks like continuously increasing background with saturation in the region of the main Mg phonon peaks. The quasilocal vibrations manifest in the d^2V/dI^2 curve in the shape of shoulder at the lower energy. After subtracting of a smoothly raised background, a maximum at about 5 mV was resolved, the position and width of which correspond to neutron scattering data.

Molybdenum-rhenium alloys with Re contents up to 20 at % were investigated by Tulina and Zaitsev (1993). By adding of Re (Re atom mass is

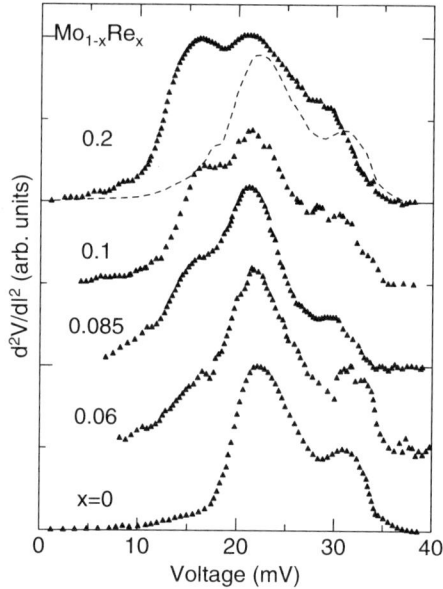

Fig. 5.27. Point-contact EPI function of the MoRe alloys with different Re concentrations indicated by each curve. Dashed curve represents theoretical phonon spectrum of the $Mo_{0.8}Re_{0.2}$ alloy. The curves are offset vertically for clarity. Data taken from Tulina and Zaitsev (1993).

twice heavier than Mo one), a new peculiarity in Mo spectra at $15\,\mathrm{mV}$ appears, the intensity of which increases with the rise of the Re concentration (see Fig. 5.27). The comparison of the EPI spectra with the phonon density of states calculated for the $Mo_{0.8}Re_{0.2}$ alloy shows that the EPI quasilocal mode is softened compared with the phonon one, and it is several times as intense. It was explained by the increase of the density of states at the Fermi level because Re donates one electron more to the conduction band compared with Mo. This decreases the force constants and renormalizes EPI. This means that the resonance quasilocal mode of the heavy impurity contributes strongly to the EPI spectrum of the alloy. Indeed, the superconducting critical temperature rises sufficiently with the Re contents [Tulina and Zaitsev (1993)].

The low-energy maximum around 7–8 meV was resolved by Tulina (1983) in the point-contact spectra of Re with the adding of a few percents of W. The maximum intensity increases with increasing of W contents. Re and W have almost the same atomic mass, what discards the appearance of the quasilocal vibrations because of the difference in the mass. The resonant vibrations in the spectrum most likely appear here as a result of the decreasing in the force constants between an impurity atom and the host ones. The characteristic feature of the ReW alloys is the appreciable increase of the superconducting transition temperature accompanying the rise in the W concentration. This allows us to expect a significant change of the EPI spectra in this system, which was observed by Tulina (1983).

5.5 Heterocontacts

In the case of heterocontacts, the contacts between two different metals, the point-contact spectrum reflects the EPI of each metal. For contacts between metals with the equal Fermi momentum, the spectrum presents itself as a sum of the spectra of both metals according to (3.27). In the event of a

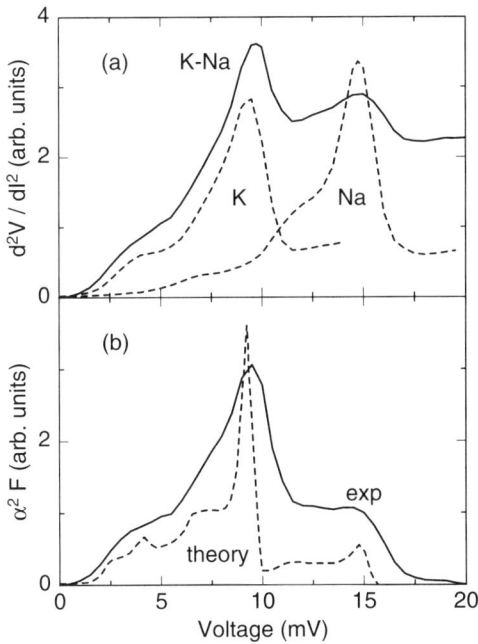

Fig. 5.28. (a) Point-contact spectrum for a K-Na heterocontact (solid curve) and for K and Na homocontacts (dashed curves) from Naidyuk (1990). (b) EPI function (solid curve) received for K-Na spectrum from (a) after subtracting of a background along with the calculation (dashed curve) by Baranger et al. (1985).

remarkable difference in the Fermi momenta, the processes of the electron trajectories reflection and refraction at the boundary start play a role along with "focusing" of electron flow in the metal with larger momentum (see Section 3.6). All of these effects lead to modification of the spectrum of a metal with the larger momentum. As it was calculated by Baranger et al. (1985) for

K–Na heterocontact, the spectrum for metal with larger p_F (in this case, Na) is not only suppressed but also its shape is modified (Fig. 5.28) because of depressing of the main LA phonon peak. The theoretical prediction is in qualitative consent with the experimental results, as shown in Fig. 5.28(b). Sato et al. (1987) have observed depressing of Al features for In–Al heterocontacts, what is also in line with the larger Fermi momentum in Al.

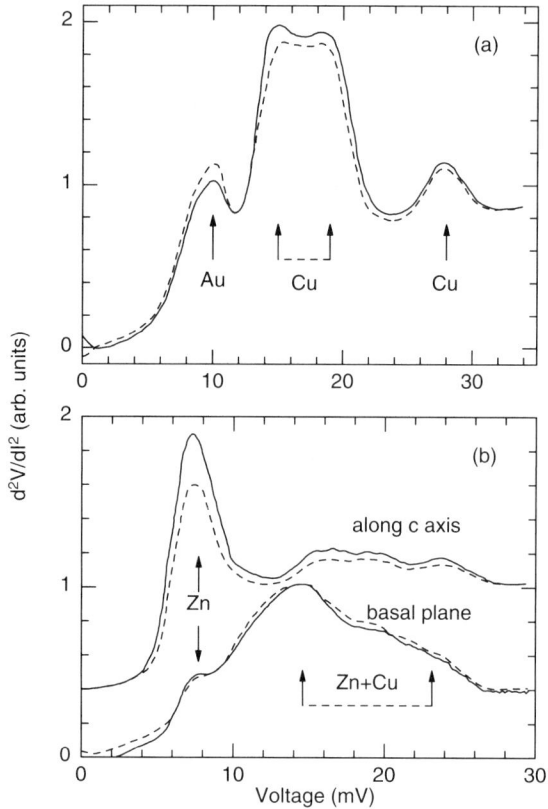

Fig. 5.29. (a) Point-contact spectra for Cu–Au heterocontact for both polarities of an applied voltage. (b) Point-contact spectra for Cu–Zn heterocontacts along c axis and in perpendicular direction in Zn for both polarities of an applied voltage. The solid curves correspond to the positive polarity in Cu. Data taken from Naidyuk and Yanson (1988, 1991).

Another effect in heterocontacts is connected with highly nonequilibrium phonon system (see Section 3.6). In the case of a small elastic phonon mean free path l_r ($l_r \ll d$), nonequilibrium phonons are accumulated in the contact and then drift isotropically from the contact center to the periphery. This leads to the drag of electrons by phonons. In a homocontact, this effect cancels because of identical phonon characteristics of the metals, but in the case of heterocontacts, this results in the asymmetry of the spectra proportional to l_r with respect to the polarity of the applied voltage [Itskovich and Shekhter (1985)].

Such asymmetry of point-contact spectra was experimentally observed for Cu–Au heterocontacts [see Fig. 5.29(a)]. It was also found that the sign of the effect for Cu–Zn heterocontacts depends on the crystallographic direction [see Fig. 5.29(b)]. The latter effect, as well as the anisotropy (sign reversal) of the drag thermo-e.m.f. in Zn, seems to be related to the predominance of the electronic-type conduction along the direction of the hexagonal c axis compared with the hole-type conduction in the basal plane. The asymmetry magnitude by Itskovich and Shekhter (1985) is directly proportional to the phonon mean free path or more specifically to the ratio l_r/d. It allows us to estimate l_r using the experimental data to be only a few nanometers. Thus, EPI spectra asymmetry in heterocontacts is caused by drift drag of charge carriers by nonequilibrium phonons and can be used to obtain specific information regarding them. Summarizing, by means of heterocontacts, it is possible to receive some additional knowledge about the more delicate features of the transport of the nonequilibrium electrons and phonons in the constriction between dissimilar metals.

5.6 Two-phonon processes

According to the right panel in Fig. 3.3, the contribution of the second-order correction processes to the backflow current corresponds to two subsequent collisions of electron in the contact region. These processes were considered by van Gelder (1978) to explain the background signal above the Debye energy. The functional dependence of two phonon processes is given by

$$B(V) = \Gamma(k,r)\frac{\int_0^V dV' \, \alpha^2 F(V')\alpha^2 F(V-V')}{\int_0^V dV' \, \alpha^2 F(V')}. \qquad (5.5)$$

Here $\alpha^2 F \equiv g$ is the point-contact EPI function, and $\Gamma(k,r)$ is the dimensionless parameter equal to

$$\Gamma(k) \simeq -0.73k \ln k - 0.16k - 0.71r - 0.34k^2 \ln k +$$
$$0.29k^2 \ln^2 k + 0.033k^3 \ln^2 k - 0.07rk \ln k + \cdots, \qquad (5.6)$$

for $k \ll 1$ and $r \ll 1$, where $k = d/l$ is the Knudsen number with practical expression:

$$k = 1.178r(TD)^{-1}$$

and

$$r = \int_0^V dV' \, \alpha^2 F(V') \simeq R_0^{-1} \left(\frac{dV}{dI}(V) - \frac{dV}{dI}(0) \right), \qquad (5.7)$$

T is the probability for electrons to pass through the orifice, and D is a factor close to unity, accounting for collision losses. It should be mentioned that only the energy conservation law is taken into account in the van Gelder theory. A more rigorous theory of $I - V$ characteristics in the second order in d/l_{in} for point contacts was developed by Kulik et al. (1984). This theory yields the van Gelder result under the assumption that the EPI matrix element α^2 is independent of the phonon momentum and the phonon spectrum is Einstein-like.

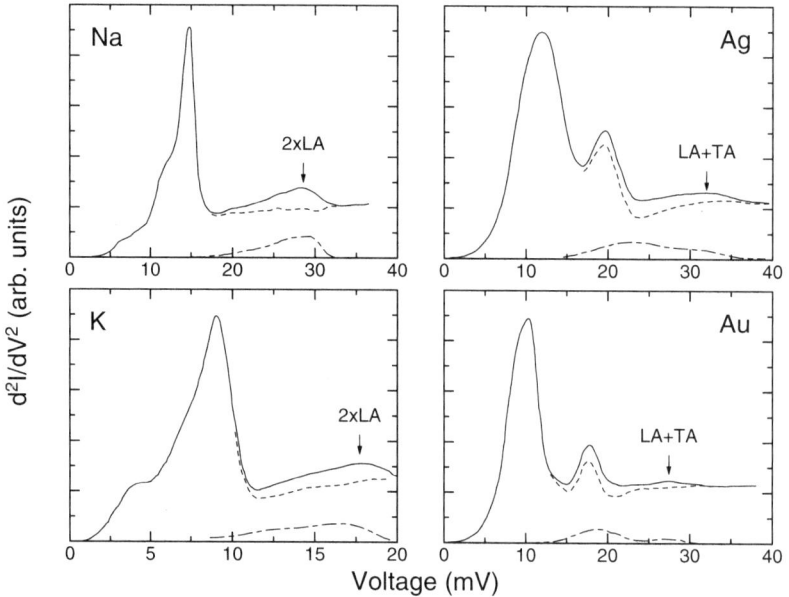

Fig. 5.30. Point-contact spectra for Na, K, Ag, and Au after transformation of the d^2V/dI^2 derivative into $-d^2I/dV^2$ (solid curves). Arrows show the position of the two-phonon maxima in the spectra. Dash-dotted curves represent calculation of the double-phonon scattering contribution to the spectra according to (5.5) with $\Gamma = 0.25$ for Na, 0.2 for K, and 0.15 for Ag and Au. Dashed curves shows behavior of background after subtracting of the calculated two-phonon contribution. Data taken from Yanson et al. (1982).

Let us turn to the experimental results. The features corresponding to the double-phonon structure are clearly seen for spectra of the simple metals presented before. They are very expressive in the alkali metals Na and K, which have simple spectra with one well-defined sharp maximum. The calculations by virtue of (5.5) give the contribution of the double-phonon processes in the spectra of the metals mentioned as shown in Fig. 5.30. For calculation, the point-contact EPI function $\alpha^2 F$ was extracted from the experimental curve after transformation of the d^2V/dI^2 derivative into d^2I/dV^2 and the background subtracting. The $\Gamma_{\exp}$ coefficient was taken to fit better experimental data in the region of the double-phonon structure. In this case, an almost linear (structureless) background appears above the maximal phonon energy after subtraction of the double-phonon contribution from the spectra (see Fig. 5.30).

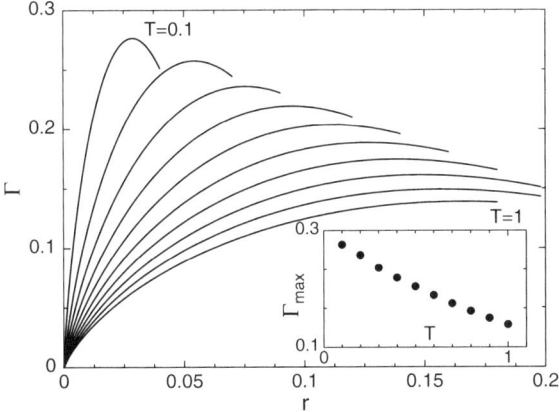

Fig. 5.31. Dependence of the Γ coefficient according to (5.6) versus parameter r (5.7). The calculations are made for $D=1$ and different transmission coefficient values: T is 0.1 for the upper left curve and increases by 0.1 for each next curve to the value $T=1$ for the bottom curve. Inset shows the dependence of the maximal Γ value on T.

The comparison of $\Gamma_{\exp}$ with the calculated one in Fig. 5.31 allows us to determine transmission coefficient T, which can be used for absolute intensity correction of the EPI function and the EPI parameter λ as well [Yanson et al. (1982)].

Let us discuss the role of the impurity scattering and point-contact structure, which can influence double-collision events. It was experimentally established that the two-phonon maxima vanish for dirty contacts. The exception can occur in the case when impurities or defects are concentrated at the interface between two electrodes and form a semitransparent barrier for electrons, which might be described by transparency T corresponding to so-called "T-model" [van Gelder (1978)]. In this case, the phonon maxima intensity decreases with the lowering of the transparency T, whereas two phonon intensity diminish slowly because Γ increases as T decreases (Fig. 5.31). On the other hand, the presence of several pure point contacts in parallel also result in decreasing of the absolute spectra intensity $N^{1/2}$ times in the case of

N identical contacts; see van Gelder (1978), but the relative intensity of the
two-phonon singularities remains here the same as for the pure single contact. Therefore, the small absolute intensity of the spectrum together with
the relative intensity of two-phonon maxima corresponding to $T \simeq 1$ indicate
the presence of multicontacts. In such a manner, the analyzing of the double-
phonon features can give information about contamination and distribution
of the impurity in the contact region, as well as it can be a criterion for the
presence of multicontacts.

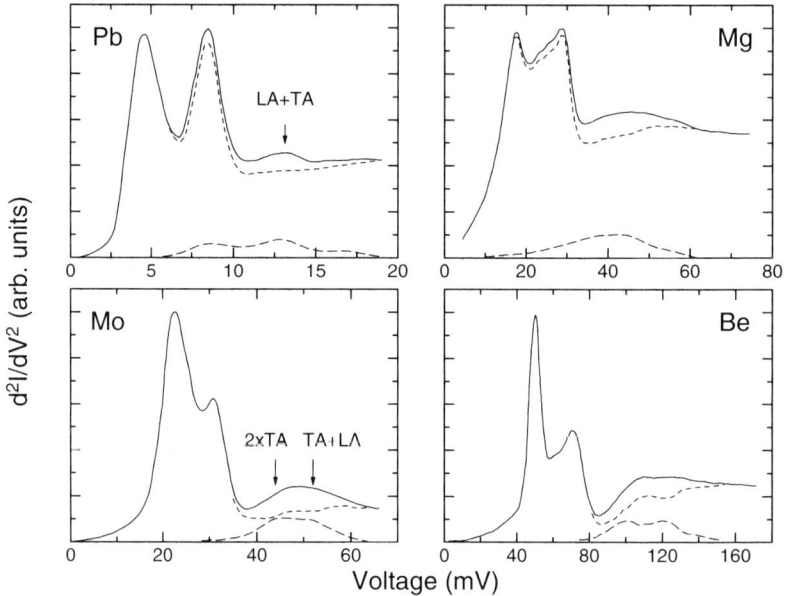

Fig. 5.32. Point-contact spectra for Pb, Mo, Mg, and Be after transformation of the
d^2V/dI^2 derivative into - d^2I/dV^2 (solid curves). Long-dash curves are calculation
by (5.5) of double-phonon scattering contribution to the spectra with $\Gamma = 0.17$
for Pb, 0.2 for Mo, and 0.25 for Mg and Be. Short-dash curves are behavior of
background after subtracting of the calculated curves. Data taken from Yanson
et al. (1982).

Two-phonon contribution to the spectra of noble metals has a more complex form (Fig. 5.30). Additionally, the most intensive part of the two-phonon
maxima falls into the region of the LA peak. In this case, the experimental
curve and the curve calculated were compared for the LA+TA maximum. The
resultant signal for Au, obtained by subtracting of the two-phonon features,
represents a straight line, with the exception of a shallow minimum at 21 mV
and a weak maximum near 31 mV [see Fig. from Yanson et al. (1982)]. The
latter corresponds well to the position of 3×TA feature. Ag has more broader

TA maximum compared with Au, and a worse coincidence with calculations is achieved.

The calculations for Pb and Mo, both having the spectra with a clear double-phonon structure, correspond well to the measured curves (Fig. 5.32). For different Mo contacts, even a correlation between determined from two-phonon feature parameter T and absolute intensity of the spectra was observed. On the other hand, transmission coefficient T is close to unity for the lead spectrum, whereas the latter has an absolute intensity much lower compared with the maximal one. This testifies, apparently, multicontacts structure of this samples prepared by thin film technology [Khotkevich and Yanson (1981)].

The two-phonon contribution to the spectra of HCP metals Mg and Be (Fig. 5.32) describes worse the experimental curves compared with all above-mentioned metals. It is likely because the van Gelder theory is valid for an Einstein-type spectrum and fails to describe more complex curves with a few peaks not well separated and transformed in a wide band.

It is interesting that a gentle linear increase of the background is still seen for some metals after subtracting of the two-phonon contribution. Yanson et al. (1982) discussed this phenomenon for K and Na from the point of view of the electron–electron interaction, which gives T^2 term in the resistivity or V^2 in the contact resistance.

It should be noted as a final remark to this chapter that in many papers, point-contact junctions are claimed as being in ballistic regime and no direct experimental proof of spectroscopic regime is exposed. In our opinion, only observation of electron–quasiparticle spectra described in this chapter can serve as an unambiguous evidence of ballistic or diffusive spectroscopic regime in constriction.

References

Akimenko A. I. and Yanson I. K. (1980) JETP Lett. **31** 191.

Akimenko A. I., Verkin A. B., Ponomarenko N. M. and Yanson I. K. (1982a) Sov. J. Low Temp. Phys. **8** 130.

Akimenko A. I., Verkin A. B., Ponomarenko N. M. and Yanson I. K. (1982b) Sov. J. Low Temp. Phys. **8** 547.

Akimenko A. I., Ponomarenko N. M., Yanson I. K., Samuely P. and Reiffers M. (1990) Z. Phys. B **79** 191.

Al-Lehaibi A., Swihart J. C., Butler W. H. and Pinski F. J. (1987) Phys. Rev. B **36** 4103.

Antonov V. N., Zhalko-Titarenko A. V., Milman V. Yu., Khotkevich A. V. and Krainyukov S. N. (1991) J. Phys.: Condens. Matter **3** 6523.

Ashraf M. and Swihart J. C. (1982) Phys. Rev. B **25** 2094.

Balkashin O. P., Jansen A. G. M., Gottlieb U., Laborde O. and Madar R. (1996) Solid State Commun. **100** 293.

Baranger H. U., MacDonald A. H. and Leavens C. R. (1985) Phys. Rev. B. **31** 6197.

Bobrov N. L., Rybaltchenko L. F., Yanson I. K. and Fisun V. V. (1987) Sov. J. Low Temp. Phys. **13** 344.

Bogomolov V. N., Klushin N. A. and Okuneva N. M. (1971) Sov. Phys. Solid State **13** 1256.

Brück E., Nowack A., Hohn N., Paulus E. and Freimuth A. (1986) Z. Phys. B. **63** 155.

Bulat I. A. (1979) Sov. Phys. Solid State **21** 583.

Burnell D. M. and Wolf E. L. (1982) Phys. Lett. A **90** 471.

Caro J. (1983) Ph.-D Thesis, Amsterdam:Rodopi. (unpublished)

Caro J., Coehoorn R. and de Groot D. G. (1981) Solid State Commun. **39** 267.

Chubov P. N., Yanson I. K. and Akimenko A. I. (1982) Sov. J. Low Temp. Phys. **8** 32.

Cracknell A. P. and Wong K. C. (1973) *Fermi Surface*, Oxford.

Dynes R. C. (1970) Phys. Rev. B. **2** 644.

Eremeev I. P., Sadikov I. P. and Chernyshov A. A. (1976) Sov. Phys. Solid State **18** 960.

Frankowski I. and Wachter P. (1981) Sol. State Commun. **40** 885.

Frankowski I. and Wachter P. (1982) Sol. State Commun. **41** 577.

Gorbachev B. I., Ivanitskii P. G., Korotenko V. T. and Pasechnik M. V. (1973) Ukrainian Phys. Journal **18** 1384.

Gorbachev B. I., Morozov S. I., Parfenov V. A. and Pasechnik M. V. (1976) Sov. Phys. Solid State **18** 2157.

Gribov N. N. (1984) Sov. J. Low Temp. Phys. **10** 168.

Gribov N. N., Caro J., Oosterlaken T. G. M. and Radelaar S. (1996) Physica B **218** 101.

Grimvall G. (1976) Phys. Scripta **14** 63.

Grimvall G. (1981) *The Electron–phonon Interaction in Metals* (North-Holland Publ. Co., Amst., N.-Y., Oxf.).

Itskovich I. F. and Shekhter R. I. (1985) Sov. J. Low Temp. Phys. **11** 649.

Jansen A. G. M., van den Bosch J. H., van Kempen H., Ribot J. H. J. M., Smeets P. H. H. and Wyder P. (1980) J. Phys. F **10** 265.

Kagan Yu., Iosilevsky Ya. (1963) Zh. Exper. Teor. Fiz. **44** 1375.

Khlyustikov I. N., and Buzdin A. I. (1988) Uspekhi Fizicheskikh Nauk **155** 47.

Khotkevich A. V. (1996) Physica B **218** 31.

Khotkevich A. V., Elenskii V. A., Kovtun G. P. and Yanson I. K. (1984) Sov. J. Low Temp. Phys. **10** 194.

Khotkevich A. V. and Khotkevich V. V. (1985) Sov. J. Low Temp. Phys. **11** 267.

Khotkevich A. V. and Krainyukov S. N. (1991) Sov. J. Low Temp. Phys. **17** 173.

Khotkevich A. V. and Yanson I. K. (1995) *Atlas of Point Contact Spectra of Electron–Phonon Interaction in Metals* Kluwer Academic Publisher, Boston.

Khotkevich A. V., Yanson I. K., Lazareva M. B., Sokolenko V. I. and Starodubov Ya. D. (1990) Physica B **165-166** 1589.

Kittel C. (1986) *Introduction to Solid State Physics* John Wiley & Sons, New York.

Koehler W. C., Child H. R., Nicklow R. M., Smith H. G., Moon R. M. and Cable J. W. (1970) Phys. Rev. Lett. **24** 16.

Krainyukov S. N., Khotkevich A. V., Yanson I. K., Zhalko-Titarenko A. V., Antonov V. N. and Nemoshkalenko V. N. (1988) Sov. J. Low Temp. Phys. **14** 127.

Krainyukov S. N., Khotkevich A. V., Yanson I. K., Zhalko-Titarenko A. V., Antonov V. N., Nemoshkalenko V. N., Mil'man Yu. V., Shitikov Yu. L. and Khlopkin M. N. (1989) Sov. Phys. Solid State **31** 419.

Kulik I. O., Omelyanchouk A. N. and Tuluzov I. G. (1984) Sov. J. Low Temp. Phys. **10** 484.

Kulik I. O. and Shekhter R. I. (1980) Sov. J. Low Temp. Phys. **6** 88.

Lee M. J. G., Caro J., de Groot D. G., and Grissen R. (1985) Phys. Rev. B **31** 8244.

Lifshitz I. M. (1965) Sov. Phys. Uspechi **7** 549.

Lynn J. W., Smith H. G. and Nicklow R. M. (1973) Phys. Rev. B **8** 3493.

Lysykh A. A., Yanson I. K., Shklyarevskii O. I. and Naidyuk Yu. G. (1980a) Sov. J. Low Temp. Phys. **6** 224.

Lysykh A. A., Yanson I. K., Shklyarevskii O. I. and Naidyuk Yu. G. (1980b) Solid State Commun. **35** 987.

MacDonald A. H. and Leavens C. R. (1982) Phys. Rev. B **26** 4293.

Meekes H. (1988) Phys. Rev. B **38** 5924.

Moser M., Hulliger F. and Wachter P. (1985a) Physica B **130** 21.

Naidyuk Yu. G. (1990) Sov. Phys.-Solid State **32** 268.

Naidyuk Yu. G., Chernoplekov N. A., Shitikov Yu. L., Shklyarevskii O. I. and Yanson I. K. (1982a) Sov. Phys. - JETP **56** 671.

Naidyuk Yu. G., Gribov N. N., Lysykh A. A., Yanson I. K., Brandt N. B. and Moshchalkov V. V. (1985) Sov. Phys. Solid State **27** 2153.

Naidyuk Yu. G., Kvitnitskaya O. E., Yanson I. K., Drechsler S.-L., Behr G., Otani S. (2002) Phys. Rev. B **66** 140301(R).

Naidyuk Yu. G. and Shklyarevskii O. I. (1982) Sov. Phys. Solid State **24** 1491.

Naidyuk Yu. G. and Yanson I. K. (1988) Sov. Phys.-Solid State **30** 888.

Naidyuk Yu. G. and Yanson I. K. (1991) Physica B **169** 479.

Naidyuk Yu. G., Yanson I. K., Lysykh A. A. and Shitikov Yu. L. (1984) Sov. Phys. Solid State **26** 1656.

Naidyuk Yu. G., Yanson I. K., Lysykh A. A. and Shklyarevskii O. I. (1980) Sov. Phys. Solid State **22** 2145.

Naidyuk Yu. G., Yanson I. K., Lysykh A. A. and Shklyarevskii O. I. (1982b) Sov. J. Low Temp. Phys. **8** 464.

Naidyuk Yu. G., Yanson I. K. and Shklyarevskii O. I. (1981) Sov. J. Low Temp. Phys. **7** 157.

Rao R. R. and Ramanand A. (1977) J. Phys. Chem. Solids **38** 831.

Rao R. R. and Ramanand A. (1979) Phys. Rev. B **19** 1972.

Reiffers M. (1992) Sov. J. Low Temp. Phys. **18** 339.

Rowell J. M., McMillan W. L. and Feldmann W. L. (1969) Phys. Rev. **178** 897.

Rowell J. M., McMillan W. L. and Feldmann W. L. (1971) Phys. Rev. B **3** 4062.

Rybaltchenko L. F., Yanson I. K., Bobrov N. L. and Fisun V. V. (1981) Sov. J. Low Temp. Phys. **7** 82.

Rybaltchenko L. F., Yanson I. K. and Fisun V. V. (1980) Sov. Phys.-Solid State **22** 1182.

Samuely P., Reiffers M., Flachbart K., Akimenko A. I., Yanson I. K., Ponomarenko N. M. and Paderno Yu. B. (1988) J. Low Temp. Phys. **71** 49.

Sato H., Okimoto H., and Yonemitsu K. (1985) Solid State Commun. **56** 141.

Sato H., Yonemitsu K., and Bass J. (1987) Phys. Rev. B **35** 2484.

Savrasov S. and Savrasov D. (1996) Phys. Rev. B **54** 16487.

Shen L. Y. Z. (1970) Phys. Rev. Lett. **24** 1104.

Shitikov Yu. L., Zemlyanov M. G., Syrykh G. F. and Chernoplekov N. A. (1980) Sov. Phys. - JETP **51** 752.

Shklyarevskii O. I., Naidyuk Yu. G. and Yanson I. K. (1982) Sov. J. Low Temp. Phys. **8** 541.

Shklyarevskii O. I., Gribov N. N. and Naidyuk Yu. G. (1983) Sov. J. Low Temp. Phys. **9** 553.

Szabo P., Reiffers M., Flachbart K., Pobell F. (1995) J. Magn. and Magn. Mat. **140-144** 847.

Tulina N. A. (1982) Solid State Commun. **41** 313.

Tulina N. A. (1983) Sov. J. Low Temp. Phys. **9** 252.

Tulina N. A. and Zaitsev S. V. (1993) Solid State Commun. **86** 55.

van Gelder A. P. (1978) Solid State Commun. **25** 1097.

Wolf E. L., Zasadzinski J. and Osmun J. W. (1980) J. Low Temp. Phys. **40** 19.

Yanson I. K. and Batrak A. G. (1979) Sov. Phys. - JETP **49** 166.

Yanson I. K., Bobrov N. L., Rybaltchenko L. F. and Fisun V. V. (1983) Sov. J. Low Temp. Phys. **9** 596.

Yanson I. K., Naidyuk Yu. G. and Shklyarevskii O. I. (1982) Sov. J. Low Temp. Phys. **8** 595.

Yanson I. K. and Shklyarevskii O. I. (1986) Sov. J. Low Temp. Phys. **12** 509.

Zasadzinski J., Burnell D. M., Wolf E. L. and Arnold G. B. (1982) Phys. Rev. B **25** 1622.

Zakharov A. A, Zemlyanov M. G. and Mikheeva M. N. (1984) Zh. Exp. Teor. Phys. **88** 1402.

Zhalko-Titarenko A. V., Mil'man Yu. V., Antonov V. N., Nemoshkalenko V. N., Khotkevich A. V. and Krainyukov S. N. (1989) Metallofizika **11** 23.

Zhernov A. P. and Kulagina T. N. (1992) Sov. J. Low Temp. Phys. **18** 312; *ibid.* 323.

Zhernov A. P., Naidyuk Yu. G., Yanson I. K. and Kulagina T. N. (1982) Sov. J. Low Temp. Phys. **8** 355.

6 PCS of nonphononic scattering mechanisms

6.1 Paramagnetic impurities

6.1.1 Kondo scattering

The impurities that retain their magnetic properties when added to a normal metal strongly scatter the conduction electrons. The scattering of this sort in a bulk conductor causes at low temperatures an abnormal rise of resistance for metals known as the Kondo effect [Kondo (1964)]. The classic Kondo alloys consist of noble metals Cu or Au with a small amount ($\ll 1\%$) of impurities as Mn or Fe. Point contacts are proved to be a very sensitive probe for detecting of the magnetic impurities in a metal. This was already mentioned in the first point-contact measurements by Lysykh et al. (1980) with Fe embedded in the Cu host. In reality, when adding only 0.01% of iron or manganese atoms to gold or copper, the point-contact spectrum displays distinct additional features at low voltages around $1\,\mathrm{meV}$ (see Fig. 6.1), which are connected with the electron scattering by localized magnetic moments as was shown by Jansen et al. (1981) and Naidyuk et al. (1982). The number of impurity atoms getting into the contact area can be evaluated at the above-mentioned concentration $c = 10^{-4}$. The total number of atoms in a constriction with the size d will be d^3/v, where v is the unit cell volume. Then the number of impurity atoms n_i is estimated as cd^3/v. Taking for a copper $v \simeq 10^{-23}\,\mathrm{cm}^3$ and the contact size $d \simeq 10^{-6}\,\mathrm{cm}$ for a typical resistance of $10\,\Omega$, we obtain $n_i \simeq 10$ atoms. It is of interest that a point contact contains a meager number of atoms, the effect of which is, however, noticeable in the measured characteristics. Thus, we can say that PCS is a unique method for studying of electron scattering by a few microscopic centers.

The theoretical explanation of the Kondo effect based on the general expression for the s-d exchange Hamiltonian

$$H = -2J\mathbf{S}\,\mathbf{s},$$

where $\mathbf{s}$ is the electron spin, $\mathbf{S}$ is the local spin, and J is the exchange integral or the coupling constant, was done by Kondo (1964) by treating of a third-order perturbation approximation. From Kondo calculation at antiferromagnetic coupling $J < 0$ follows a logarithmic increase of the resistivity by lowering temperature:

$$\rho(T) = \rho_0(1 + 2\,N(\epsilon_F)J\,\ln(T/T_F)). \tag{6.1}$$

However, the ln-term diverges at $T \to 0$. This and others problems were overcome in the further treatment of the s-d model by Andrei (1980) and Wiegmann (1980) who succeeded in exact solution of the s-d Hamiltonian. Here we concentrate on the experimental evidence of this Kondo scattering on the diluted paramagnetic impurities in the point contacts. Kondo systems with a high concentration of magnetic impurities and more complicated behavior are discussed in Section 14.1.

Omelyanchouk and Tuluzov (1980, 1985) used both the Boltzmann kinetic equation, containing the collision integral for electrons and magnetic impurities, and the method of quasiclassic Green's functions to calculate $I - V$ curves of point contacts. The corresponding correction to the current by the account of the electron scattering on disordered magnetic impurities is given by

$$I_i(V) = -\frac{1}{eR_i}\left(eV - \frac{3J}{\epsilon_F}F_T(eV)\right), \tag{6.2}$$

where i stands for impurity,

$$F_T(x) = \int_{-\epsilon_0/2}^{\epsilon_0/2} d\epsilon \int_{-\infty}^{\infty} \frac{d\epsilon'}{\epsilon - \epsilon'}\left(\tanh\frac{\epsilon}{2k_BT}\left(\tanh\frac{\epsilon' - x}{2k_BT} - \tanh\frac{\epsilon' + x}{2k_BT}\right)\right), \tag{6.3}$$

with $\epsilon_0 \sim \epsilon_F$ being the width of the band. Subsequently, $d^2I/dV^2(V)$ contains features at zero-bias voltage caused by this scattering with asymptotic expression:

$$-\frac{d^2I}{dV^2} = \frac{3}{2R_i}\frac{J}{\epsilon_F}\begin{cases} 2/V, & \text{if } eV \gg k_BT, \\ (e/k_BT)^2V, & \text{if } eV \ll k_BT, \end{cases} \tag{6.4}$$

where

$$\frac{1}{R_i} = \frac{8k_Fd}{R_0}c\,S(S+1)\left(\frac{J}{\epsilon_F}\right)^2 \langle K \rangle,$$

and R_0 is the resistance of a clean contact, c is the impurity concentration, k_F is the Fermi wavenumber, and $\langle K \rangle$ is averaged geometrical form-factor (see Table 3.2). Measurements carried out on CuMn and CuFe alloys by Naidyuk et al. (1982) proved these formulas both qualitatively and quantitatively (see Fig. 6.1). Using the point-contact spectra, the value of the exchange constant J was estimated for CuFe with 0.01% Fe alloy as $J/\epsilon_F = 0.13 \pm 0.02$; besides the geometrical form-factor $\langle K \rangle$ was found for CuMn (Mn=0.012%) alloy, taking into account the known ratio $J/\epsilon_F=0.0815$.

Magnetic field results in the Zeeman splitting of degenerate spin states with the energy difference $2\Delta = 2g\mu_B H$, where μ_B is the Bohr magneton. This depresses spin-flip scattering, which now costs energy 2Δ and decreases Kondo resistance. In other words, in the absence of an external magnetic field,

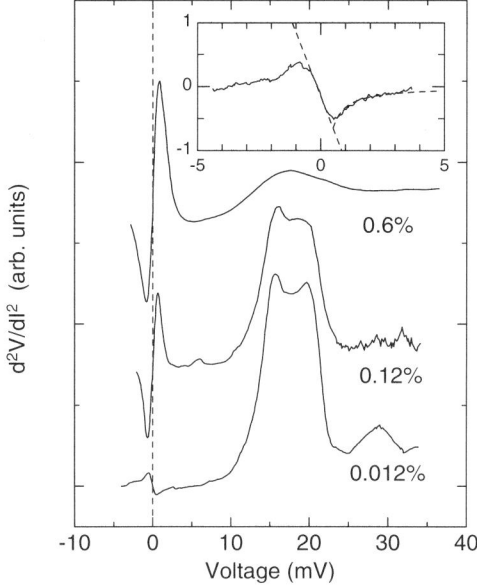

Fig. 6.1. Point-contact spectra $d^2V/dI^2(V)$ of CuMn alloys at 1.7 K with the increasing of Mn contents marked by each curve. Maxima between 10 and 35 mV correspond to the Cu phonons. The inset shows zero-bias feature for the lowest concentration. Dashed lines show linear and V^{-1} behavior. Data taken from Naidyuk et al. (1982).

the spin-flip scattering of electron on magnetic impurity is elastic, whereas in the magnetic field, it is accompanied by the energy change by 2Δ. In point contact, the inelastic spin-flip processes are forbidden until $eV < \Delta$, and for $eV \geq \Delta$, they become possible. This leads to decreasing of the contact resistance at $|eV| < \Delta$ that is a minimum develops on the $dV/dI(V)$ characteristic at zero-bias in magnetic field (Fig. 6.2).

The microscopic calculations carried out by Omelyanchouk and Tuluzov (1985) yields $-d^2I/dV^2(V)$ curve with a minimum at $eV \sim k_B T$ in zero-field and maximum at $eV \sim \mu_B H = Q$ caused by inelastic scattering in a magnetic field (Fig. 6.3). At $T \to 0$, this maximum tends to a δ-function shape. The differential resistance $R_d(V) = dV/dI(V)$ in the case of nonzero magnetic field is given by

$$R_i(R_0^{-1}(0) - R_d^{-1}(V)) = 1 + \Gamma_T(eV + Q) - \Gamma_T(eV - Q) \qquad (6.5)$$

$$- \frac{3J}{2\epsilon_F} \left\{ \left[1 - \frac{<M^2>}{S(S+1)} + \Gamma_T(eV + Q) - \Gamma_T(eV - Q) \right] \frac{\partial F_T(x)}{\partial x}\Big|_{x=eV} \right.$$

$$+ \left[\frac{1}{2} + \frac{<M^2>}{2S(S+1)} + \Gamma_T(eV + Q) \right] \frac{\partial F_T(x)}{\partial x}\Big|_{x=eV+Q}$$

$$\left. + \left[\frac{1}{2} + \frac{<M^2>}{2S(S+1)} - \Gamma_T(eV - Q) \right] \frac{\partial F_T(x)}{\partial x}\Big|_{x=eV-Q} \right\},$$

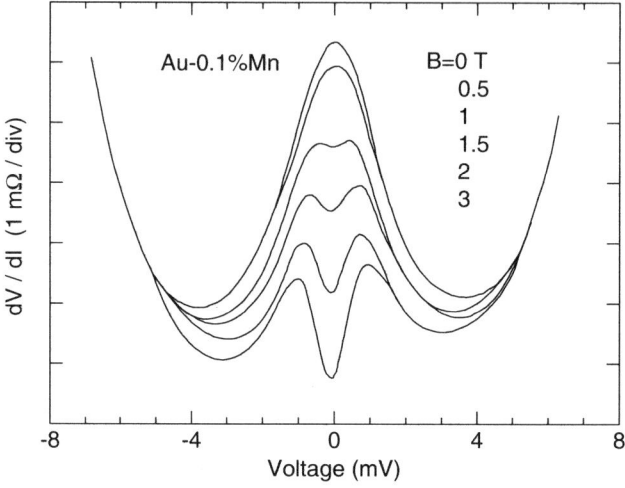

Fig. 6.2. $dV/dI(V)$ curves of Au-0.1%Mn point contact at different magnetic fields and $T=1.2$ K. Data taken from Duif et al. (1987).

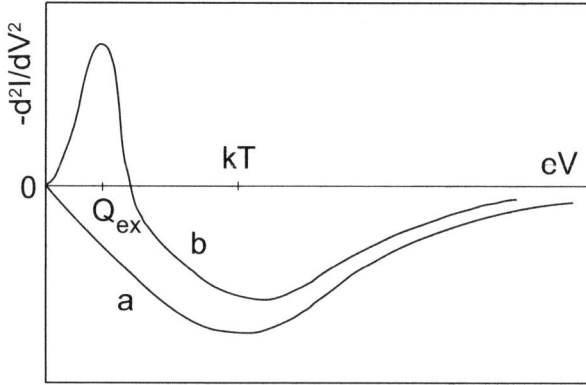

Fig. 6.3. Theoretical point-contact spectra $d^2I/dV^2(V)$ of a dilute magnetic alloy at the external magnetic field (a) $H=0$ and (b) $H \neq 0$. Here $Q_{ex} = \mu_\mathrm{B} H$ and T is the bath temperature. After Omelyanchouk and Tuluzov (1985).

where

$$\Gamma_T(x) = \frac{\langle M \rangle}{2S(S+1)} \tanh \frac{x}{2k_\mathrm{B}T}$$

and

$$\langle M \rangle = \frac{1}{2} \coth \frac{Q}{2k_\mathrm{B}T} - \left(S + \frac{1}{2}\right) \coth \left(S + \frac{1}{2}\right) \frac{Q}{k_\mathrm{B}T},$$

$$\langle M^2 \rangle = \langle M \rangle^2 - \left(S + \frac{1}{2}\right)^2 \mathrm{csch}^2 \left(S + \frac{1}{2}\right) \frac{Q}{k_\mathrm{B}T} + \frac{1}{2} \mathrm{csch}^2 \frac{Q}{k_\mathrm{B}T}.$$

An analogous result to the above expression was obtained by d'Ambrumenil and White (1982). They generalized expression for the tunneling current given by Appelbaum (1967), who calculated the conductance assisted by interaction of electrons with magnetic moments localized in the barrier. As follows from (6.5) the zero-bias maximum in $dV/dI(V)$ splits in a magnetic field. It was clearly observed by Jansen et al. (1981) and Naidyuk et al. (1982). The distance between the splitted maxima in $dV/dI(V)$ equals roughly $2g\mu_B H$. This allows us to determine g-factor from the $dV/dI(V)$ curves in a magnetic field easily. Finally, Duif et al. (1987) carried out complete calculation according to (6.5), which shows a perfect correspondence with the theory (Fig. 6.4).

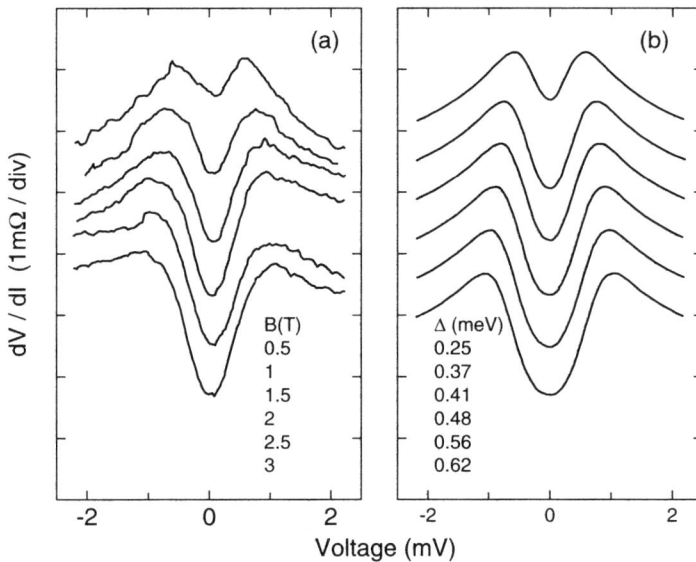

Fig. 6.4. (a) Raw $dV/dI(V)$ curves of a Au-0.1%Mn point contact with $R_0 = 1.1\,\Omega$ at $T = 1.2$ K for a different magnetic field. (b) Calculated spectra for this contact. The indicated Zeeman energies Δ have been adjusted in the calculation to fit the experimental curves. Data taken from Duif et al. (1987).

6.1.2 Spin glass

From the above-mentioned comparison between the calculated and the measured curve or for simplicity from the distance between the maxima in $dV/dI(V)$, the Zeeman energy Δ can be determined as a function of magnetic field. Usually, extrapolation of the Zeeman energy to zero magnetic field gives a finite Δ for alloys with a concentration of magnetic impurities above 0.1% (Fig. 6.5). This can be explained by the presence of internal

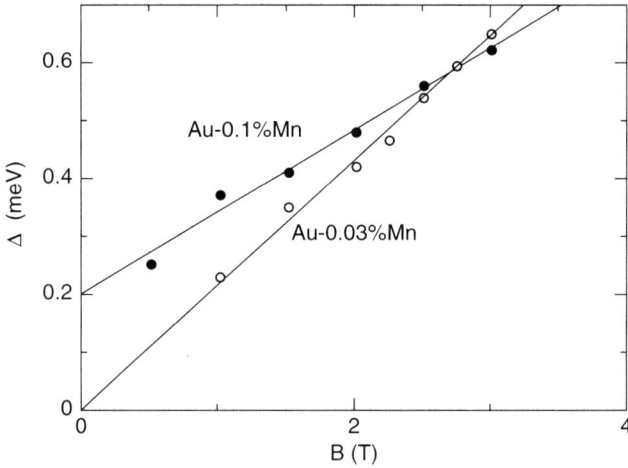

Fig. 6.5. Zeeman energy Δ calculated from the $dV/dI(V)$ curves of Au-0.1%Mn and Au-0.03%Mn point contacts plotted as a function of the magnetic field. The g-value can be obtained from the slope of the lines, using $\Delta = g\mu_B H$, and it is equal to 2.4±0.3 and 3.7±0.3 correspondingly. Data for Au-0.1%Mn samples show $\Delta \neq 0$ at zero field caused by the presence of the internal field (see text). Data taken from Duif et al. (1987).

field, which originates from an interaction between impurities assumed to be present in the spin glass state. The internal field caused by the indirect exchange RKKY interaction results in randomly fixed spatial direction of spins instead of indefinite spin direction as for the Kondo alloys. Figure 6.6 shows that the concentration increase leads to the appearance of zero-bias minimum in $dV/dI(V)$. This is analogously as by the external magnetic field increase. At the same time, as impurity amount rises, $d^2V/dI^2(V)$ shows transformation of "negative" zero-bias anomaly into a "positive" one (see, e.g., Fig. 6.1).

The influence of impurities can be described well by (6.5), including the internal field Q_{int} in the calculations. An expression for the inelastic current in a point contact based on alloy with ordered magnetic impurities was obtained also by Omelyanchouk and Tuluzov (1985). It turned out that $-d^2I/dV^2(V)$ has a maximum at a voltage $V \sim \mu_B H_{int}/e$, the width of which is of the order of $k_B T$ (Fig. 6.7). The nonuniform distribution of an internal field in the contact leads to an additional smearing of this maximum by an amount of magnitude of the internal field nonuniformity, i.e., by $\mu_B(\delta H_{int})$. Such

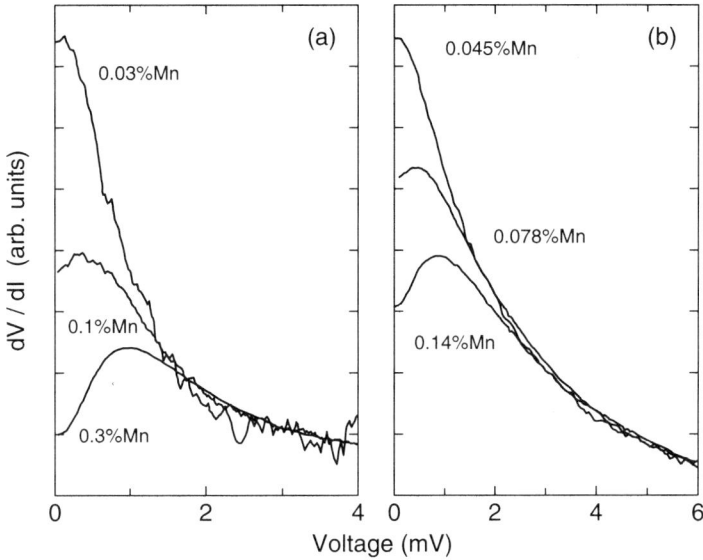

Fig. 6.6. Raw $dV/dI(V)$ curves for three AuMn (a) and three CuMn (b) point contacts with different Mn contents, showing the effect of internal field with increasing of the impurity concentration. The point-contact resistances are between $0.6\,\Omega$ (a) and $2.4\,\Omega$ (b). Data taken from Duif et al. (1987).

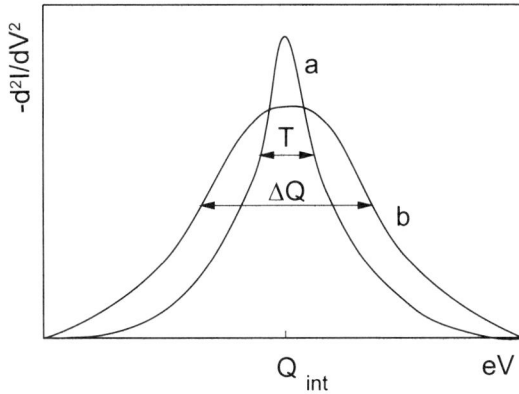

Fig. 6.7. Theoretical point-contact spectra $-d^2I/dV^2(V)$ of the ferromagnetic phase of the spin glass in the case of uniform (a) and nonuniform (b) internal magnetic field. Here $Q_{\mathrm{int}} = \mu_B H_{\mathrm{int}}$ and ΔQ is the nonuniformity magnitude of the internal magnetic field. After Omelyanchouk and Tuluzov (1985).

maxima in $\mathrm{d}^2V/\mathrm{d}I^2(V)$ have been observed in the CuMn alloys with the increasing of impurity concentration (Fig. 6.1).

6.1.3 Size effect

The variation of the impurity concentration in a microscopic scale may play a role by point-contact study of alloys, especially with decreasing of the contact dimension or increasing of its resistance. As was mentioned above a meager

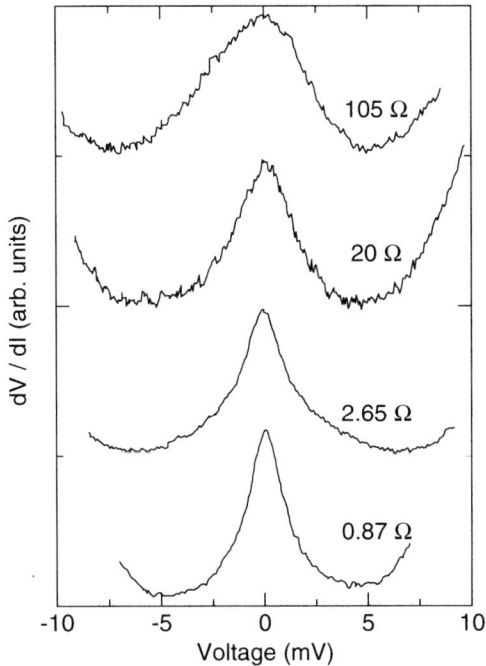

Fig. 6.8. Kondo maximum in $\mathrm{d}V/\mathrm{d}I(V)$ curves of a Cu-0.1%Mn mechanically controllable break junction with the resistance increasing from $0.87\,\Omega$ to $105\,\Omega$, which corresponds to a contact diameter from 32 nm to 2.9 nm. The relative value of zero-bias maxima is between 0.3% and 0.6% of the R_0. The curves are shifted vertically for clarity. The experimental resolution is better as 1 mV for all curves. Data taken from Yanson et al. (1995).

amount of impurities can exist in the constriction, and therefore, statistical fluctuations may be noticeable. Investigation of the contact size effect on the magnetic-impurity scattering was started by Yanson et al. (1995) by using mechanically controllable break junctions. It was found that the zero-bias maximum in $\mathrm{d}V/\mathrm{d}I(V)$ caused by the Kondo scattering broadens and its relative amplitude increases as d decreases (Fig. 6.8). The authors used (6.4) for the case $eV \gg k_{\mathrm{B}}T$ and found a significant enhancement of the exchange coupling parameter J. This also leads to a few orders of increasing the Kondo temperature defined in a standard way [Daybell (1973)]:

$$T_{\mathrm{K}} = \frac{\epsilon_{\mathrm{F}}}{k_{\mathrm{B}}} \exp\left(-\frac{2\epsilon_{\mathrm{F}}}{3J}\right). \qquad (6.6)$$

The experiments prove the theoretical conjecture of size-dependent Kondo scattering. Zarand and Udvardi (1996) reported that spatial- and energy-dependent fluctuations of the local density of states, becoming more pronounced with decreasing of the contact size, can quantitatively explain the size dependence of the Kondo temperature.

The further investigations by Yanson et al. (1996) showed that the assumption of the Omelyanchouk and Tuluzov (1985) theory that the electron–impurity scattering may occur essentially at any point in the metal is not more applicable in this case. The following theory by Kolesnichenko et al. (1996) takes into account the discreteness of the impurity position around the contact. This causes an enormous increase of the scattering at $d/r_0 < 1$, where r_0 is the average interimpurity distance. Moreover, regarding two types of impurities, the mesoscopic effects result in dependence of the correction for the contact resistance versus bias voltage if one of them is only magnetic. In this instance, as was shown by Kolesnichenko et al. (1997), impurity distribution in the constriction can affect both the intensity and position of the Kondo maximum in dV/dI (V).

6.2 Two-level systems

The departure of the point-contact characteristics $dV/dI(V)$ or $d^2V/dI^2(V)$ from a smooth monotonic behavior at zero bias belongs to a class of features generally named zero-bias anomalies. These anomalies are well known also in the tunneling spectroscopy (see Wolf (1985) in Chapter 2); however, they are not often fully comprehensible. Following from the previous paragraph, the Kondo effect also causes typical zero-bias anomalies, i. e., minimum or maximum on dV/dI at zero bias. Nevertheless, analogous features can be often observed for point contacts between clean normal metals. In principle, it may occur by incident contamination of the contact region by surface dirties or oxides, when the contact is established mechanically. However, the zero-bias anomalies were clearly resolved for constrictions prepared by electron beam lithography at ultra high vacuum conditions [Ralph and Burmann (1992)] as well as for mechanically controllable break junctions with breaking of a sample at the low temperatures [Keijsers et al. (1995)], what discards an incident pollution.

A more universal mechanism to explain the origin of the zero-bias anomalies was proposed by Kozub (1984). It is associated with the existence of a

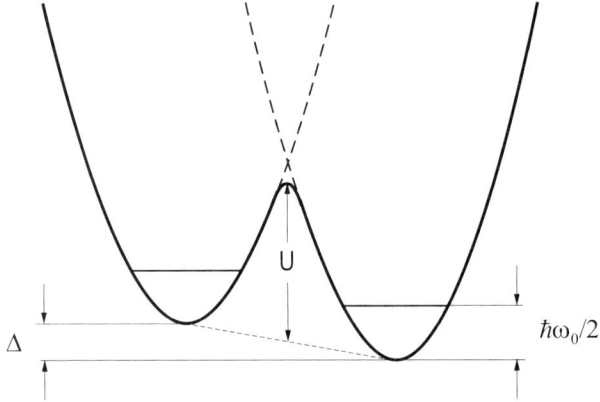

Fig. 6.9. Double-well potential characterized by a barrier height U and asymmetry energy Δ describing the model of a two-level system.

special class of arrangements in strongly disordered or amorphous materials – two-level systems. Nowadays, it is commonly believed (even though the microscopic nature is still under elucidation) that two-level systems or two-level fluctuators are a local rearrangement of a small group of atoms being in the metastable configuration, which can occupy two slightly different configurations states. Two-level systems can be described by a double-well potential presented in Fig. 6.9. At low temperatures ($k_B T \ll U$), transitions between these states are possible only by quantum-mechanical tunneling. This leads to the energy splitting $E = \sqrt{\Delta^2 + \Delta_0^2}$ of the two eigenstates of a two-level system, where Δ_0 is the tunnelling energy, which is a result of overlapping of the ground state wave functions (right and left). Now this picture is commonly used for the description of the physical properties of a glassy or amorphous system [Black (1981)]. Establishing or readjusting of the point contacts at the low temperatures can cause a strong metal deformation or a lattice distortion that leads to the appearance of that kind of fluctuators in the constriction region. Two-level resistance fluctuations have been observed for the first time by Ralls and Buhrman (1988) for small nanobridges between two metallic films prepared by the electron-beam lithography. This unambiguously proves the possibility of observing two-level fluctuators in the point contacts.

Let us return to the theoretical side of the problem. Kozub (1984) considered inelastic processes at excitation of a two-level system by electrons to an upper level. This contributes to the point-contact spectra as

$$\frac{1}{R}\frac{dR}{dV} \simeq \frac{1}{V}\sum_j C_j \sigma_j \gamma (eV - E_j) \, , \qquad (6.7)$$

where $C_j \approx (d/r_j^4)\min(l,d)$ is a kinetic factor, r_j is the distance of the two-level system from the contact center, σ_j is the inelastic scattering cross-section, and E_j is the excitation energy of the two level system. γ describes a profile of the spectra and contains a sharp peak with a height $\sim E_j/k_BT$ and a width $\sim k_BT$ at $eV = E_j$. An estimation of the inelastic contribution to the current given by Kozub (1984) is about 10^{-3} of the net current.

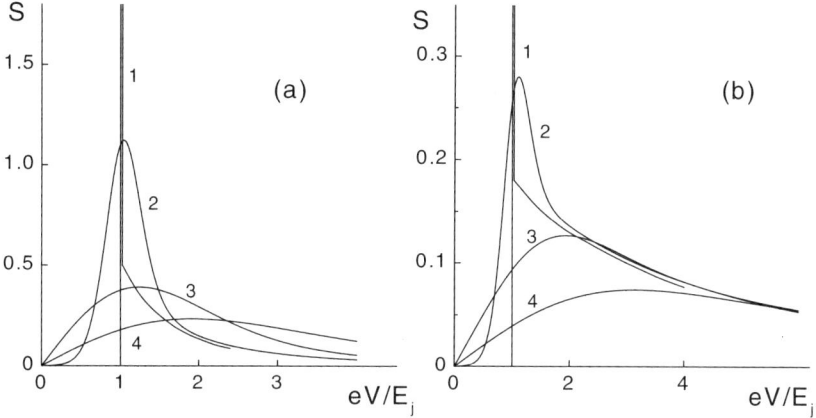

Fig. 6.10. Function $S(v)$ determines a shape of the point-contact spectrum for $q = 0.5$ (a) and $q = 0.1$ (b) with $t = 0, 0.1, 0.5,$ and 1 for curves 1–4, respectively. After Kozub and Kulik (1986).

Kozub and Kulik (1986) analyzed an elastic scattering of electrons on the two-level fluctuators, which is governed by the voltage and temperature dependence of population of the two-level system in their upper level. This leads to the point-contact spectrum:

$$\frac{1}{R}\frac{dR}{dV} = \sum_j \frac{eC_j}{2E_j}(\sigma_j^+ - \sigma_j^-)\tanh\left(\frac{1}{2t}\right) S(v, t, q), \tag{6.8}$$

where $v = eV/E_j$ and $t = k_BT/E_j$ are the reduced voltage and temperature, and $\sigma_j^\pm$ is the scattering cross-section of the upper (+) and lower (-) levels. The parameter q is determined by the electrostatic potential distribution in the contact and is equal to 0.5 for a fluctuator located in the contact center and vanishes on the periphery. The function $S(v, t, q)$ is given in Fig. 6.10. It is interesting that the singularity at $eV = E_j$ contains a δ-function contribution with a wing at larger energies; i. e., $S(v)$ is strongly asymmetric. Equation (6.8) tolerably describes the experimental data in Fig. 6.11 [see also Akimenko and Gudimenko (1993)]. The shape of the elastic line is different from that

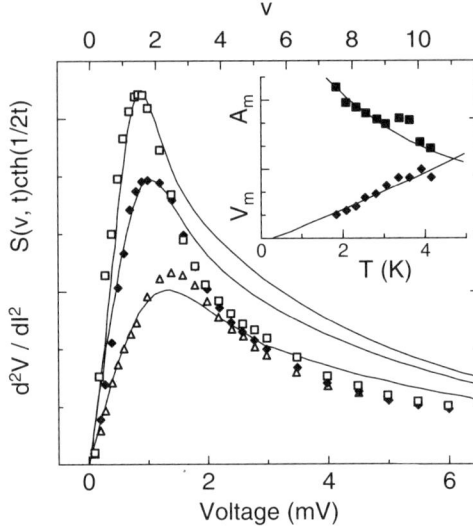

Fig. 6.11. Zero-bias anomaly in $\mathrm{d}^2V/\mathrm{d}I^2$ for a $11.3\,\Omega$ Cu thin-film break junction at different temperatures (symbols) along with theoretical fit by means of (6.8) at $E_j = 0.56\,\mathrm{meV}$, $q = 0.06$ (solid curves). Effective temperatures from the upper curve to the bottom one are 1.84, 2.56, and 3.9 K. Inset shows temperature dependence of the peak position (diamonds) and its height (squares) along with theoretical calculations (solid curves). Data taken from Keijsers et al. (1995).

as in the case of inelastic two-level fluctuators spectroscopy where a peak at $eV = E_j$ is symmetric (6.7). Moreover, the inelastic scattering of electrons by two-level systems is very weak as they represent a two-well potential with a small tunneling probability between wells, i.e., for slowly relaxed two-level systems. On the other hand, the elastic scattering crucially depends on a local position of the two-level system and on the difference in the scattering cross-section as seen from (6.8). The latter defines also a sign of the point-contact features, i.e., both positive and negative value of $\mathrm{d}R/\mathrm{d}V$ can be observed; that is, maximum or minimum on $\mathrm{d}V/\mathrm{d}I$ at zero bias manifest this type of scattering.

Kozub and Rubin (1996) considered also an adiabatic electron "dressing" of defects caused by the static electron response to their potential with renormalization of the fluctuator energy splitting. This leads to an increasing of conductance for the fluctuator in its upper level and, according to (6.8), produces a negative signal as in the case of the Kondo scattering. Temperature rise magnifies occupation numbers of the upper levels and increases the conductance. The same holds for an increase of the applied bias, which further imitates Kondo-like behavior.

Another theoretical concept by Vladar and Zawadowski (1983) suggests that a strong-coupling between conduction electrons and two-level systems results in something analogous to the Kondo condensate regime as in the case of magnetic impurity. This leads to describing of the above-mentioned coupling in the frame of the two-channel Kondo model, taking into account spin degeneracy. Ralph and Buhrman (1992) studied small metallic constrictions at very low temperatures and found both V and T dependence of its conductivity to be logarithmic (Fig. 6.12). They claimed that conductivity has mag-

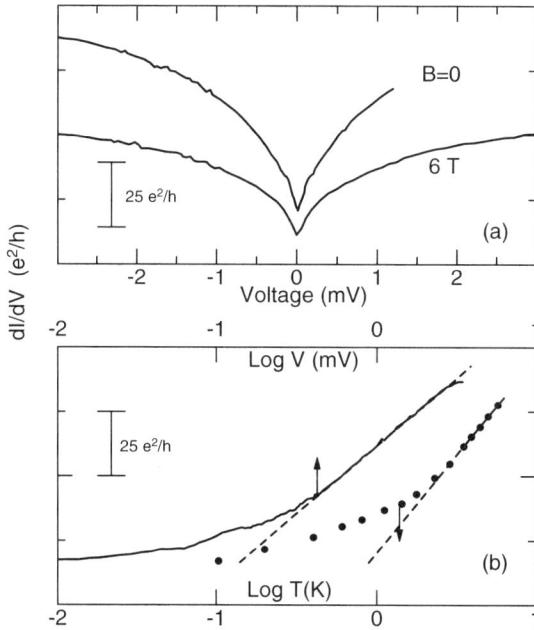

Fig. 6.12. (a) dI/dV of a Cu nanofabricated constriction with $R_0 = 6.5\,\Omega$ at $T = 0.1$ K and modulating signal $V_1 = 20\mu$V. The curve in magnetic field is shifted down by $20\ e^2/h$ for clarity. (b) Voltage (solid curve) and temperature (symbols) dependence of conductivity. Straight dashed lines show logarithmic V and T behavior. Data taken from Ralph and Buhrman (1992).

netic field dependence, which is not characteristic for the Kondo effect, and their behavior is caused by the strong-coupling regime between conduction electrons and the two-level systems. Which model, Kozub-Kulik or Vladar-Zawadowski, corresponds to the real situation is not simple to distinguish. Both models yield slightly different power-law or ln-term for temperature or voltage dependencies. It is also difficult to discriminate this difference properly in the experimental situation, as was stated by Keijsers et al. (1995, 1996) by measuring break junctions both for simple metals and for metallic glasses.

However, as found by Ralph et al. (1994), the scaling prediction, which fits
the two-channel Kondo model, favors the Vladar–Zawadowski approach. It
is worth noting that the Kozub and Rubin (1996) model predicts some special features of the resistance at higher temperatures and biases, beyond the
two-channel Kondo scaling region and holds also for any sort of mobile defects, whereas the two-channel Kondo model describes two-level systems of
more special type with a small asymmetry and large tunneling probability.
Independently, on the model, two-level fluctuators are still believed to be
responsible for some kind of zero-bias anomalies in the point-contact characteristics, and further confirmation for this comes from the measurements of
$1/f$-noise given in Chapter 9.

6.3 Crystal field

Continuously filled from La to Lu, the $4f$-electron shell determines a variety
of physical properties of rare-earth intermetallic compounds. This shell is
pushed deep into the closed $5s$ and $5p$ shells, which is why $4f$ electrons behave
atomic-like. Thus, one can find localized magnetic moments by virtue of the
Hund's rules. In this case, the ground state is $2J+1$ times degenerate, what is
defined by the total angular momentum J. In the crystal environment, which
breaks a spherical symmetry, the degeneration of the ground state is lifted.
This effect influences many transport and thermodynamics properties of the
rare-earth compounds

According to the developed theory by Kulik et al. (1988), a nonlinear contribution to the $I-V$ characteristic of the point contact caused by excitation
of the rare-earth ions to the higher levels can be written as

$$-\frac{\mathrm{d}^2 I}{\mathrm{d}V^2} = \frac{4eL}{3\hbar v_F R_0}\left(g_f(eV) + g_b(eV)\right), \qquad (6.9)$$

where L is the characteristic contact size, the function g_f describes inelastic
excitation of the rare-earth ions by electrons, and g_b represents the contribution of reabsorption of exited ions by electrons as well as elastic scattering of
electrons on ions. The last function represents a so-called background component, which has a smooth dependence compared with g_f. At zero temperature, g_f is given by the expression:

$$g_f(eV) = N(\epsilon_F)n_f \sum_{i>k}(N_k - N_i)\langle|W_{\mathbf{pp'}}^{ik}|^2 K(\mathbf{p}, \mathbf{p'})\rangle\delta(eV - \Delta_{ik}), \qquad (6.10)$$

where n_f is the concentration of the rare-earth ions, $N_{i,k}$ is the occupation
numbers of the levels i and k, $\langle...\rangle$ is the averaging over directions of the
electron momenta $\mathbf{p}$ and $\mathbf{p'}$ on the Fermi surface, $W_{\mathbf{pp'}}^{ik}$, is the matrix element for scattering of an electron from a state $\mathbf{p}$ to a state $\mathbf{p'}$ along with

simultaneous excitation of an ion from a state k to a state i, K is the point-contact geometrical form factor, and Δ_{ik} is the energy of a $k \rightarrow i$ transition. In the case of the three-level system, the calculated g_f and g_b are displayed in Fig. 6.13. The spectra at $T \rightarrow 0$ represent δ-function-like peaks at energies

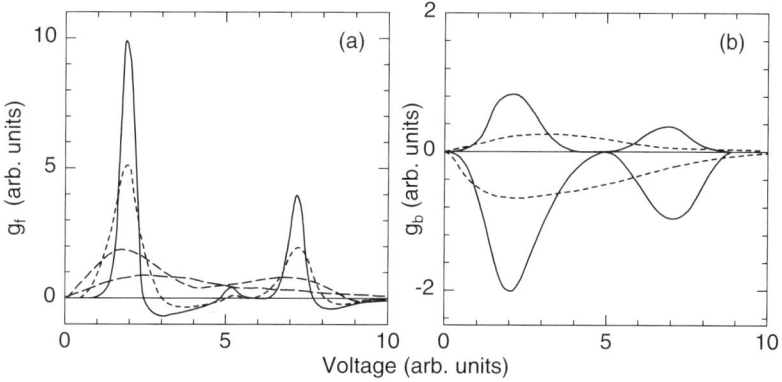

Fig. 6.13. Functions g_f (a) and g_b (b) determining a shape of the point-contact spectra of the three-level system with the distance between levels in dimensionless units $E_1 - E_0 = \Delta_1 = 2$, $E_2 - E_1 = \Delta_2 = 5$ and $|W^{11}|^2 = 1$, $|W^{22}|^2 = 0.3$, $|W^{33}|^2 = 0.2$, $|W^{21}|^2 = 0.5$, $|W^{31}|^2 = 0.2$, $|W^{32}|^2 = 0.3$, $|W^{11} - W^{22}|^2 = 0.6$, $|W^{22} - W^{33}|^2 = 0.6$, $|W^{11} - W^{33}|^2 = 0.8$. Positive sign of curves in (b) corresponds to f-impurities model, whereas a negative sign to the regular lattice of f-centers. The temperature in the same units from top to bottom in (a) is 0.1, 0.2, 0.5, and 1 and for solid and dashed curves in (b) is 0.2 and 1, respectively. After Kulik et al. (1988).

corresponding to the excited ions levels. Moreover, the peaks are asymmetric and have behind the maximum a small negative contribution to the spectra. As seen from Fig. 6.13(b), the behavior of g_b is more flowing with a sign that is dependent on whether the f-ions build a regular lattice or are placed as random impurities. With increasing temperature typical for PCS, smearing of the spectra occurs.

In the intermetallic compound PrNi$_5$, the nine-fold degenerated ground state of Pr^{3+} ion is splitted by the intracrystalline electric field of the hexagonal lattice into three doublets and three singlets (see Fig. 6.14). At low temperatures, the Pr^{3+} ion is in the singlet ground state Γ_4. When the ion is excited, it can move to the level Γ_{5A} as well as to the higher energy levels Γ_3 and Γ_{5B}. Other transitions to the levels Γ_1 and Γ_6 are forbidden by the selection rules. However, in an external magnetic field, this exclusion is lifted (it remains only for the high-symmetrical direction along the hexagonal c axis) and the Zeeman splitting of the spin-degenerated levels occurs (see Fig. 6.14).

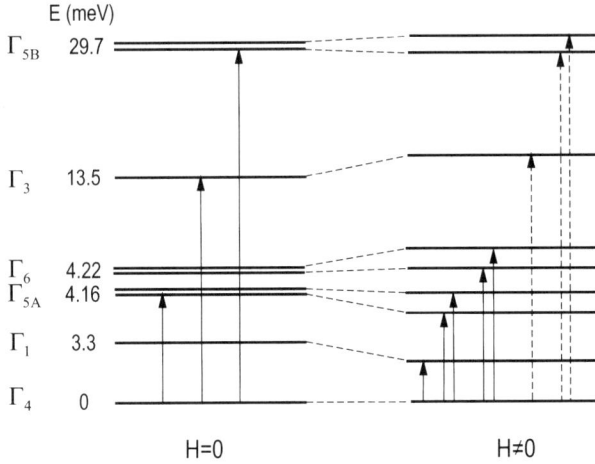

Fig. 6.14. Schematic representation of the degeneracy lifting of the Pr^{+3} ion ground state in the crystal field of the $PrNi_5$ compound and additional Zeeman splitting in an external magnetic field, which is not parallel to the hexagonal c axis. The arrows show the allowed transitions. The dashed arrows show the transition probability that is negligible compared with other ones. The zero-field level schema is according to Table I given by Reiffers et al. (1989).

The processes of the Pr^{3+} ion excitation by electrons, because of which the ion moves to a higher lying level, reveal themselves in the point-contact spectrum as peaks at the corresponding energies. In a zero field, the $PrNi_5$ spectrum [see Fig. 6.15 and Akimenko et al. (1984)] resolves three maxima associated with the transitions shown in Fig. 6.14. The predominate peak in the spectrum around 4 meV corresponds to the excitation of the Pr^{3+} ion to the state Γ_{5A}. Evidently, these processes will give a main contribution into the transport properties of this substance. The probability of the other transitions producing a gentle maxima near 13 and 30 mV [Akimenko et al. (1986)], is appreciably smaller. Respectively, their influence on the electric conductivity and other kinetic and thermodynamic properties of $PrNi_5$ is much weaker. These are the simplest qualitative conclusions that can be made directly from the shape of the $PrNi_5$ point-contact spectrum.

The appearing of transitions to previously forbidden levels in the magnetic field along with the Zeeman splitting of the remaining degenerated levels was observed by Reiffers et al. (1989) and is shown in Fig. 6.15 for $B \neq 0$. The measurements carried out on single-crystalline samples of $PrNi_5$ for three principal crystallographic directions show an anisotropy of the Zeeman splitting. The authors concentrated on the behavior of dominant maximum at 4 meV. The $\Gamma_4 \to \Gamma_3$ transition occurs in the spectra as a gentle shoulder at about 13.5 meV, the intensity of which reduces in a magnetic field because of decreasing of the transition probability, whereas the $\Gamma_4 \to \Gamma_{5B}$ transition

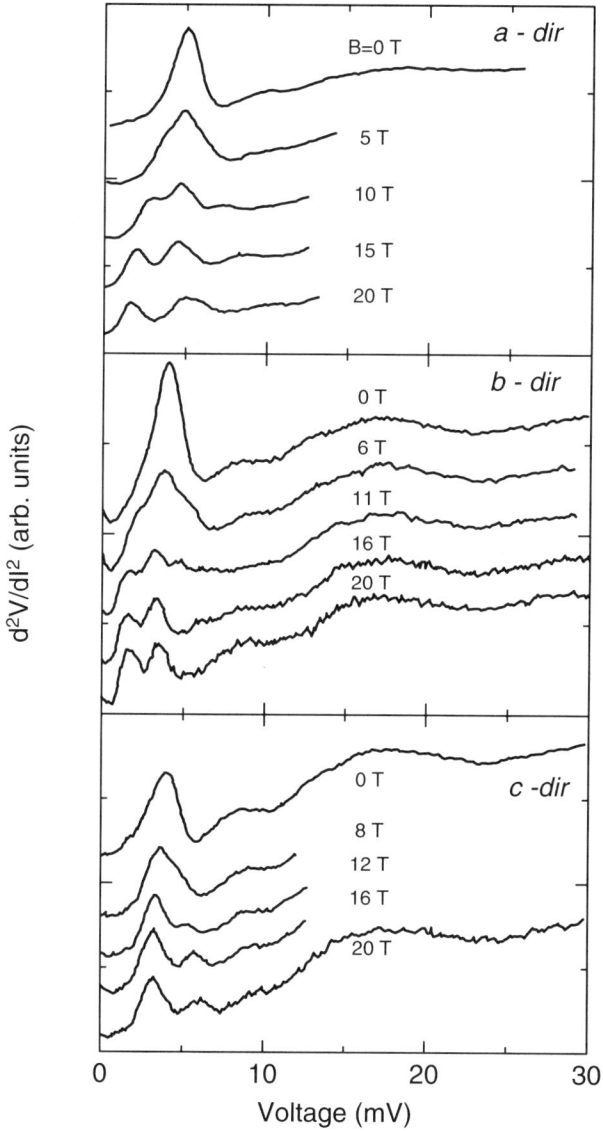

Fig. 6.15. Measured in a magnetic field at $T=1.5\,\mathrm{K}$ $\mathrm{d}^2V/\mathrm{d}I^2$ dependence of PrNi$_5$-Cu point contacts with contact axis and magnetic field along the a, b, and c directions of the hexagonal lattice. The light maxima at about $8\,\mathrm{mV}$ and between 15 and $20\,\mathrm{mV}$ are caused by the phonons. Data taken from Reiffers et al. (1989).

at about 30 meV is hardly distinguishable from the longitudinal Cu phonons. At the beginning, a broadening of the maximum at 4 mV is observed for small magnetic fields below 5 T. This implies that the splitting energy evaluated by the formula $2\Delta = 2g\mu_B B$ (where g is the Lande factor, and μ_B is the Bohr magneton) is smaller for fields below 5 T than the resolution of this technique, which is about 1 mV. A distinct splitting of a 4-meV peak into two maxima occurs at higher fields $B \geq 10$ T parallel to the c axis. It is caused by splitting of the Γ_{5A} doublet. With increasing magnetic field for the a and b directions, a few additional peaks are clearly resolved with even a distinct difference between mentioned directions in the basal plane. Here the forbidden in zero magnetic field $\Gamma_4 \rightarrow \Gamma_1$ transition is clearly observed. The positions of the peaks versus a magnetic field for different contacts and three magnetic field orientations along the main crystallographic directions in PrNi$_5$ are shown in Fig. 6.16. The fit of the data allows us to define the crystal-field level scheme presented in Fig. 6.14 more precisely and to determine a set of the crystal-field parameters [Reiffers et al. (1989)], which also accounts satisfactorily for magnetic susceptibility, specific heat, and thermal expansion. The above results may be regarded as an example of how PCS enables a reconstruction of a quantitative model of the crystal-field level scheme of PrNi$_5$ that represents a useful and direct method for study of the crystal-field effects.

Naidyuk et al. (1990) observed distinct Zeeman splitting even for the polycrystalline PrNi$_5$ samples. It may be possible only if there is one single crystallite in the point-contact region. Hence, qualitatively the Zeeman splitting can also be seen in the case of polycrystal, though for an unknown orientation of single crystallite. By measuring a magnetoresistance of the same PrNi$_5$ contacts, the authors proposed a method to determine point-contact parameters, namely, a constriction diameter and an electron mean free path. They took an analogous formula to (3.19) where they used the magnetic field derivatives of the point-contact resistance and resistivity instead of the derivatives over the temperature. It turned out that all contacts under investigation were at least in the diffusion regime.

Another example is PrAl$_3$, where the lifting of degeneracy of the Pr^{+3} multiplet in hexagonal symmetry crystal field also results in three doublet and three singlet levels, where Γ_1 is the singlet ground state. In this case, only the transition $\Gamma_1 \rightarrow \Gamma_6$ is allowed from the ground state with energy 4.7 meV as follows from neutron measurements by Alekseev et al. (1982). Point-contact spectra of PrAl$_3$ measured by Ponomarenko et al. (1989) also exhibit one dominant peak at 4.5 mV, which appears to be connected with the mentioned transition.

Brück et al. (1986) reported about the observation in the point-contact spectra of PrCu$_6$ a pronounced peak at 5 mV or three features (peaks or shoulders) at about 3, 6, and 9 mV, which are likely caused by the crystal-field splitting. McLean and Lozarich (1984) resolved maximum at 4 mV for spectrum for a clean Pr. They connect the maximum with excitation of propagat-

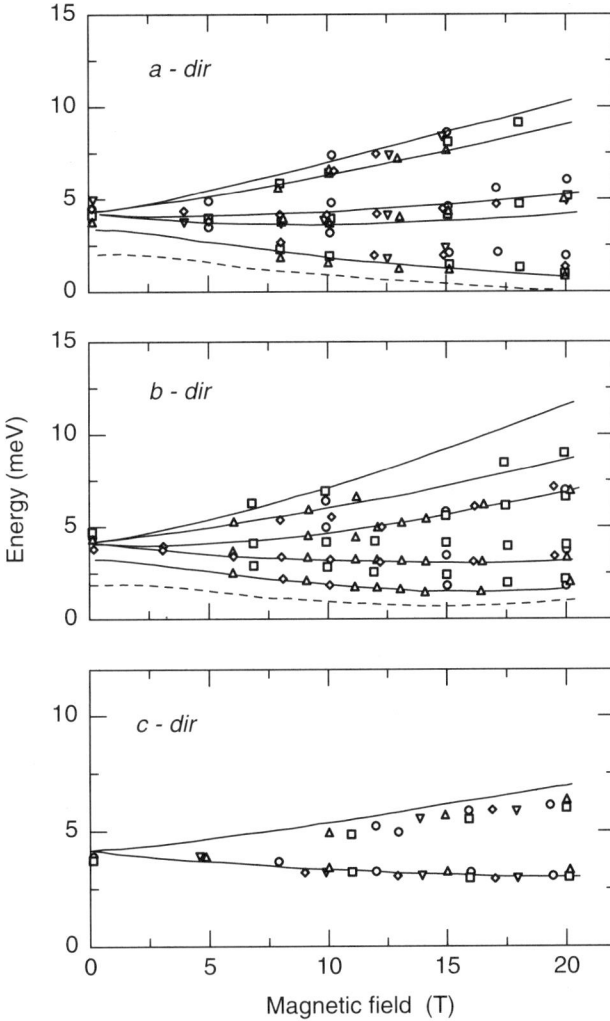

Fig. 6.16. Magnetic field dependence of the maxima or shoulders in the point-contact spectra (see Fig. 6.15) for three main crystallographic directions in PrNi$_5$ (different symbols are for different contacts). The solid curves are calculated to fit better experimental results. The same parameter set was used to reconstruct the crystal-field level scheme in Fig. 6.14. The dashed curves show the calculated energy of the Γ_1 level with previously used crystal-field parameters showing that PCS data improve the crystal-field level scheme. Data taken from Reiffers et al. (1989).

ing crystal-field excitations. It is worth noting that Frankowski and Wachter (1982) considered a low-energy peak at 5 meV in d^2V/dI^2 of $Tm_{0.87}Se$ point contacts as caused by the crystal-field excitations. Nevertheless, a large negative value of the spectra above this energy and a linear dependence of the spectra at zero-bias voltage typical for a spin glass system point out a magnetic nature of this feature. Of course, measurements in a magnetic field are very desirable to make a final conclusion in the above-mentioned cases.

Reiffers et al. (1999) communicated about splitting of the crystal-field levels in a magnetically ordered state below 7 K in $NdNi_5$ without the application of the external magnetic field. Nowack et al. (1997) found in point-contact spectra of $YbBe_{13}$ both zero-bias anomaly and maximum around 3 mV, which were connected with the Kondo scattering and the crystal-field effects, correspondingly. They also raised an issue of modification of the crystal-field scattering by the presence of the Kondo effect.

6.4 Valence and spin fluctuations

Metals in the beginning (end) of the La-series, namely, Ce (Yb), have one superfluous (one missing) electron to make the $4f$-shell empty (filled in). As the empty or filled-in f-shell is more stable, the Ce ions can be "donor" centers adding electrons in the conduction band ($Ce^{+3} \rightarrow Ce^{+4} + e^-$), whereas Yb ions vice versa can capture conducting electrons ($Yb^{+3} + e^- \rightarrow Yb^{+2}$). These processes are known as valence fluctuations and substances as intermediate-valence compounds. In this case, $4f$-level approaches in energy the Fermi level and f electrons strongly hybridize with the conduction electrons. Hence, f electrons may tend to be delocalized. The hybridization causes scattering resonance near the Fermi level leading to virtual bound states of the width $\Delta \sim 10\text{--}100$ meV and correspondingly enhanced electronic DOS proportional to Δ^{-1}. This results in the increase of the electronic specific heat, Pauli susceptibility and so on in an order of magnitude comparatively to the normal metal. On the other hand, if the hybridization is weak, i. e., $4f$-level is situated deep below the Fermi level, the magnetic RKKY interaction between rare-earth ions plays a dominate role, leading to a magnetic ordered state. It takes place by substitution of the rare-earth ions Ce and Yb by elements from the middle of the La-series where both magnetic moments of ions and interaction between ions become stronger. In principle, the same ion, i. e., Ce, depending on the lattice environment (the distance between Ce-ions), may behave in a different way. Thus, Ce-compounds both magnetically ordered like CeB_6 or $CeAl_2$ and intermediate-valence like $CePd_3$ or $CeNi$ are well known.

Intermediate-valence compounds were studied by point contacts as soon as this method was developed (see Bussian et al. (1982), for example). Nevertheless, the Joule heating effects are believed to give the main contribution to the nonlinear conductivity of point contacts in this case. One of a few examples of the nonthermal regime in point contacts of such kind of compounds is the measurements on CeNi$_5$ carried out by Akimenko et al. (1985). Along with a pronounced phonon structure in the point-contact spectra of CeNi$_5$, pointing out at least the diffusive spectral regime, some additional structures (peaks or shoulders) at energies about 2 and 4–5 mV were observed. Position and intensity of the peaks were found to be dependent on the polarity of an applied voltage in the case of CeNi$_5$-Cu heterocontacts (Fig. 6.17). Considering this

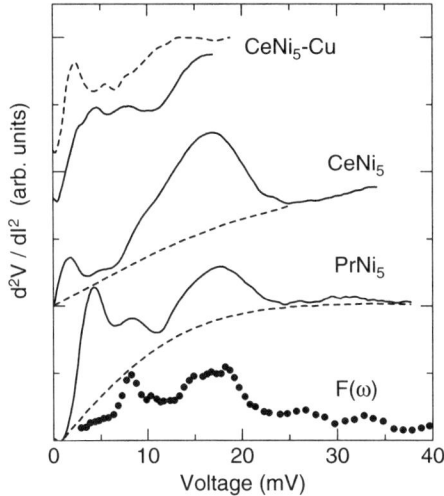

Fig. 6.17. Point-contact spectra of CeNi$_5$ and PrNi$_5$ homocontacts with resistances about 12 Ω at $T = 4.2$ K. Bottom curve (symbols) represents dependence of the phonon density of states in LaNi$_5$. Smooth dashed lines are the assumed background. Two upper curves show the asymmetry versus bias polarity of the low-energy spectral part in the case of a CeNi$_5$-Cu heterocontact. Solid curve corresponds to positive polarity on CeNi$_5$. Curves are shifted vertically for clarity. Data taken from Akimenko et al. (1985).

observation as well as a comparison with the phonon density of states of the isostructural LaNi$_5$ compound, both the crystal-field excitation of the Ce^{+3} ions and electron–phonon interaction cannot be a reason for these features. Akimenko et al. (1985) proposed the following explanation of the nature of the low-energy peaks. Conduction electrons with the excess energy eV in the point contact can be captured by Ce^{+4} ions to the $4f$-shell, which gives an additional contribution to the contact resistance. On the other hand, holes can be captured by the Ce^{+3} ions. The proposed energy scheme is the following. The energy position of two Ce-ion configurations are arranged near the Fermi level, however, only on the opposite sides: The Ce^{+3} configuration has an energy position below ($\sim$ 2 meV) and Ce^{+4} above ($\sim$ 1 meV) the Fermi level. A level width was estimated as 6–8 meV and 1.5–2 meV for the Ce^{+3} and Ce^{+4} ions, correspondingly, which leads to the averaged valence about 3.8.

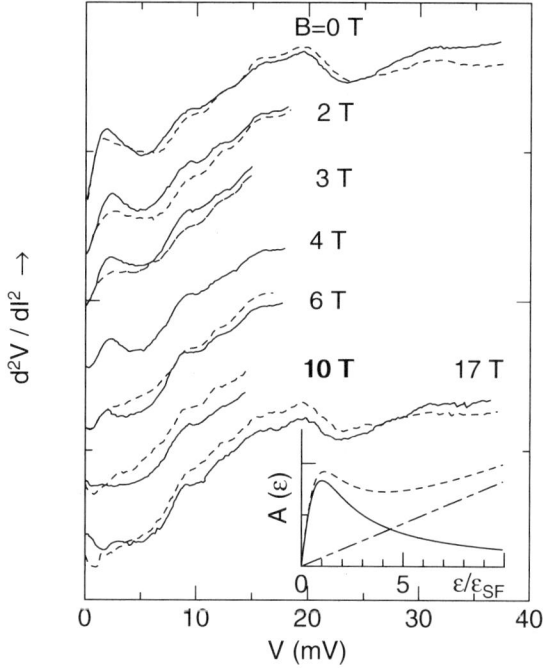

Fig. 6.18. Magnetic field dependence of the point-contact spectrum of a $CeNi_5$-Cu heterocontact at 1.5 K for two bias polarity. Solid curves correspond with positive polarity on $CeNi_5$. Note: A phonon-related contribution to the spectra above 5 mV remains unchanged in a magnetic field as high as 17 T. Curves are shifted vertically for clarity. Inset: The spectral function of paramagnons $A(\epsilon)$ (solid curve) according to (6.11). Dashed curve represents $A(\epsilon)$ superimposed on linear background marked by dash-dotted line. Data taken from Naidyuk et al. (1992).

At the same time, there is a contradiction concerning the valence state of Ce in the $CeNi_5$ among different authors. In particular, Gignoux and Gomez-Sal (1985) stated that this compound did not even belong to the intermediate-valence systems and that the Ce-ions were in the nonmagnetic state with the maximal valence Ce^{+4}. Moreover, the magnetism of Ni atoms in $CeNi_5$ should be taken into account. A broad maximum in the magnetic suscepti-bility at 100 K is caused by the thermal smearing of the $3d$ electronic DOS enhanced by the spin fluctuations. Naidyuk et al. (1992) investigated single crystal samples of $CeNi_5$ in a magnetic field and found similar features in the point-contact spectra at low biases. However, the asymmetry in position and intensity of the low-energy peaks was not so expressed here (Fig. 6.18) as should be, if the mechanism Akimenko et al. (1985) works. Additionally, only one low-energy peak was usually clearly resolved at about 2–4 mV, the position and intensity of which depend mainly on the crystallographic orientation in constriction. By applying the magnetic field, this maximum is depressed and disappears by reaching 10 T. The authors consider the interaction of electrons with spin fluctuations or paramagnons as a reason for the appearance of this peak. The spectral function of the paramagnons can be written according to Kaiser and Doniach (1970) as

$$A(\tilde{\epsilon}) = a \frac{\tilde{\epsilon}}{1 + \tilde{\epsilon}^2}, \tag{6.11}$$

where $\tilde{\epsilon} = \epsilon/(k_B T_{sf})$. Hence, $A(\tilde{\epsilon})$ increases linearly with the energy ϵ followed by a maximum at the corresponding spin-fluctuation temperature T_{sf} and then falls off at higher energies. Naidyuk et al. (1992) found a qualitative agreement with this model, e. g., a linear zero-bias dependence of d^2V/dI^2 (see Fig. 6.18 and inset). Again the expected suppression of the spin fluctuations by magnetic field is obviously seen in the spectra (Fig. 6.18). Naidyuk et al. (1994) examined also asymmetry of the point-contact characteristics in $CeNi_5$-Cu heterocontacts and concluded that it can be well explained by including thermoelectric effects.

Relating to the valence-fluctuation effect, it is worth noting that the presence of ions with different valence in the point contact corresponds to the high concentration of the impurities. This leads to a short elastic mean free path of electrons of the order of the lattice constant. Therefore, point contacts with real valence-fluctuation compounds are in the thermal limit (see Section 7.4).

References

Akimenko A. I., Ponomarenko N. M., Yanson I. K., Janos Š. and Reiffers M. (1984) Sov. Phys. Solid State **26** 1374.

Akimenko A. I., Ponomarenko N. M. and Yanson I. K. (1985) JETP Lett. **41** 286; (1986) Sov. Phys. Solid State **28** 615.

Akimenko A. I. and Gudimenko V. A. Solid State Commun. (1993) **87** 925.

Alekseev P. A., Sadikov I. P., Shitikov Yu. L. (1992) Phys. Stat. Sol. (b) **114** 161.

Andrei N. (1980) Phys. Rev. Lett **45** 379; (1982) Phys Lett. A **87** 299.

Appelbaum J. A. (1967) Phys. Rev. **154** 633.

Black J. L. (1981) in *Glassy Metals* ed. by H. J. Güntherodt and H. Back (Springer Verlag, Berlin) 167.

Brück E., Nowack A., Hohn N., Paulus E. and Freimuth A. (1986) Z. Phys. B. **63** 155.

Bussian B., Frankowski I. and Wohlleben D. (1982) Phys. Rev. Lett. **49** 1026.

d'Ambrumenil N. and White R. M. (1982) J. Appl. Phys. **52** 2052.

Daybell M. D. 1973 in *Magnetism*, ed. by G. Rado and H. Suhl (Academic Press, New York), **5** 121-147.

Duif A. M., Jansen A. G. M. and Wyder P. (1987) J. Phys. Cond. Matter **1** 3157.

Frankowski I. and Wachter P. (1982) Solid State Commun. **41** 577.

Gignoux D. and Gomez-Sal J. C. (1985) J. Appl. Phys. **57** 3125.

Jansen A. G. M., van Gelder A. P., Wyder P. and Strässler S. (1981) J. Phys. F: Metal Phys. **11** L15.

Kaiser A. B. and Doniach S. (1970) Intern. J. Magnetism **1** 11.

Keijsers R. J. P., Shklyarevskii O. I. and van Kempen H. (1995) Phys. Rev. **51** 5628.

Keijsers R. J. P., Shklyarevskii O. I. and van Kempen H. (1996) Phys. Rev. Lett. **77** 3411.

Kolesnichenko Yu. A., Omelyanchouk A. N. and Tuluzov I. G. (1996) Physica B **218** 73.

Kolesnichenko Yu. A., Omelyanchouk A. N., van der Post N. and Yanson I. K. (1997) Low Temp. Phys. **23** 934.

Kondo J. (1964) Progr. Theor. Phys. (Kyoto) **32** 37.

Kozub V. I. (1984) Sov. Phys. Solid State **26** 1186.

Kozub V. I. and Kulik I. O. (1986) Sov. Phys. - JETP **64** 1332.

Kozub V. I. and Rubin A. M. (1996) Physica B **218** 64.

Kulik I. O., Omelyanchouk A. N. and Tuluzov I. G. (1988) Sov. J. Low Temp. Phys. **14** 82.

McLean A. B. and Lozarich G. G. (1984) J. Phys. F: Met. Phys. **14** L185.

Lysykh A. A., Yanson I. K., Shklyarevskii O. I. and Naidyuk Yu. G. (1980) Solid State Commun. **35** 987.

Naidyuk Yu. G., Shklyarevskii O. I. and Yanson I. K. (1982) Sov. J. Low Temp. Phys., **8** 362.

Naidyuk Yu. G., Reiffers M., Jansen A. G. M., Wyder P. and Yanson I. K., (1990) Sov. J. Low Temp. Phys. **16** 522.

Naidyuk Yu. G., Reiffers M., Jansen A. G. M., Wyder P., Yanson I. K., Gignoux D. and Schmitt D. (1992) Int. J. Mod. Phys. **7** 222.

Naidyuk Yu. G., Reiffers M., Omelyanchouk A. N., Yanson I. K., Jansen A. G. M. and Wyder P. (1994) Physica B **194-196** 1321.

Nowack A., Wasser S., Schlabitz W., Kvitnitskaya O. E., and Fisk Z. (1997) Phys. Rev. B **56** 14964.

Omelyanchouk A. N. and Tuluzov I. G. (1980) Sov. J. Low Temp. Phys., **6** 626.

Omelyanchouk A. N. and Tuluzov I. G. (1985) Sov. J. Low Temp. Phys., **11** 211.

Ponomarenko N. M., Akimenko A. I., Yanson I. K., Burkhanov G. S., Chistyakov O. D. and Kol'chugina N. B. (1989) Sov. Phys. Solid State **31** 1970.

Ralls K. S. and Buhrman R. A. (1988) Phys. Rev. Lett. **60** 2434.

Ralph D. C. and Buhrman R. A. (1992) Phys. Rev. Lett. **69** 2118.

Ralph D. C., Ludwig A. W. W., von Delft J. and Buhrman R. A. (1994) Phys. Rev. Lett. **72** 1064.

Reiffers M., Naidyuk Yu. G., Jansen A. G. M., Wyder P., Yanson I. K., Gignoux D. and Schmitt D. (1989) Phys. Rev. Lett. **62** 1560.

Reiffers M., Salonova T., Gignoux D. and Schmitt D. (1999) Europhysics Lett. **45** 520.

Vladar K. and Zawadowski A. (1983) Phys. Rev. **28** 1564, 1582, 1596.

Wiegmann P. B. (1980) JETP Lett. **31** 364.

Yanson I. K., Fisun V. V., Hesper R., Khotkevich A. V., Krans J. M., Mydosh J. A. and van Ruitenbeek J. M. (1995) Phys. Rev. Let. **74** 302.

Zarand G. and Udvardi L. (1996) Physica B **218** 68.

7 Thermal effects in point contacts

7.1 Thermal conductivity of point contacts

It is well known that the electron contribution to the thermal conductivity dominates in the metals. As a consequence, the thermal conductivity of pure metals is one or two orders of magnitude higher than that of semiconductors or dielectrics. If the relaxation time of electrons is identical for electrical and thermal processes, then electrical and thermal conductivity is connected by the well-known Wiedemann–Franz law [Kittel (1986)]. Theoretical treatment of the heat transfer through a point contact in the ballistic regime when a temperature gradient is applied to the contact instead of (or together with) a potential difference was done by Bogachek et al. (1985b). In the zeroth-order approximation, the heat flux was calculated analogously to the ballistic injection of electrons as

$$Q(T) = \frac{\pi^2}{3} \left(\frac{k_B}{e} \right)^2 \frac{T}{R_0} \Delta T, \qquad (7.1)$$

where $\Delta T = T_2 - T_1$ is the difference between temperatures on both sides of contact T_1 and T_2, and R_0 is the Sharvin resistance. If we determine the thermal resistance of a point contact as $R_T = \Delta T / Q$, then

$$\frac{R_0}{R_T} = \frac{\pi^2}{3} \left(\frac{k_B}{e} \right)^2 T = L_0 T, \qquad (7.2)$$

where L_0 is the Lorenz number. That is the Wiedemann–Franz law is valid for point contacts as well.

The heat transport through the metallic contacts has been investigated by Shklyarevskii et al. (1986b). The measured heat flux Q as a function of the contact resistance is shown in Fig. 7.1. A good agreement between (7.1) and the experimental data was found for very low ohmic contacts, whereas the heat flux is sufficiently higher as expected for contacts with resistances above 0.01 Ω. This is connected with the contribution caused by phonon heat transfer through the current nonconducting regions of the mechanical contact. The authors found also a quadratic temperature dependence of the heat flux for low ohmic contacts in accordance with (7.1) if $T_2 \gg T_1$. From (7.2), they

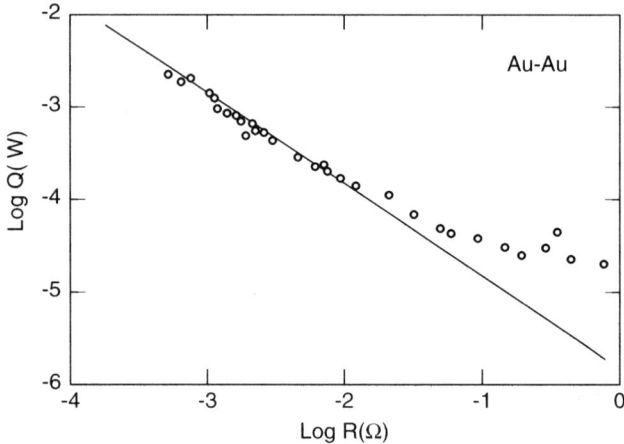

Fig. 7.1. Resistance dependence of the heat flux through an Au-Au contact. The electrodes were kept at $T_1 = 4.2\,\mathrm{K}$ and $T_2 = 10\,\mathrm{K}$. The solid line is calculated according to (7.1). Data taken from Shklyarevskii et al. (1986b).

determined a temperature dependence of the Lorenz number, which appears to be the same as for the bulk transport.

Scattering of electrons has to influence the heat flux through the contact. The calculation of the electron–phonon interaction contribution to the thermal conductivity of point contact yielded an amazing result. It is proved that the second derivative of the first-order heat flux Q^I versus voltage is directly related to the electron–phonon interaction function $\alpha^2 F$ [Bogachek et al. (1985b)]:

$$\frac{\mathrm{d}^2 Q^I}{\mathrm{d}V^2} = \frac{\pi A}{3} V e^3 d^3 N(\epsilon_F) \alpha^2 F(T, V), \tag{7.3}$$

where $A = 1$ in the ballistic regime and $A = 16\,l_{\mathrm{el}}/3\pi d \ll 1$ in the diffusive regime. It is difficult to measure this derivative, and this has not been done experimentally yet. Nevertheless, Reiffers et al. (1986) observed specific features of the heat conductivity in the point contacts. They attached one of the electrodes to the heat exchanger, while the other one was held in vacuum (Fig. 7.2). Both electrodes carried miniature pick-ups to check the electrode temperature. The temperature was found to be rising slightly with the voltage in both electrodes. However, the temperature increase was different depending on the polarity of the bias voltage. When the temperature difference or its asymmetry versus the bias voltage was plotted, the curve turn-out looked similar to the electron–phonon interaction function of the contacting metals (Fig. 7.3). The experiment provides an evidence that the electron–phonon interaction affects the heat conductivity in point contacts and confirms, although not directly, the above-mentioned theoretical prediction. It should be

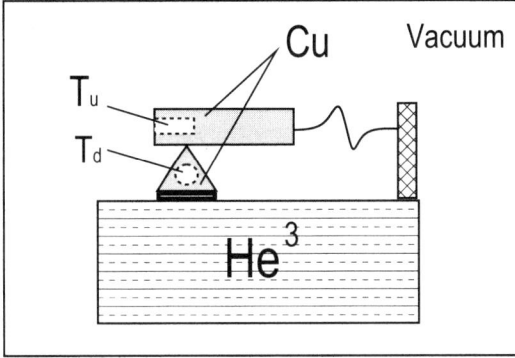

Fig. 7.2. Experimental arrangement for measuring thermal effects in point contacts with pick-up thermometers (u) and (d) inside the electrodes. Adapted from Reiffers et al. (1986).

noted that the temperature difference ΔT_{d} in Fig. 7.3 has an additional fine structure. The positions of these maxima and minima correlate well with the extremes in the noise spectra of Cu measured by Akimenko et al. (1984). This can be related to peculiarities of the electron–phonon scattering, caused by the presence of N (normal), U (umklapp), and multiphonon processes.

7.2 Thermoelectric effects in ballistic regime

Let us suppose that the contact is affected by a temperature gradient. Let one electrode be held at the helium temperature T_0 and the other at $T > T_0$. For an electrical current in a ballistic contact, Bogachek et al. (1985b) obtained the following equation:

$$I^0 = \frac{emd^2}{8\pi\hbar^2} \int_0^\infty d\epsilon \left(f\left(\frac{\epsilon - eV/2 - \mu_2}{k_B T_2} \right) - f\left(\frac{\epsilon + eV/2 - \mu_1}{k_B T_1} \right) \right), \quad (7.4)$$

with distribution functions f and chemical potential μ in the left and right electrodes taken at different temperatures. At zero current, this relation yields a thermopower caused by electron diffusion from the hot electrode to the cold one. Additionally, phonon flow from the hot part to the cold one "drags" electrons. This causes, so-called, phonon drag thermopower. Consider an electrical circuit as shown in Fig. 7.4, where the electrodes (1) and (2) and connecting wires are made of the same material. The voltage measured by the voltmeter in this case is caused by the difference between thermovoltages of the point contact $\mathcal{E}_{\mathrm{pc}}$ and the bulk material:

$$V_{\mathrm{ab}} = \mathcal{E}_{\mathrm{PC}} - \int_{T_1}^{T_2} S(T)\, dT, \quad (7.5)$$

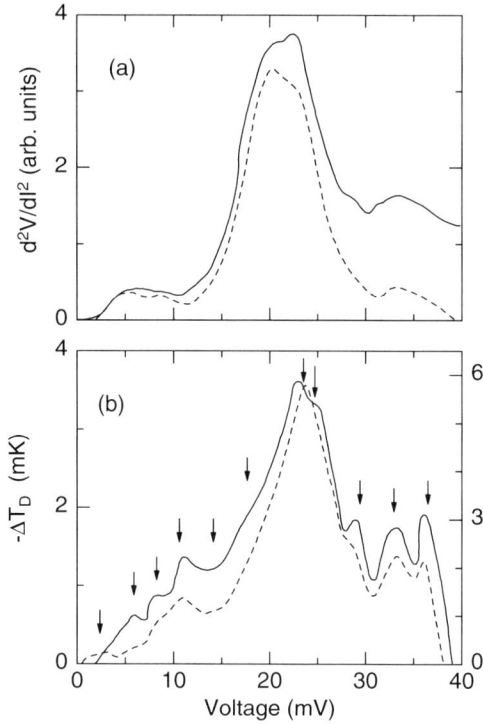

Fig. 7.3. (a) Point-contact spectrum of Cu (solid curve) for a contact with $R_0 = 0.84\,\Omega$ at $T_0 = 0.72\,\mathrm{K}$. Dashed curve shows the same spectra after subtracting of background. (b) Temperature asymmetry $\Delta T_\mathrm{d} = T_\mathrm{d}(+V) - T_\mathrm{d}(-V)$ versus bias voltage V (see Fig. 7.2). Dashed curve depicts $T_\mathrm{d}(-V)$ with subtracted contribution from Joule heating $\propto V^2/R_0$. The arrows show peculiarities in the noise spectra of Cu from Akimenko et al. (1984). Data taken from Reiffers et al. (1986).

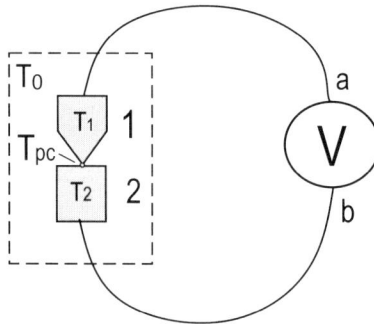

Fig. 7.4. Electrical circuit for a point contact between two metals (1) and (2) kept at temperatures T_1 and T_2 inside the bath with the temperature T_0.

where $S(T)$ is the Seebeck coefficient of a material building the circuit. As was shown by Bogachek et al. (1985a), the thermopower of the point contact is not the same as in the bulk. Namely, drag-related thermopower in the ballistic constriction has an additional small term $d/l_{\rm in}$. Therefore, the thermopower is suppressed in comparison with that of the bulk metal. As the diffusive thermopower of the point contact does not change compared with the bulk[1], voltage $V_{\rm ab}$ measured in the circuit is proportional to the drag-related thermopower.

Hence, for a homocontact where the electrodes are maintained at different temperatures, a potential difference will occur. This happens because an electric circuit contains a "bottle-neck" caused by the point contact, where kinetic processes are specific. It is "hard" for the phonons to drag the electrons through the constriction, which results in the drag thermopower suppression. This, probably, offers a unique possibility of measuring the absolute thermopower in a material-homogeneous electrical circuit. Such experiments

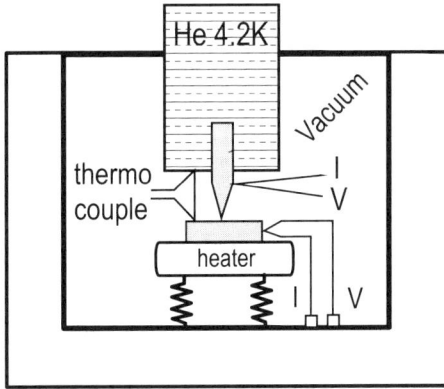

Fig. 7.5. Experimental arrangement for measuring thermoelectric effects in point contacts. The circuit between V and I leads is made of the same material. Adapted from Shklyarevskii et al. (1986a).

were done by Shklyarevskii et al. (1986a). The experimental setup is shown in Fig. 7.5. One part of a contact—the sharp Ag needle—is at a liquid helium temperature. The another part—Ag slab—is thermally attached to the heater and could be heated up to the room temperature. The current and voltage leads are made of the same starting material. The measured thermopower versus temperature is given for an Ag point contact in Fig. 7.6. It is seen that the thermopower resembles the bulk one even in absolute values. This occurs because of the prevalence of the phonon-drag thermopower for

---

[1] As was shown theoretically by Bogachek et al. (1989), the diffusion part of thermopower should also be suppressed in the point contact, but less than the part associated with the drag.

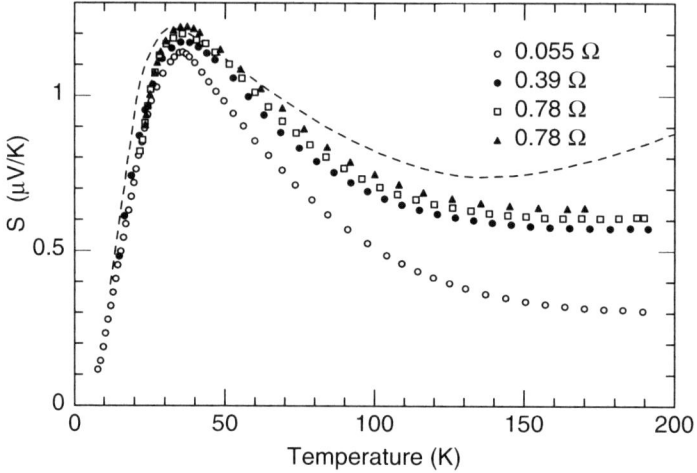

Fig. 7.6. Measured thermopower between two voltage leads versus temperature for Ag point contacts with different resistance. The dashed curve represents the thermopower of bulk Ag. Data taken from Shklyarevskii et al. (1986a).

the noble metals at low temperatures. The latter is suppressed in the contact, which leads to $\mathcal{E}_{PC} \to 0$ in (7.5) so that V_{ab} behavior is determined by the bulk thermopower. Further extensive studies of thermoeffects in homogeneous circuits with point contact for noble and transition metals as well as Kondo alloy AuMn have been conducted by Shklyarevskii et al. (1989).

7.3 Thermal regime in clean metals

The thermal regime in the point contacts was examined by Verkin et al. (1979) during the study of the clean ferromagnetic metals. Along with the electron–phonon scattering, a strong magnetic (electron–magnon) scattering occurs in these metals. For example, the electrical resistivity of an ferromagnetic Fe at the room temperature (close to the Debye temperature) exceeds resistivity of Cu about six times [Kittel (1986)]. Correspondingly, the electron mean free path decreases what favors the thermal regime. Above the Curie temperature T_C, transition of a ferromagnet into paramagnetic state results in a kink in the resistivity, because magnons and associated scattering disappear. As a consequence, the contact resistance at the voltage corresponding to T_C has features similar to that of $\rho(T)$ originated from the Maxwell term. This peculiarity in $R_{PC}=dV/dI$ is more expressed in the second derivative $d^2V/dI^2 \approx dR_{PC}/dV$ of the $I-V$ characteristic. For point contacts in ferromagnetic metals Ni, Fe, and Co (see Fig. 7.7), a maximum in the dV/dI curve and a N-type feature in d^2V/dI^2 appears. The voltage position of the

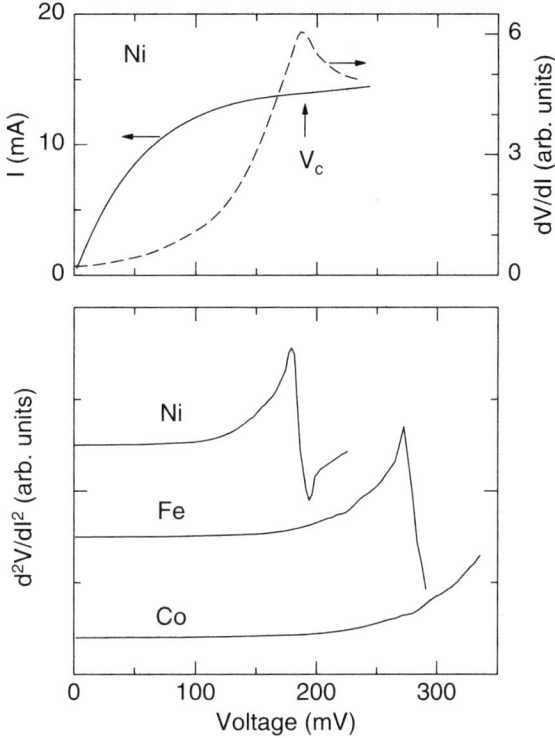

Fig. 7.7. Upper panel: $I - V$ characteristic (solid curve) and its derivative $\mathrm{d}V/\mathrm{d}I$ (dashed curve) of a Ni point contact at $T = 4.2\,\mathrm{K}$. Bottom panel: $\mathrm{d}^2V/\mathrm{d}I^2$ of the same Ni contact and of Fe and Co point contacts. The EPI features at low biases on $\mathrm{d}^2V/\mathrm{d}I^2$ (see Fig. 5.9) are not seen because of their negligible intensity compared with the thermal peculiarities. Data taken from Verkin et al. (1979).

peculiarities (maximum in $\mathrm{d}V/\mathrm{d}I$ or zero for $\mathrm{d}^2V/\mathrm{d}I^2$) corresponds well to the Curie temperatures T_C by virtue of the formula (3.22): $eV_\mathrm{c} \simeq 3.63\,k_\mathrm{B}T_\mathrm{C}$. By substitution of nickel by copper or beryllium atoms, T_C decreases and the feature in $\mathrm{d}^2V/\mathrm{d}I^2$ also shifts to the lower voltages (Fig. 7.8), which is in agreement with the thermal model. Some rare-earth metals are ordered magnetically below the room temperature. Correspondingly, thermal features in $\mathrm{d}^2V/\mathrm{d}I^2$ were observed in Gd, Tb, Dy, and Ho at voltages below $100\,\mathrm{mV}$ [Akimenko et al. (1982)]. All of this testifies to the fact that actually the peculiarities are caused by the metal transition from the magnetic to the nonmagnetic state when Curie or Néel temperature in the contact is reached.

The observation of the critical voltage V_s may be considered as one of the straightforward and easy-to-grasp examples of the temperature increasing inside the contact. This voltage cannot be exceeded in the point contact, no

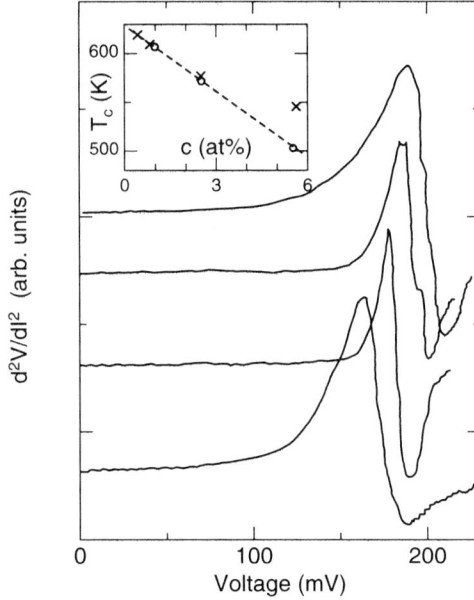

Fig. 7.8. $\mathrm{d}^2V/\mathrm{d}I^2\,(V)$ characteristics of $\mathrm{Ni}_{1-x}\mathrm{Be}_x$ point contacts at $T = 4.2$ K (x = 0.004, 0.04, 0.025, and 0.057 from the top curve to the bottom one). Inset: Curie temperature determined by magnetic (open circles) and point-contact measurements (crosses) for different x. Data taken from Naidyuk et al. (1984).

matter how much the current rises. The nature of this phenomenon is that each metal has a so-called "softening" temperature, which is about one-third of the melting one [Holm (1967)]. At this temperature, a metal readily tends to plastic deformation. When the voltage reaches a corresponding "softening" value, the contact spot will increase under the action of pressing force and the short resistance will decrease, so that even by the current buildup, V_s cannot be exceeded. This was observed for many metals with different melting temperatures, which correlates with V_s [Holm (1967)]. That is the reason why the N-type feature is not observed for Co (see Fig. 7.7). It was impossible to reach the Curie temperature in the contact, which appears to be higher than the "softening" one.

The criterion of thermal regime is also the dependence of the peculiarities position on bath temperature T_0. As follows from (3.21), the required temperature in the contact is reached at lower voltages, the higher the bath temperature is. Thus, all kinds of anomalies in $\mathrm{d}^2V/\mathrm{d}I^2$, if they are of a thermal nature, will shift over the voltage scale depending on the bath temperature T_0. At the same time, the spectral features in $\mathrm{d}^2V/\mathrm{d}I^2$ do not change their positions with the temperature, but they only broaden in accordance with (3.4). For example, EPI maxima are almost smeared above 50–100 K, but their position remains the same [van Gelder et al. (1980)]. On the contrary, thermal anomalies can be found, e. g., for the ferromagnetic metals even at room temperature, only they are shifted toward the lower voltages compared with the measured ones at helium temperature.

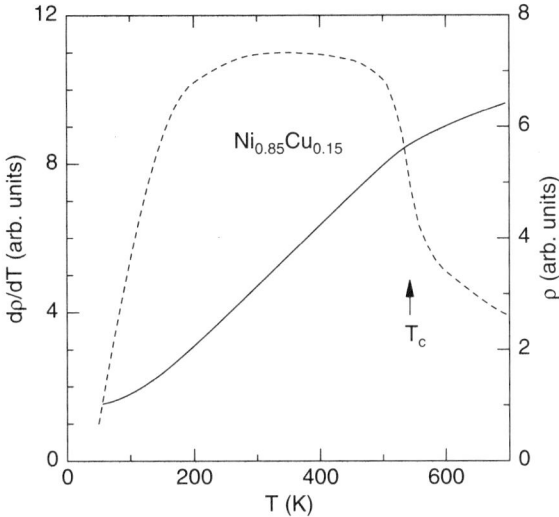

Fig. 7.9. Temperature dependence of the resistivity (solid curve) and its derivative (dashed curve) of the $Ni_{0.85}Cu_{0.15}$ alloy calculated from d^2V/dI^2 (V) of the point contact. Data taken from Verkin et al. (1980).

Using (3.23), the $\rho(T)$ dependence can be reconstructed from the dV/dI or d^2V/dI^2 characteristics as it was done first by Verkin et al. (1980) for NiCu alloys (see Fig. 7.9). Here a kink in the calculated $\rho(T)$ and drop in $d\rho(T)/dT$ are clearly resolved at T_C. Conversely, using $\rho(T)$, an $I - V$ curve and its derivatives for point contact can be found in the thermal limit. Such calculations were performed, and they showed a good agreement with experimental curves for the clean metal Cr [Khotkevich and Krainyukov (1991)], for the intermediate-valence compounds [Paulus (1985)], and for the heavy-fermion systems [Jansen et al. (1987)]. The above-mentioned compounds have usually a high resistivity and thermal effects play here a crucial role (see Chapter 14 for details). Thus, such calculations prove the thermal model for point contacts both qualitatively and quantitatively.

A modified "local heating model" based on the idea developed by Tinkham et al. (1977) was proposed by Negishi et al. (1993) during a study of magnetic metals Ni and α-Mn by point contacts. They stated that when a constant additive correction to the bulk resistivity near a contact interface

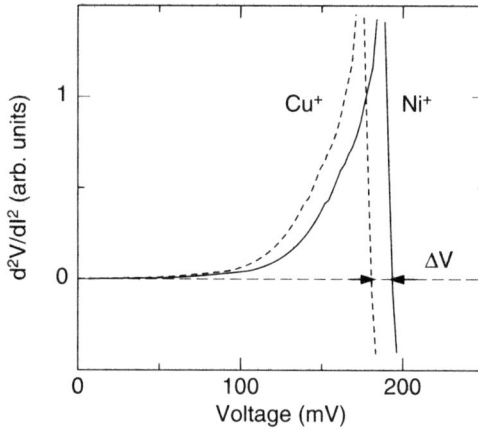

Fig. 7.10. d^2V/dI^2 (V) characteristics of a Ni-Cu heterocontact measured in the thermal regime for two bias polarities at $T = 4.2\,\text{K}$. Data taken from Naidyuk et al. (1985).

is included, the calculated $I - V$ curves and their derivatives are in good agreement with the measured experimentally.

A thermal equilibrium in constriction is established with a virtually small delay of about 10^{-9}–10^{-10} s because of the small size of the contact as shown by Balkashin and Yanson (1982) by study of nonstationery phenomena in point contacts. Therefore, using an alternating signal for detecting of derivatives, a temperature in the point contact can also be modulated. This can be considered as a temperature-modulating spectroscopy [Verkin et al. (1980)]. The advantages of this method are that: (1) a simple measurement of dV/dI or dR/dV reflects the main features in $\rho(T)$, or in $d\rho(T)/dT$ (2) the absence of a system for controlling and stabilizing the sample temperature and, (3) possibility of using small (even submilimeter size) quantities of the substance.

7.4 Heterocontacts in thermal regime

Another consequence of heating in the thermal regime is the asymmetry of $I-V$ characteristics of heterocontacts depending on the polarity of an applied voltage [Naidyuk et al. (1985), Paulus and Voss (1985)]. The heating of the contact results in appearance of the Seebeck voltage over the contact because of the different Seebeck coefficients of contacting materials. Considering a circuit as presented in Fig. 7.4 and supposing $T_0 = T_1 = T_2 < T_{\text{PC}}$, the thermoelectric voltage is

$$V_{\text{ab}} = \int_{T_0}^{T_{\text{PC}}} (S_1 - S_2)\mathrm{d}T, \tag{7.6}$$

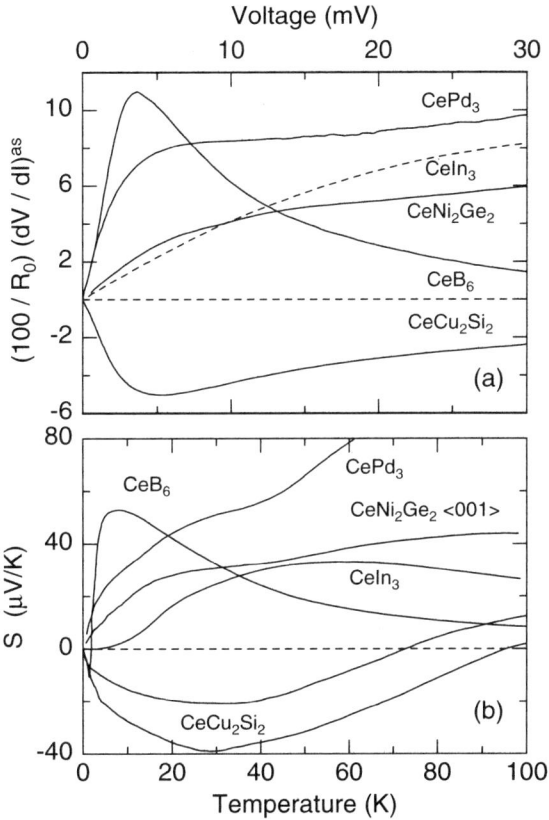

Fig. 7.11. (a) Asymmetric part of dV/dI of heterocontacts based on different Ce compounds. (b) Seebeck coefficients of the same compounds. Data taken from Paulus (1985).

where S_1 and S_2 are the Seebeck coefficients for metals (1) and (2). This voltage adds to the bias voltage in one polarity and subtracts in another one, which results in the shift between thermal peculiarities in d^2V/dI^2 curves measured for two polarities of biases on the value $\Delta V = 2V_{ab}$. It was found by Naidyuk et al. (1985) (see Fig. 7.10) that $2V_{ab} \simeq 10\,\mathrm{mV}$ for Cu-Ni heterocontacts, and calculation of $2V_{ab}$ by (7.6) yields $22\,\mathrm{mV}$. The agreement is satisfactorily taking into account simplicity by evaluation. Besides the Peltier effect will cause an extra heating or cooling of the contact, depending on the direction of the current flow. Moreover, the Thompson effect has to be considered as mentioned by Naidyuk et al. (1985) because of the presence of both temperature and electric-field gradients.

Basically, thermoelectric effects lead to asymmetry of the $I - V$ characteristic in the thermal limit. The asymmetric part of the differential resistance

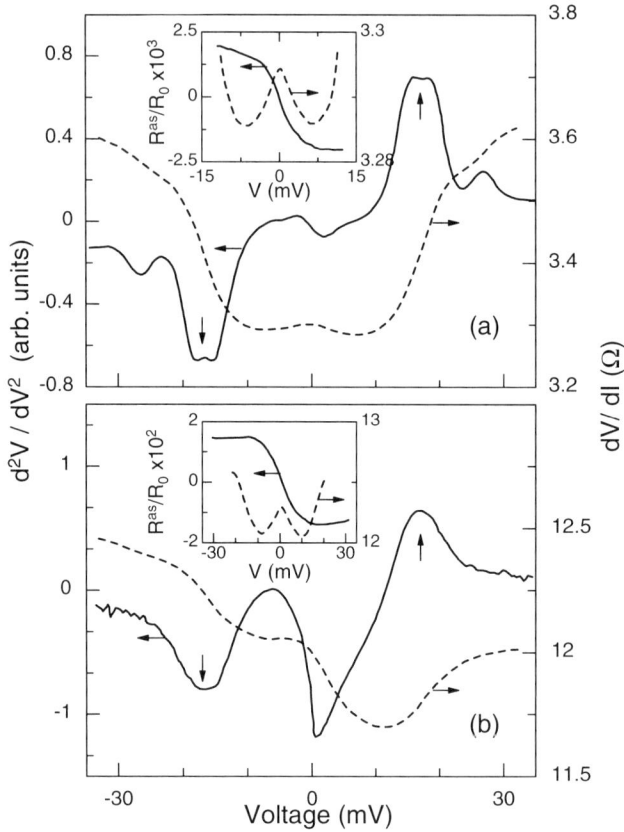

Fig. 7.12. (a) d^2V/dI^2 (V) (solid curve) and dV/dI (V) (dashed curve) characteristics for a Cu-CuFe (Fe=0.01%) heterocontact at T=4.2 K. Inset: symmetric (dashed curve) and asymmetric (solid curve) part of dV/dI. (b) The same for a Cu-CuFe (Fe=1%) heterocontact. Note: The asymmetry in dV/dI is here an order of magnitude larger than in CuFe (Fe=0.01%). Vertical arrows show the TA phonons in Cu. Data taken from Naidyuk et al. (1985).

calculated by Itskovich et al. (1985) is given by (3.28). Thus, the asymmetry is proportional to the difference of the Seebeck coefficients of the contacting metals. This asymmetry is especially expressive in substances with a short mean free path of electrons and large $S(T)$. Such systems include heavy-fermion compounds, Kondo lattices, and intermediate-valence systems (see Chapter 14). The study of the asymmetry of dV/dI curves in such conductors yields direct information about the thermopower behavior therein. For this purpose, one should use a normal metal as a counterelectrode, the Seebeck coefficient of which is usually much smaller than that of the above-mentioned compounds, i. e., $S_1 \gg S_2$. Hence, the dV/dI asymmetry will be directly proportional [see (3.28)] to $S(T)$ of the substance being studied. Figure 7.11 displays an asymmetric part of dV/dI in comparison with the Seebeck coefficient for some Ce compounds. The correspondence between both dependencies is obvious. Such measurements can be used for qualitative analyzing of $S(T)$ behavior, especially in samples, when substances under study are available in small quantities or if the traditional methods for determining of the Seebeck coefficient are hard to apply.

Naidyuk et al. (1985) observed asymmetry in d^2V/dI^2 of Cu-CuFe and Cu-CuMn heterocontacts, where structures related both to the Kondo effect and the electron–phonon interaction are clearly seen (Fig. 7.12). The pronounced phonon features prove that the contacts are not in the real thermal regime. The authors explain the asymmetry by relative large contribution to the Seebeck coefficient caused by magnetic scattering along with nonuniform temperature distribution in the constriction. In this case, already small, namely, by a few degree temperature increase in the point contact can lead to remarkable asymmetry of the conductivity according to (3.28). This temperature increase is not enough to smear the phonon maxima of Cu for 0.01% Fe concentration sufficiently, whereas for a sample with 1% Fe the transverse phonon peak of Cu is sufficiently broadened and LA maximum around 30 mV is absent. This corresponds to the higher temperature increase in this case. It looks like the temperature growth in the constriction is a precursor of the transition to the thermal regime at the voltage increase.

References

Akimenko A. I., Verkin A. B., Ponomarenko N. M. and Yanson I. K. (1982) Sov. J. Low Temp. Phys. **8** 547.

Akimenko A. I., Verkin A. B. and Yanson I. K. (1984) Sov. J. Low Temp. Phys. **10** 605; J. Low Temp. Phys. **54** 247.

Balkashin O. P. and Yanson I. K. (1982) Sov. Phys. Tech. Phys. **27** 522.

Bogachek E. N., Kulik I. O., Omelyanchouk A. N. and Shkorbatov A. G.(1985a) JETP Lett **41** 633.

Bogachek E. N., Kulik I. O. and Shkorbatov A. G. (1985b) Sov. J. Low Temp. Phys. **11** 656.

Bogachek E. N., Kulik I. O. and Shkorbatov A. G. (1989) Sov. J. Low Temp. Phys. **15** 156.

Holm R H (1967) *Electric Contacts* Springer Verlag, Berlin.

Itskovich I. F. and Shekhter R. I. (1985) Sov. J. Low Temp. Phys. **11** 649.

Jansen A. G. M., de Visser A., Duif A. M., France J. J. M and Perenboom J. A. A. J. (1987) J. Magn. Magn. Mat. **63 & 64** 670.

Khotkevich A. V. and Krainyukov S. N. (1991) Sov. J. Low Temp. Phys. **17** 173.

Kittel C. (1986) *Introduction to Solid State Physics* John Wiley & Sons, New York.

Naidyuk Yu. G., Gribov N. N., Shklyarevskii O. I., Jansen A. G. M. and Yanson I. K. (1985) Sov. J. Low Temp. Phys. **11** 580.

Naidyuk Yu. G., Yanson I. K., Lysykh A. A. and Shitikov Yu. L. (1984) Sov. Phys. - Solid State **26** 1656.

Negishi H., Takase K., Funaki K., Furuta K. and Inoue M. (1993) J. Low Temp. Phys., **91** 391.

Paulus E. (1985) Ph. D. Thesis, Köln. (unpublished)

Paulus E. and Voss G. (1985) J. Magn. Magn. Mat. **47 & 48** 539.

Reiffers M., Flachbart K. and Janoš Š (1986) Sov. Phys. - JETP **41** 633.

Shklyarevskii O. I., Jansen A. G. M., Hermsen J. G. H. and Wyder P. (1986a) Phys. Rev. Lett. **57** 1374.

Shklyarevskii O. I., Jansen A. G. M. and Wyder P. (1986b) Sov. J. Low Temp. Phys. **12** 536.

Shklyarevskii O. I., Jansen A. G. M. and Wyder P. (1989) Sov. J. Low Temp. Phys. **15** 96.

Tinkham M., Octavio M. and Skocpol W. J. (1977) J. Appl. Phys., **48** 1311.

van Gelder A. P., Jansen A. G. M. and Wyder P. (1980) Phys. Rev. **22** 1515.

Verkin B. I., Yanson I. K., Kulik I. O., Shklyarevskii O. I., Lysykh A. A. and Naidyuk Yu. G. (1979) Solid State Commun., **30** 215.

Verkin B. I., Yanson I. K., Kulik I. O., Shklyarevskii O. I., Lysykh A. A. and Naidyuk Yu. G. (1980) Izv. Akad. Nauk SSSR, Ser. Fiz. **44** 1330.

8 Point contacts in the magnetic field

Point-contact experiments in the magnetic field may be used both to study the material properties and to search the influence of the restricted geometry on the constriction conductivity (trajectory effects). The mesoscopic phenomena in the magnetoconductivity also relate to the latter issues. Sharvin and Maxwell contributions to the contact resistance are altered by a magnetic field in a different way. Generally, the Maxwell part of the point-contact resistance behaves in accordance with the bulk magnetoresistance, whereas the ballistic resistance is sensitive to a magnetic field because of the appearance of Landau quantization levels on the Fermi surface. In the other words, a magnetic field influence on the contact conductance can be separated into classic trajectory or quantum-mechanical effects.

8.1 Magnetoresistance

The character of a current spreading in a point contact will be radically changed because of a strong magnetic field at $\omega_c \tau \gg 1$, where $\omega_c = eB/m^*$ is the cyclotron frequency, m^* is the effective cyclotron mass, and $\tau = l_{el}/v_F$ is the electron relaxation time. Motion of electrons becomes quasi-one-dimensional in a field parallel to the contact axis z and a spatial spreading of the current can only be caused by the scattering of electrons by impurities (Fig. 8.1).

As shown by Bogachek et al. (1987), the effective potential drop length L at the orifice increases in a high magnetic field from the value of the contact size $\sim d$ at $B = 0$ to $\sim d(l_{el}/r_B)$, where $r_B = v_F/\omega_c$ is the cyclotron radius (Fig. 8.2). The contact magnetoresistance at $r_B \ll d$ behaves as

$$R(B) = R_0 \sqrt{1 + (l_{el}/r_B)^2}. \tag{8.1}$$

For a strongly contaminated contact $l_{el} \ll r_B$, this leads to the quadratic field dependence of $R(B)/R_0 \simeq 1 + 1/2\,(l_{el}/r_B)^2$. In the case $l_{el} \gg r_B$, the contact resistance increases linearly with a magnetic field: $R(B)/R_0 \simeq l_{el}/r_B$. It is worth noting that both quadratic and linear dependencies of the point-contact magnetoresistance are determined by the same ratio l_{el}/r_B. This behavior was observed by Ass and Gribov (1987) on bismuth point contacts.

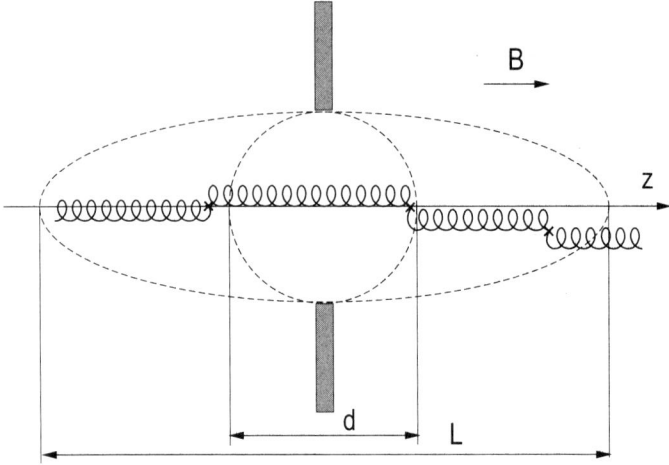

Fig. 8.1. Trajectory of an electron in the point contact in a strong magnetic field along the contact axis z. The current spreads only because of elastic scattering processes marked by crosses. The effective region of high current density is stretched out from d at zero field to $L \simeq d(l_{el}/r_B)$ in a high magnetic field. Adapted from Bogachek et al. (1987).

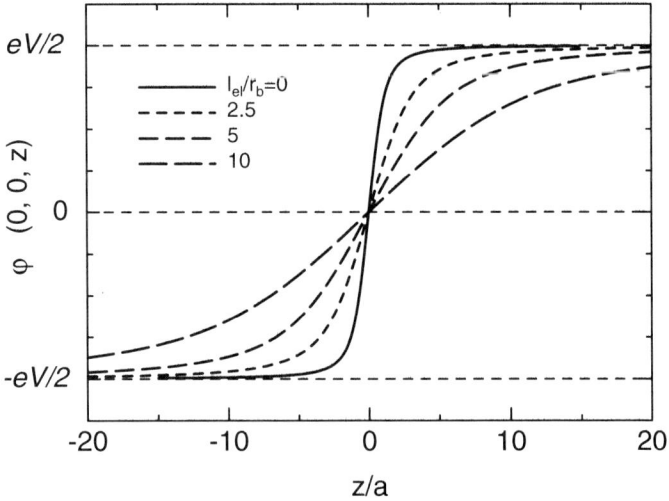

Fig. 8.2. Electrostatic potential along z axis for ballistic point contact with radius a at different ratio of l_{el}/r_B value calculated according to the formula (34) from Bogachek et al. (1987).

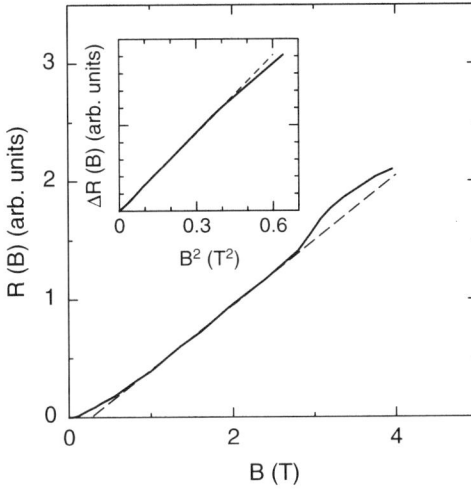

Fig. 8.3. Longitudinal magnetoresistance of Bi-Cu point contact with zero-field resistance $2\,\Omega$ at $4.2\,\mathrm{K}$. Dashed line shows linear dependence. At higher fields, the deviation from the linear dependence is caused by the appearance of Shubnikov-de Haas oscillations. Inset shows the initial part of the curves versus B^2. Dashed line shows quadratic dependence. Data taken from Ass and Gribov (1987).

In the region of weak fields $l_{\mathrm{el}} \ll r_{\mathrm{B}}$, they found a quadratic field dependence of $R(B)$, whereas $R(B)$ was linear in strong fields ($l_{\mathrm{el}} \gg r_{\mathrm{B}}$) (Fig. 8.3) in agreement with the theory. The measurements yielded l_{el}, which value, determined both from quadratic and linear parts of $R(B)$, coincide. As was shown by Moskalets et al. (1989), the nonuniform distribution of impurities in the contact depresses, quadratic term at weak fields and leads to a linear, quadratic, and again linear dependence of $R(B)$ with a field increase. During the investigation of the point-contact magnetoresistance, it is possible to determine a set of contact parameters, both traditional ones, such as contact diameter, elastic mean free path, and the size connected with the spatial localization of the contaminated region in the constriction.

The magnetoresistance of a point contact can be connected also with a field-dependent bulk resistivity contributing to the diffusive Maxwell term of the contact resistance. It is well known from the solid-state textbooks that the longitudinal magnetoresistivity of the bulk material is not affected, in general, by a magnetic field, whereas the transverse magnetoresistivity of compensated metals, where the number of electrons and holes is equal, increases quadratically with a magnetic field:

$$\rho_\perp(B) = \rho_0(1 + (\omega_{\mathrm{c}}\tau)^2). \tag{8.2}$$

In high magnetic fields and for pure samples, $\omega_{\mathrm{c}}\tau$ can exceed easily 10^2, resulting in strong anisotropy of the resistivity. It leads to the sufficient increase of the Maxwell component in the contact resistance, which can be observed even for a ballistic contact. Moreover, in the case of high $\omega_c\tau$, the sample size will be important. Resistance of a point contact when it is placed on a thin plate with the thickness t and the length (diameter) D is defined as [Swartjes et al. (1988)]

$$R(B) = \frac{\rho}{2\pi t}(1 + (\omega_c\tau)^2)\ln\left(\frac{D}{2d}\right),\qquad(8.3)$$

in the magnetic field perpendicular to the contact area. In the case of uncompensated metals, the magnetoresistance vanishes on account of the Hall voltage, which compensates the deflection of electrons in the magnetic field. However, in the Corbino geometry with a contact in the center of a circular sample, this compensation fails and a magnetoresistance is again quadratic versus a field. Experiments done by Swartjes et al. (1988) for compensated metal, namely, bismuth and uncompensated aluminum, are explained by the mentioned approach. In both cases, the magnetoresistance comes, apparently, from the resistivity-dependent Maxwell part of the point-contact resistance.

8.2 Magnetooscillations

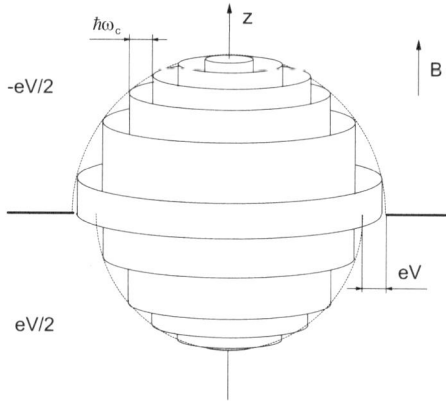

Fig. 8.4. Electron distribution function for a ballistic point contact in a quantizing magnetic field parallel to the contact axis z and at bias voltage eV. At $T \to 0$, the electron states situated on the Landau tubes are separated by the energy $\hbar\omega_c$. In this case, the z component of the electron momentum for the states passing the Fermi level is approximately zero. Diffraction or elastic scattering to different Landau levels may increase this component. Adapted from Kokkedee (1992).

Let a magnetic field be applied along the z axis. Then the electron trajectory in the x–y plane perpendicular to the field direction is a circular orbit with the cyclotron radius $r_B = v_F/\omega_c$. The requirement of a periodic electron wave function along this orbit leads to the quantization of the electron energy with the quantum $\hbar\omega_c$. As a result, the electronic states are redistributed on the tubes (Landau levels) inside the Fermi surface separated one from another by the energy $\hbar\omega_c$ (Fig. 8.4). A magnetic field rise results in the increase of a distance between tubes, and as a consequence, a part of them should escape from the Fermi surface. This leads to the oscillating density of states at the Fermi level with a period $\sim 1/B$ as well as to oscillation of a number of physical quantities known as de Haas–van Alphen or Shubnikov–de Haas effects. The latter describes quantum oscillations of a resistivity connected with oscillations of the scattering time τ. The relative change in the resistivity is believed to be proportional to the relative change of the density of states on the Fermi level: $\Delta\rho/\rho \approx (\hbar\omega_c/\epsilon_F)^{1/2}$.

The Maxwell resistance of the point contact is directly proportional to the resistivity and hence should exhibit the quantum oscillations mentioned. However, this effect will be pronounced in the diffusive regime, whereas the Maxwell resistance is negligible for the ballistic contact. On the other hand, the Sharvin resistance is determined by the Fermi surface and, in principle, can oscillate as well. Oscillations of the point-contact resistance in a magnetic field were investigated by Gribov et al. (1987) in Bi, by Swartjes et al. (1988) in Bi, Ga, and Al, by Shklyarevskii et al. (1988) in Sb, and by Bobrov et al. (1995, 1996) in Al and Be. The theoretical models were developed by Bogachek et al. (1985), Bogachek and Shekhter (1988), and van Gelder [Swartjes et al. (1989), van Gelder (1991)]. In the van Gelder model, oscillations of the Sharvin resistance have a relative amplitude $\Delta R/R \sim 1/q \ll 1, q = \epsilon_F/\hbar\omega_c$, which is even smaller than that of the Maxwell part. The effect can be enhanced by a factor $q^{1/2}$, taking into account the diffraction of electrons between quantum states when the magnetic length $\Lambda_B = (2\hbar/eB)^{1/2}$ characterizing "size" of the electron states on the Landau tube is of the order of a point-contact size. This length for a field of a few Tesla is about 10–20 nm that is comparable with the typical contact size.

Bogachek et al. (1985) calculated a point-contact magnetoresistance in the ballistic case $l \gg r_B \gg d$ at quantizing magnetic fields by neglecting diffraction effects. The electrical conductivity was analyzed by the Wigner distribution function of electrons when the trajectory description of the electron gas is feasible. The magnetoresistance is found to be

$$\Delta R/R_0 = q^{-3/2}(j_1(q) + j_2(q)), \tag{8.4}$$

where j_1, j_2 describes resistance oscillations because of renormalization of the chemical potential and the electron energy, respectively. At $T = 0$,

$$j_{1,2} = \gamma_{1,2} \sum_{k=1}^{\infty} (-1)^k k^{-3/2} \sin(2\pi k q - \pi/4), \tag{8.5}$$

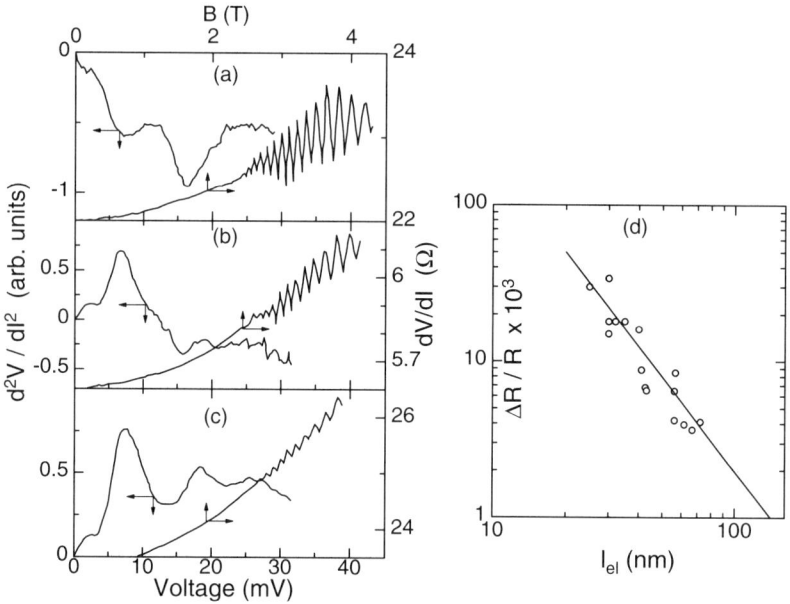

Fig. 8.5. Magnetoresistance of Sb contacts by increasing an electron mean free path from 10 nm for the top curve (a) to 42 nm for the lower one (c). Also, EPI spectra (d^2V/dI^2) are displayed. They are transformed from negative to positive in the region of the phonon peaks by increasing an electron mean free pass (see Section 11.1 for explanation). Right panel shows a reduced amplitude of the quantum oscillations at 3 T for different contacts as a function of the electron mean free path. Data taken from Shklyarevskii et al. (1988).

with $\gamma_1 = \sqrt{2}/4$ and $\gamma_2 \simeq \ln q$. The oscillation amplitude is here $\sim q^{-3/2}$, which is even smaller than for the previous van Gelder model. In further calculations, Bogachek and Shekhter (1988) showed that the scattering-induced isotropism of the electron distribution function will increase a contribution of the quantum oscillations of the density of states to an electron transport in a contact. For a dirty long channel $r_B \gg L \gg d \gg l_{el}$ at $T = 0$,

$$\Delta R/R_0 \simeq \sqrt{2}q^{-1/2}\sum_{k=1}^{\infty}(-1)^k k^{-1/2} \cos(2\pi kq - \pi/4). \qquad (8.6)$$

The oscillation amplitude increases here by a factor q compared with the clean limit (8.5). That is why electrons on the extreme orbits giving the main contribution to the oscillations of the density of states have the velocity component v_z along the contact axis being nearly zero, and hence, they contribute slightly to the contact resistance. The isotropic elastic scattering increases this component, which enhances both the resistance and its oscillations. Indeed, Shklyarevskii et al. (1988) observed the increase of the quantum oscillations with decreasing of the elastic mean free path in the constriction by investigation of Sb point contacts (Fig. 8.5).

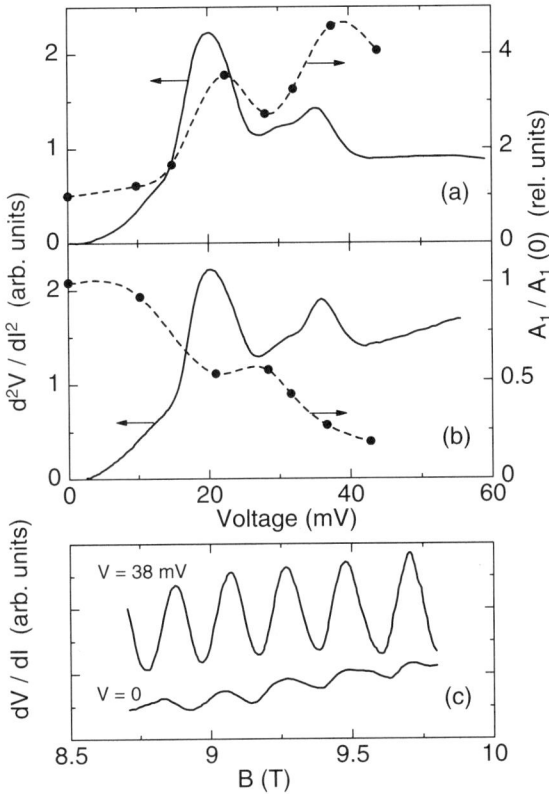

Fig. 8.6. Point-contact spectra of Al contacts [solid curves in (a) and (b)] show both an increase and a decrease of the relative amplitude A of the quantum oscillations (symbols) at 9.5 T with increasing of the bias voltage. For contact (a), $R_0 = 1.41\,\Omega$ and $\Delta R/R_0 \approx 5 \times 10^{-4}$, and for (b), $R_0 = 8.73\,\Omega$ and $\Delta R/R_0 \approx 3 \times 10^{-3}$. (c) shows a fragment of the oscillations at zero and at the finite bias voltages. Data taken from Bobrov et al. (1995, 1996).

Bobrov et al. (1995, 1996) studied the energy dependence of the oscillation amplitude in voltage-biased Al and Be contacts. They found both nonmonotonous increasing and decreasing of the oscillation amplitude by voltage sweep, with structures similar to the point-contact spectrum for the electron–phonon interaction (Fig. 8.6).

The possible reason of the phenomena observed could be an additional scattering of electrons on nonequilibrium phonons and phonon–phonon collisions, which modify the distribution of nonequilibrium phonons. To summarize, it is worth it to show analysis of the magnetoquantum oscillations in the Be point contact at $V = 0$ (Fig. 8.7). This yields results analogous to the standard de Haas–van Alphen experiments leading to frequency 985 T, which corresponds to the third-band electron (cigar) Fermi-surface of Be.

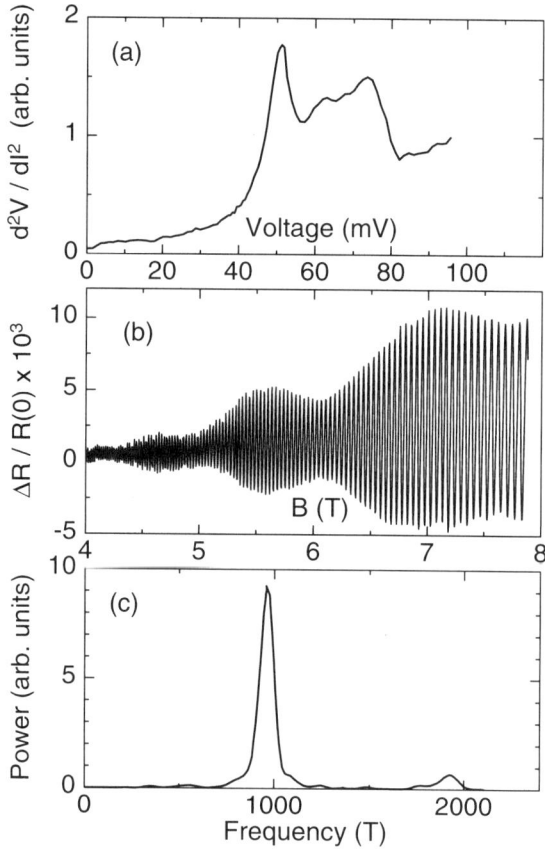

Fig. 8.7. (a) Point-contact spectrum of Be at $T = 1.3$ K. (b) Magnetoresistance of the same point contact at $V = 0$ ($R_0 = 16.24 \, \Omega$). Here magnetic field and contact axis are along the c axis. (c) Fourier transform of the curve from (b). Data taken from Bobrov et al. (1995).

8.3 EPI in a magnetic field

A quasi-one-dimensional character of the current spreading in the magnetic field modifies the scattering processes in a constriction affecting also the electron–phonon interaction. Following Bogachek et al. (1987), in strong magnetic fields $r_B \ll d$, a geometrical form-factor K for a point contact entering in the electron–phonon interaction function (3.7) is

$$K(\mathbf{p}, \mathbf{p}') = \frac{9\pi}{128} \frac{l_{el}}{d} \left(2(n_z - n_z')^2 + \frac{1 - (\omega_c \tau)^2}{1 + (\omega_c \tau)^2} (n_\perp - n_\perp')^2 \right), \qquad (8.7)$$

where $\mathbf{n} = (n_\perp, n_z) = \mathbf{v}/v_F$. Equation (8.7) will be valid if at least one of the parameters r_B or l_{el} is less than a contact diameter. At $B = 0$, (8.7) is similar to the expression for the K factor in the diffusive limit (see Table 3.2), whereas at $B \to \infty$, its value vanishes. The first term in (8.7) gives a positive contribution to the spectrum describing the scattering along the contact axis, whereas the second term becomes negative in high magnetic fields. For a weak magnetic field $r_B \gg d$, the averaged K-factor is given by

$$\langle K(\mathbf{p}, \mathbf{p}') \rangle \approx \langle K_0 \rangle (1 + \eta l_{el} d / r_B^2), \qquad (8.8)$$

where $\langle K_0 \rangle = 0.25$ is the averaged "conventional" K factor in zero field, $\eta \approx 1$. As can be seen, the K factor determines the intensity of a point-contact spectrum and it increases quadratically with a field. The transition from a weak magnetic field to a strong one occurs at $r_B \sim d$; therefore, the increasing of maxima intensity in the point-contact spectrum can be replaced by their decrease. At sufficiently high magnetic fields, the current spreading region can be commensurable with the inelastic mean free path at

$$B = (c p_F / e d) \sqrt{l_{in}/l_{el}} , \qquad (8.9)$$

leading to a gradual transition from the ballistic to the thermal regime.

At a practically easily attainable magnetic field of about 10 T, a cyclotron radius for simple metals is about 500 nm, which is much larger than the typical point-contact size of around 10–100 nm. Therefore, in reality, it is possible to investigate the phenomena discussed above only for semimetals having an order of magnitude smaller cyclotron radius caused by a small effective mass of quasiparticles. A magnetic field impact on the electron–phonon interaction spectrum in point contacts was observed in antimony by Yanson et al. (1985). Later Yanson et al. (1992) investigated these effects in detail on an Sb single crystal oriented along the principal crystallographic axes, both in longitudinal and transverse magnetic fields. They found that the maxima in the point-contact spectra can both decrease and increase (Fig. 8.8) depending not only on the orientation of the contact axes with respect to the Sb crystallographic axis and magnetic field, but also on the interrelation among the contact diameter, the Larmor radius of the charge carriers, and their mean free path.

The Fermi surface of Sb consists of three distorted electronic ellipsoids and six hole quasi-ellipsoids. Therefore, both intravalley and intervalley scattering processes are feasible. Intravalley scattering can excite phonons with wave

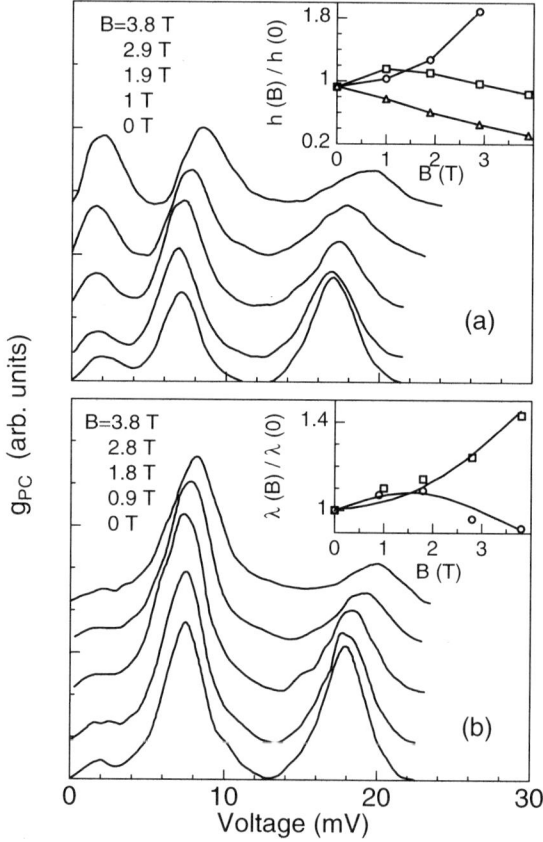

Fig. 8.8. (a) Magnetic field effect on the point-contact EPI function of Sb at $\mathbf{B}\|\mathbf{j}$ and the contact axis along the binary C_2 axis of the rhombohedral A7 structure. The curves are offset vertically. Inset shows the field dependence of the reduced amplitude of the intravalley h_1 (circles), the intervalley acoustic h_2 (squares), and the optical h_3 (triangles) peaks. The lines are guides on eyes. (b) The same for the field $\mathbf{B}\perp\mathbf{j}$. Inset shows the magnetic field dependence of the reduced EPI constant λ for $\mathbf{B}\|\mathbf{j}$ (squares) and $\mathbf{B}\perp\mathbf{j}$ (circles). The lines are guides on eyes. Data taken from Yanson et al. (1992).

vector $0 < q < k_F$, and the electron–phonon interaction has a considerable dispersion, which results, according to Bogachek et al. (1987), in the intensity increase of the spectra in a magnetic field even if the $r_B \gg d$ condition no longer holds. At intervalley scattering, the phonon momentum change is large $\Delta q \sim q \gg k_F$ but the phonon wave number dispersion is weak and an intensity of corresponding maxima in spectrum can both increase and decrease depending on the ratio l_{el}/d, r_B/d and k_F/k_D. Experimental data in Fig. 8.8 generally confirm this picture. Qualitatively, an analogous magnetic field dependence of the point-contact spectra for another semimetal As was measured also by Gribov et al. (1991) (Fig. 8.9). They mentioned that the

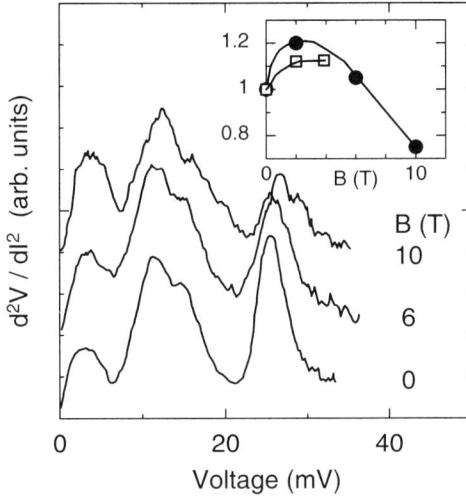

Fig. 8.9. Magnetic field effect on the point-contact spectrum of As-Cu contact with **B** along the contact axis and perpendicular to the C_3 axis of As single crystal. Inset shows the field dependence of the reduced integral intensity of the spectra for two contacts with different resistance. The curves are offset vertically. Data taken from Gribov et al. (1991).

localization phenomenon should be suggested as well for a full account of the magnetic field dependence of point-contact spectra for a dirty constriction (see Chapter 11).

8.4 Magnetofingerprints

Universal conductance fluctuations (UCF) or magnetofingerprints phenomena are characteristic for the mesoscopic systems and originate from quantum interference of electrons scattered at the impurities. In contrast to the Aharonov–Bohm (1959) effect for a disorder metal loop, UCF are descended from the large number of oscillating contributions from a number of interference loops in a bulk conductor. A restriction to this effect is that the size of a system should be smaller than the inelastic diffusion length $l_\varphi \approx \sqrt{l_{el} l_{in}}$. In polycrystalline metal film, l_φ can be about $1\,\mu$m at $T < 1$ K, whereas $l_{el} \sim 10$–50 nm. A magnetic field alters the UCF changing phase of the electron wave function, and it has generally the same effect as modification of the impurity configuration. Lee et al. (1987) found that magnetic field $B_c = \Phi_0/S$ corresponding to one flux quantum $\Phi_0 = h/e$ through the sample area S alters UCF in the same way as a complete change of an impurity configuration and acts as a typical period of fluctuations. At $T \to 0$, UCF amplitude is of the order of $e^2/h = 3.88 \times 10^{-5}\,\Omega^{-1}$ independent on a sample size, dimensionality of a system, and degree of disorder.

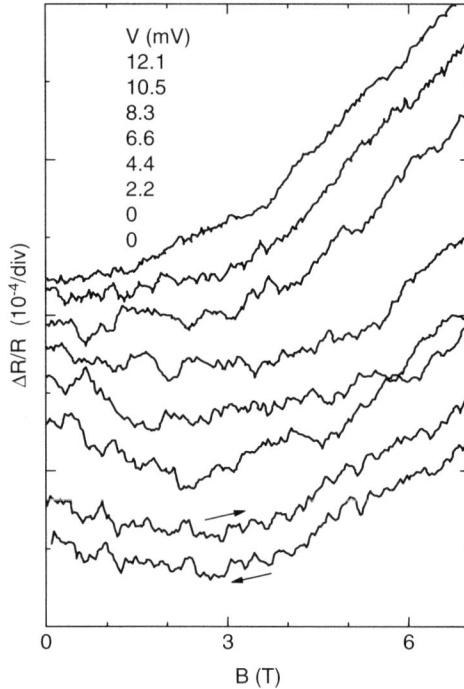

Fig. 8.10. Magnetofingerprints of $11\,\Omega$ Ag point contact at $T = 0.4\,\text{K}$ at a different bias voltage from $12.1\,\text{mV}$ (for the top curve) to $0\,\text{mV}$ (for two bottom curves). The two lower curves demonstrate the reproducibility of the fluctuations for the magnetic field sweep up and down (direction is shown by arrows) at $V = 0$. Data taken from Holweg et al. (1991).

Magnetoconductance fluctuations with a reproducible aperiodic structure were observed in the nanofabricated (see Section 4.1.4) point contacts by Holweg et al. (1991). By applying a bias voltage, the fluctuation patterns evolve gradually, but completely different magnetofingerprints appear above $2\,\text{mV}$ (Fig. 8.10). With the bias voltage rise, a typical period of fluctuations increases while their amplitude decreases. At zero bias, the measured amplitude of the fluctuations is about two orders of magnitude lower than for UCF. After initial increase with bias, the amplitude of fluctuations becomes roughly constant. Further on it starts to decrease above the voltage, corresponding to the energy of the first phonon peak (Fig. 8.11). As an explanation, the

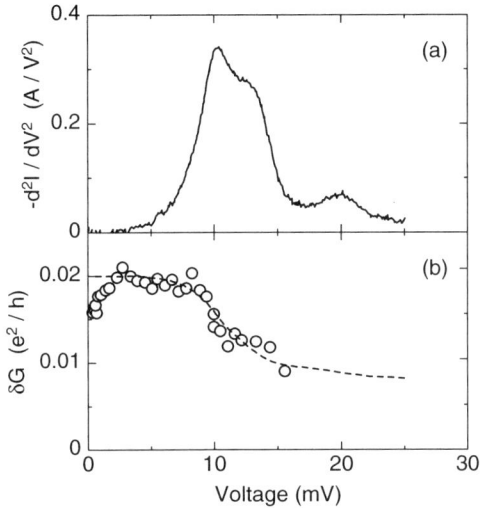

Fig. 8.11. Point-contact spectrum (a) and rms amplitude (b) of the conductance fluctuations for the point contact in Fig. 8.10 along with calculated voltage dependence $\exp(-l_{\mathrm{el}}/l_{\mathrm{in}})$ (dashed line). Here $l_{\mathrm{in}} \sim \tau^{-1}$ follows to (2.5). Data taken from Holweg et al. (1991).

authors proposed a model in terms of the quantum interference of electrons near the orifice over loop sizes of the order of the elastic mean free path. They found that the amplitude of the conductance fluctuations depends on the effective size and degree of disorder

$$\delta G \approx (e^2/h)d/l_{\mathrm{el}}. \tag{8.10}$$

The decrease of δG above 10 mV was explained by phonon emission by energized electrons, resulting in a reduced inelastic mean free path. Hence, the conductance fluctuations in the ballistic point contacts can be referred to as non-UCF. Kozub et al. (1994, 1996) mentioned that the situation in the ballistic point contact is different from the bulk, whereas only a few acts of scattering provide a dominant contribution to the backscattering. Therefore, the "local" instead of the "global" interference effects should be considered along with a finite number of scattering. In this case, if both of the interfering trajectories include two or more scattering centers, remote regions from the contact will produce the main contribution to δG, although with an additional term d/l_{el} in (8.10). The authors received a better correlation with the experimental findings, assuming that the remote regions have a larger mean free path. With decreasing of the electronic mean free path, the diffusive motion along remote backscattered trajectories in the electrodes should be considered as well [Caro et al. (1996)].

References

Aharonov Y. and Bohm D. (1959) Phys. Rev. **115** 485.

Ass E. I. and Gribov N. N. (1987) Sov. J. Low Temp. Phys. **13** 365.

Bobrov N. L., Kokkedee J. A., Gribov N. N., Yanson I. K. and Jansen A. G. M. (1995) Physica B **204** 83.

Bobrov N. L., Kokkedee J. A., Gribov N. N., Yanson I. K., Jansen A. G. M. and Wyder P. (1996) Physica B **218** 42.

Bogachek E. N., Kulik I. O. and Shekhter R. I. (1985) Solid State Commun. **56** 999.

Bogachek E. N., Kulik I. O. and Shekhter R. I. (1987) Sov. Phys.- JETP **65** 411.

Bogachek E. N. and Shekhter R. I. (1988) Sov. J. Low Temp. Phys. **14** 445.

Caro J., Kozub V. I. and Holweg P. A. M. (1996) Physica B **218** 81.

Gribov N. N., Shklyarevskii O. I., Ass E. I. and Andrievskii V. V. (1987) Sov. J. Low Temp. Phys. **13** 363.

Gribov N. N., Samuely P., Kokkedee J. A., Jansen A. G. M., Wyder P. and Yanson I. K. (1991) Phys. Rev. Lett. **66** 786.

Holweg P. Λ. M., Kokkedee J. A., Caro J., Verbruggen A. H., Radelaar S., Jansen A. G. M. and Wyder P. (1991) Phys. Rev. Lett. **67** 2549.

Kokkedee J. A. (1992) Ph. D. Thesis, Grenoble. (unpublished)

Kozub V. I., Caro J. and Holweg P. A. M. (1994) Phys. Rev. B **50** 15126; (1996) Physica B **218** 89.

Lee P. A., Stone A. D. and Fukuyama H. (1987) Phys. Rev. B **35** 1039.

Moskalets M. V., Ass E. I., Gribov N. N. and Koshkin I. V. (1989) Sov. J. Low Temp. Phys. **15** 578.

Shklyarevskii O. I., Gribov N. N. and Yanson I. K. (1988) Sov. J. Low Temp. Phys. **14** 229.

Swartjes H. M., Jansen A. G. M. and Wyder P. (1988) Phys. Rev. B. **38** 8114.

Swartjes H. M., van Gelder A. P., Jansen A. G. M. and Wyder P. (1989) Phys. Rev. B. **39** 3086.

van Gelder A. P. (1991) Physica B **175** 68.

Yanson I. K., Gribov N. N., and Shklyarevskii O. I. (1992) JETP Lett. **42** 195.

Yanson I. K., Shklyarevskii O. I. and Gribov N. N. (1992) J. Low Temp. Phys. **88** 135.

9 Electrical fluctuations in point contacts

9.1 Physics of noise

Noise is a great problem in science and technology because it dictates sensitivity of existing electronic devices limiting the magnitude and accuracy of measurements, and the quality of processing various signals. Noise is an unwanted voltage or current of fluctuation nature generated by the electronic device or some circuit component. Because the current carriers in a conductor are in permanent motion, this will cause a rise in fluctuation voltage V_{n} appearing at the conductor edges. At each particular moment of time, this voltage has a definite value, but when it is averaged over a long time, it becomes equal to zero. This type of noise is called Nyquist or thermal noise, and its average square value $\langle V_{\mathrm{n}}^2 \rangle$ originating on the resistance R in the frequency range Δf is proportional to the temperature T:

$$\langle V_{\mathrm{n}}^2 \rangle = 4k_B T R \Delta f. \tag{9.1}$$

Another type of noise is connected with the discreteness of the charge of the current carriers. Thus, for example, the events of electron flying out of the cathode (emitter) of electronic tube (transistor) form a sequence of independent events that occured at random times that lead to the so-called shot noise. The same takes place in any conductor because of the charge carriers motion under applied bias voltage. The spectral density of the shot noise is given by

$$S_{\mathrm{V}} = \langle V_{\mathrm{n}}^2 \rangle / \Delta f \propto eVR; \tag{9.2}$$

i.e., the shot noise increases with a voltage rise. Electrons and holes can appear and disappear at random in an impurity-free semiconductor because of generation and recombination processes. This also causes fluctuation voltage in the sample called a generation-recombination noise. Another source of noise was observed at first in the electronic vacuum tubes. It was attributed to the average emission rate alteration causing microscopic current fluctuations. The emission rate alteration can be treated as an effect of collective interaction of electrons, as a consequence of which such oscillations or flicker of current occur. This noise is known as a flicker noise. It is characterized by the fact that the spectral density of the flicker noise is frequency dependent close to the $1/f$ law, that is why it is also called $1/f$ noise. According to the empirical

Hooge's (1969) formula, a spectral density of the flicker noise can be written as

$$S_{\mathrm{R}} = S_{\mathrm{V}}/I^2 = \gamma R^2/N_{\mathrm{e}} f^{\alpha}, \qquad (9.3)$$

where $\alpha \approx 1$ and γ is a constant. $1/f$ noise dominates for tiny systems (samples) with the small number of charge carriers N_{e}. That is, it will overlord in thin and narrow films, tiny constrictions, point contacts, and so on. The microscopic nature of this noise is not yet clear. Different models are suggested that explain the flicker noise in solids; in particular, recently two-level systems (TLS) that present as elementary disorder carriers in the crystal lattice have been considered.

9.2 Two-level systems

Nowadays the most exploiting model of the $1/f$ noise in metals is concerned with atomic size defects described by double-well potential (Fig. 6.9). The tunneling of atoms (or group of atoms) between two states of a defect with a different scattering cross-section result in resistance fluctuations. The short screening length in metals leads to a small scattering cross-section of an individual defect; therefore, small (tiny) enough systems (samples) like contact or nanobridge are preferable to resolve the microscopic nature of the $1/f$ noise, as mentioned above. Additionally, the small-size devices allow studying, in principle, an individual two-level fluctuator.

Machlup (1954) showed that a system switched between two discrete resistance states with exponential distribution time spent in the upper or lower state

$$\tau = \tau_0 \exp(\epsilon/k_B T_{\mathrm{d}}), \qquad (9.4)$$

where τ_0 is the attempt time, ϵ is the activation energy, and T_{d} is the defect temperature, has a Lorentz shaped spectral density:

$$S_{\mathrm{V}}(f) \sim \frac{\tau}{1 + (2\pi f \tau)^2}, \qquad (9.5)$$

where τ is the characteristic mean fluctuation time. The superposition of many Lorentzian spectrum noise sources with a broad distribution of the fluctuation rate τ_i gives rise to generic $1/f$ noise [Dutta et al. (1979)]. The time τ_i depends both on temperature and voltage applied to the device. In the latter case, the defect can be heated to an effective temperature well above the lattice temperature because of accelerated electrons.

For the first time, defect-motion-induced two-level fluctuations of the resistance were observed by Ralls and Buhrman (1988) in wide cooper nanobridges prepared by electron-beam lithography 3–40 nm. As seen from Fig. 9.1, the sample resistance noise is dominated by random switching between two values. The distribution of times spent in two states was found to be exponential, resulting in the Lorentzian noise spectrum. The authors

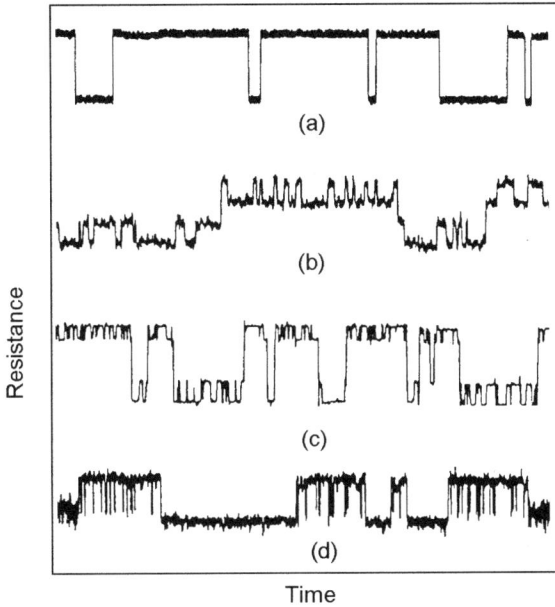

Fig. 9.1. Resistance fluctuations in Cu the nanobridges for $T < 150$ K with different types of behavior: (a) a single fluctuator, (b) two independent fluctuators, (c) amplitude modulation and (d) frequency modulation of one fluctuator by another. Resistance fluctuations are from 0.005% to 0.2% of the total resistance. Time scales are somewhat arbitrary, as they depend on the bath temperature. After Ralls and Buhrman (1988).

estimated changes in the scattering cross-section to be 1–30 Å, which is of the order of atomic dimension, i.e., reasonable for the defect motion. Analogous fluctuations of resistance were registered by Holweg et al. (1992) (Fig. 9.2) at helium temperature and biases 50–100 mV in several gold nanoconstrictions with a relatively high resistance ($> 83\,\Omega$). They found that the bias dependence of fluctuations differs from device to device, but it is reproducible for a specific device, illustrating the mesoscopic character of phenomena. Holweg et al. (1992) measured a shape of the spectral density of a single two-level fluctuator, which is indeed close to the Lorentz function (9.5) and is constant for the low frequencies and behaves like $1/f^2$ for the higher frequencies (Fig. 9.2). Many of the fluctuators in constriction are activated, giving rise to a typical $1/f$ power spectrum at room temperature (Fig. 9.3). This demonstrates that visible at the helium temperature, two-level fluctuations represent a fundamental mechanism of the $1/f$ noise. Ralls and Buhrman (1988) and Holweg et al. (1992) estimated attempt time, activation energy, and electromigration parameter, which correspond to that for the defects in metals. The τ_0^{-1} value is in the range of the typical phonon frequencies (10^{10}–10^{13} s^{-1}) and $\epsilon \sim 200$–

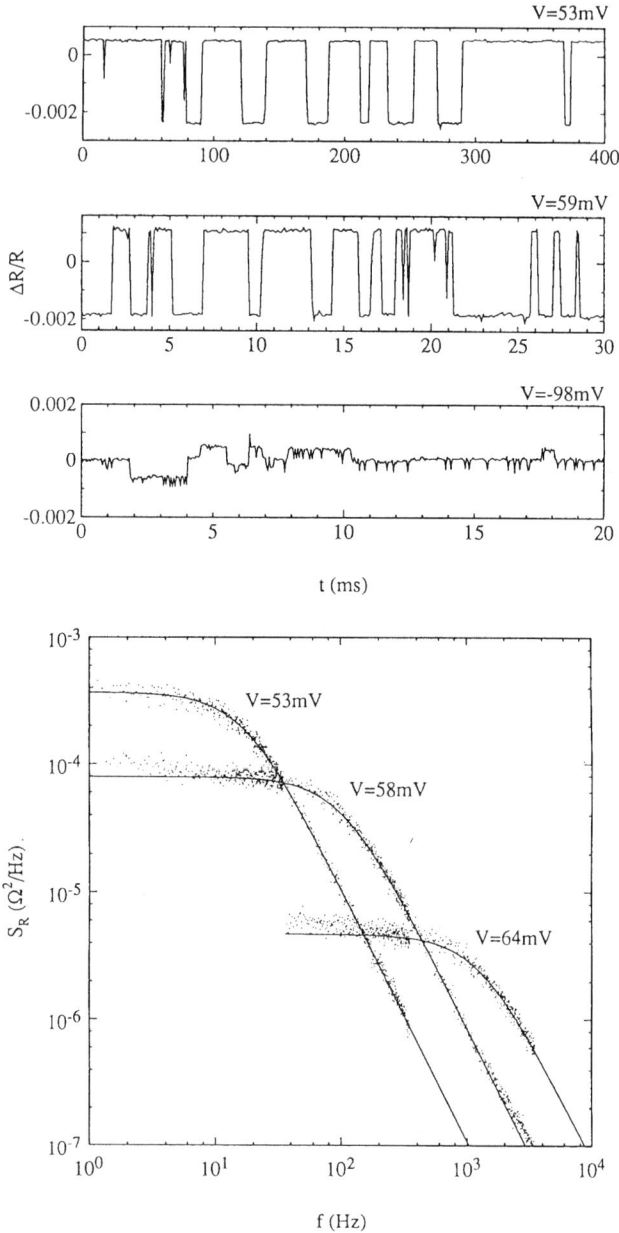

Fig. 9.2. Resistance fluctuations of the $83\,\Omega$ Au nanobridge at $T = 4.2\,\mathrm{K}$ versus time for three bias voltages (three upper panels). The switching rate of the two-level fluctuator increases with a bias rise from 53 to 59 mV. More fluctuators are visible simultaneously at higher voltage (-98 mV). The spectral density of the same two-level fluctuator at a few biases indicating the Lorentz line shape is shown on the bottom panel. Because of defect heating for larger biases the knee of the Lorentzians shifts to the higher frequency. The fit (solid curves) was done using Lorentzian spectra (9.5) and an $1/f$ term to account for the small residual $1/f$ noise. After Holweg et al. (1992).

400 mV as expected for the defect motion in metals. The analysis assumes that complex defects are involved in the fluctuation phenomena.

The voltage applied to a constriction tilts the fluctuator potential (see Fig. 6.9) resulting in the lowering or raising of an activation energy depending on the bias polarity. With increasing of the bias voltage, the fluctuation rate increases, suggesting a thermal activated behavior [see (9.4)] described by a rise of the defect temperature T_d. However, as mentioned by Holweg et al. (1992), the voltage dependence of fluctuations can be better understood by local heating of the defect by inelastic scattering with electrons rather than overall sample heating by an applied voltage. They found that the defect temperature rises linearly with the bias voltage and is independent on the specific defect parameters. On the other hand, nonequilibrium phonons created by electrons and accumulated in the constriction may also influence the defect temperature, although interaction of TLS with phonons in metals at low temperatures is less efficient [Black (1981)].

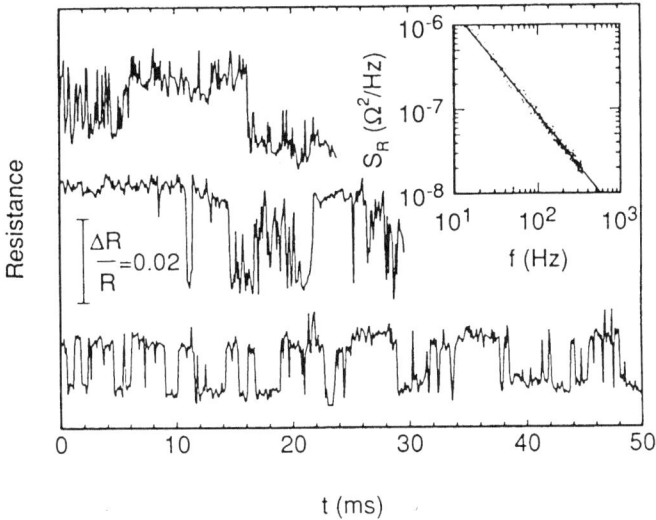

Fig. 9.3. Resistance fluctuations of a $1.2\,\mathrm{k}\Omega$ Au nanobridge at $300\,\mathrm{K}$ and at 12-mV bias. The interval between time traces, taken under identical condition, is a few seconds. At $300\,\mathrm{K}$, a switching between the discrete resistance levels is still visible, but the character of the noise changes in time. Inset shows a typical spectral density of fluctuations of $54\,\Omega$ nanobridge at $300\,\mathrm{K}$ and 16-mV bias, illustrating the $1/f$ character of the noise. After Holweg et al. (1992).

9.3 Point-contact noise spectroscopy

The $1/f$ noise has been investigated by Hooge and Hoppenbrouwers (1969) for ordinary metallic contacts. They showed that the $1/f$ noise of contacts can be described by an extension of (9.3), which is valid for the bulk samples. Small contact size, as follows from (9.3), makes a contribution of the $1/f$ noise dominant over the thermal Nyquist noise (9.1) as well as over the shot noise (9.2). The most fascinating features of the $1/f$ noise in contacts was found when studying the bias dependence of the noise voltage $\langle V_n^2 \rangle$. Yanson et al. (1982) observed that $\langle V_n^2 \rangle$ of a ballistic point contact increases nonmonotonically with the bias voltage showing both maxima and minima (Fig. 9.4). It turns out that the position of the minima corresponds well to the multiple phonon energies. This phenomenon was attributed to the emission of coherent phonons by nonequilibrium electrons in the constriction. The correlation in the emission processes leads to the noise decrease.

During further investigations, Akimenko et al. (1984) studied the noise in the energy range of characteristic phonon frequencies at different regimes of electron flow. Although for contacts in the thermal regime spectral density of noise monotonically increases with a voltage showing a power law V^n with n between 2 and 5, for ballistic contacts in the above energy region, they found a reproducible fine structure (minima and maxima) in the voltage dependence of noise (Fig. 9.5). The position of peculiarities in the noise spectra correlates also with the calculated EPI function in the case of Na, taking into account Umklapp processes only. For other metals being studied, namely, Cu and Sn, the maxima of the noise spectrum correspond to the energies at which the Umklapp scattering is efficient, and the minima are located at the energies where the normal processes are expected to be dominant. The authors came to a conclusion that the noise reducing occurs because of the correlation between the normal electron–phonon scattering events as a result of the stimulated emission of the coherent phonons with small group velocities. On the other hand, the noise spectra of contacts in the diffusive or in the thermal limit monotonically (nearly quadratically) increase at low biases without any peculiarities. Such sensitivity of the noise spectra even to the elastic scattering on impurities provides an opportunity to obtain direct information about the regime of the electron flow through the contact. This can be referred to as point-contact noise spectroscopy.

Akimenko et al. (1985) measured both the point-contact spectrum and noise spectrum for the three symmetric directions in copper. They demonstrated a close relationship between the noise spectrum and the background of the EPI spectrum. This leads to the suggestion that fluctuations in the number of nonequilibrium phonons generated in the contact region could be also responsible for the spectral low-frequency fluctuations of the $1/f$ type.

Let us consider the theoretical examination of the noise in contacts. The theory of the flicker noise in the diffusive and ballistic regime for point contacts was proposed by Zagoskin et al. (1987). Although in the diffusive regime

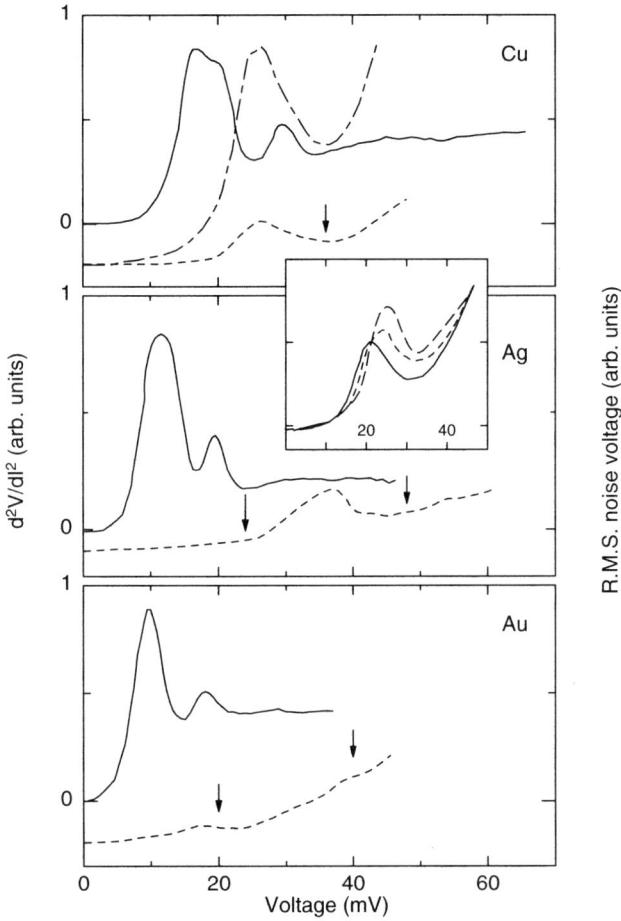

Fig. 9.4. Point-contact spectra d^2V/dI^2 (solid curves, left scale) and the dependence of the rms noise voltage (dashed curves, right scale) versus bias for noble metal contacts with resistance between 0.57 and 0.8 Ω at 4.2 K and $f = 50$ kHz. For Cu the dash-dotted curve is five times amplified the dashed one. The arrows indicate multiplied by factor 2 and 4 position of the main transverse (TA) peak. Inset shows the noise voltage for one of the Cu contacts at different temperatures: 10, 20, and 30 K from the top curve to the bottom one. Data taken from Yanson et al. (1982).

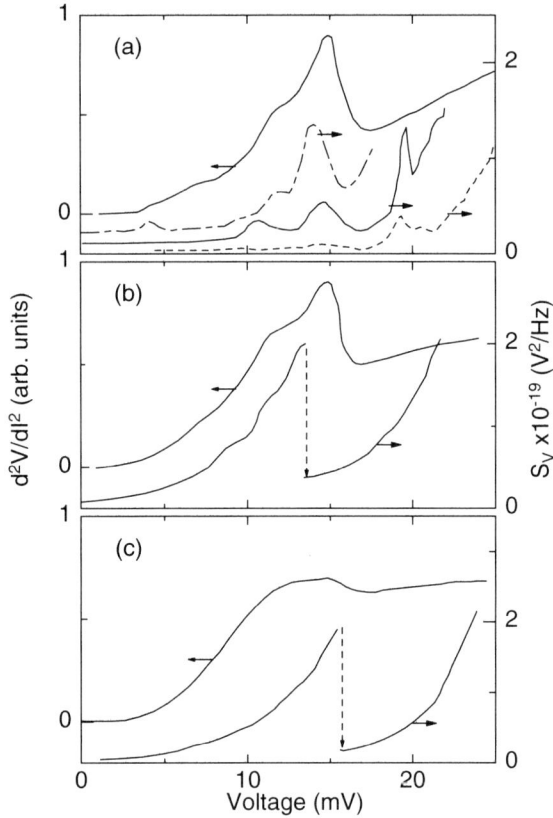

Fig. 9.5. Point-contact spectra $\mathrm{d}^2V/\mathrm{d}I^2$ and the noise spectral density S_V versus bias voltage for sodium contacts with different regimes of electron flow. (a) Ballistic regime: $R = 0.5\,\Omega$, $T = 1.7\,\mathrm{K}$. Dashed and dash-dotted curves were measured after the resistance of contact dropped to $0.37\,\Omega$. For these curves, the ordinate axis scale should be multiplied by 5 and 0.2, respectively. (b) Diffusive regime: $R = 0.66\,\Omega$, $T = 1.7\,\mathrm{K}$. (c) Thermal regime: $R = 1.33\,\Omega$, $T = 1.7\,\mathrm{K}$. The vertical dashed arrows mark the decreasing of sensitivity by factor 10. Data taken from Akimenko et al. (1984).

the intensity of the noise is proportional to $1/N_e$, it behaves like $N_e^{-1/3}$ in the ballistic limit, where N_e is the number of charge carriers in the contact region. Exploiting a model of excitation of both phonons and TLS by accelerated electrons, the authors showed the weakening of an electron–TLS exchange when a bias voltage reaches characteristic phonon energies. Thus, the transfer of the excess energy of electrons to phonons intensifies. Since the average inelastic interaction of TLS with electrons determines the average occupation number of the upper fluctuator level; i. e., it is responsible for the fluctuations of the resistance, and the EPI contribution to the contact noise is expected. This effect is manifested in the appearance of smooth features (kinks) in the S_V dependence on the bias voltage at the characteristic phonon energies (Fig. 9.6). This theory partially explains the experiments mentioned

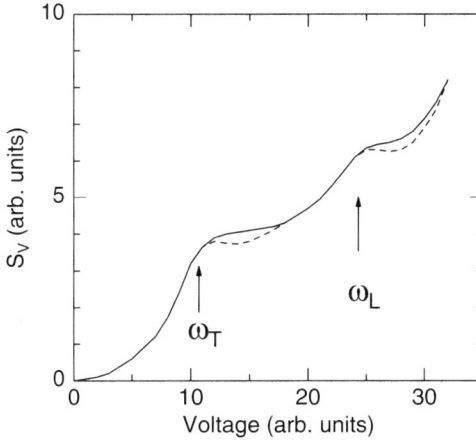

Fig. 9.6. Schematic dependence of the flicker noise spectral density versus energy in the case of two characteristic phonon frequencies ω_T and ω_L. The effect is determined by the parameter d/l_{in} (l_{in} is the electron–phonon mean free path) that is small for the ballistic contact. The dashed curve shows the desired behavior of S_V for $d/l_{in} \simeq 1$. After Zagoskin et al. (1992).

above, where distinct maxima in S_V are observed instead of the predicted kinks. The other problem to be explained is the shifting of the extremes in the noise spectrum versus voltage with tuning a filtering frequency, mentioned by Matej (1993), as well as their temperature dependence (see inset in Fig. 9.4).

Let us turn to the Nyquist and to the shot noise for contacts, as discussed at the beginning of this chapter. Despite the fact that the $1/f$ noise in a contact dominates, its intensity goes down with the frequency; hence, the other noise mechanisms may be observable at least at high frequencies. Kulik and Omelyanchouk (1984) showed that only Nyquist noise is present with the intensity in accordance with (9.1) in the current carrying state of a contact in the collision-free ballistic regime, whereas the shot noise is reduced by the factor d/l_{in}:

$$S_V \propto 2eVR\frac{d}{l}, \tag{9.6}$$

where l is the phase-breaking electron mean free path $l \gg d$. Decreasing of the shot noise was observed experimentally and is shown in Fig. 9.7. EPI gives rise to the electrical fluctuations that are analogous to the shot noise because of discreteness of absorption and emission events of the nonequilibrium phonons. It is curious that the shot noise should exhibit peculiarities at the characteristic phonon energies in the Kulik and Omelyanchouk (1984)

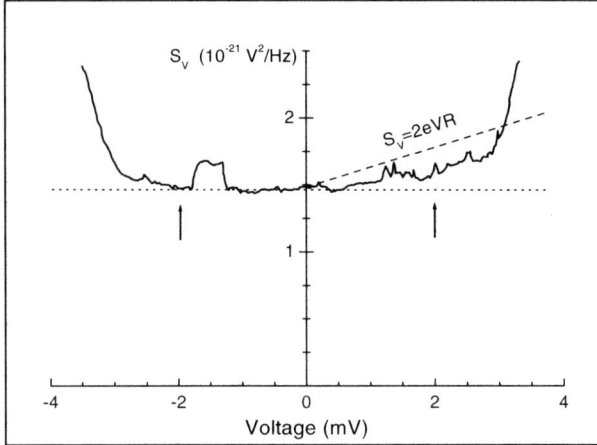

Fig. 9.7. Initial part of point-contact noise spectrum for sodium contacts with R $= 0.56\,\Omega$ at $T = 1.7\,\mathrm{K}$. The horizontal dashed line marks the amplifier noise level. The inclined dashed line corresponds to the expected shot noise. Arrows at $\pm 2\,\mathrm{mV}$ mark approximate position of incipience of the steeper rise of the noise. Data taken from Akimenko et al. (1984).

model so that the second derivative of the noise spectral density with respect to the voltage is proportional to the EPI function.

References

Akimenko A. I., Verkin A. B. and Yanson I. K. (1984) J. Low Temp. Phys. **54** 247.

Akimenko A. I., Verkin A. B. and Yanson I. K. (1985) Sov. J. Low Temp. Phys. **11** 391.

Black J. L. (1981) in *Glassy Metals* ed. by H. J. Güntherodt and H. Back (Springer Verlag, Berlin) 167.

Dutta P., Diman P. and Horn P. M. (1979) Phys. Rev. Let. **43** 646.

Dutta P. and Horn P. M. (1981) Rev. Mod. Phys. **53** 497.

Holweg P. A. M., Caro J., Verbruggen A. H. and Radelaar S. (1992) Phys. Rev. B **45** 9311.

Hooge F. N. (1969) Phys. Lett. A **29** 139.

Hooge F. N. and Hoppenbrouwers A. M. H. (1969) Phys. Lett. A **29** 642.

Hooge F. N. (1976) Physica B **83** 14.

Kulik I. O. and Omelyanchouk A. N. (1984) Sov. J. Low Temp. Phys. **10** 158.

Machlup S. (1954) J. Appl. Phys. **25** 341.

Matej V. (1993) Solid State Commun. **86** 179.

Ralls K. S. and Buhrman R. A. (1988) Phys. Rev. Lett. **44** 2434.

Ralls K. S. and Buhrman R. A. (1991) Phys. Rev. B **44** 5800.

Yanson I. K., Akimenko A. I. and Verkin A. B. (1982) Solid State Commun. **43** 765.

Zagoskin A. M., Kulik I. O. and Omelyanchouk A. N. (1987) Sov. J. Low Temp. Phys. **13** 332.

Zagoskin A. M., Kulik I. O. and Omelyanchouk A. N. (1992) Sov. J. Low Temp. Phys. **18** 475.

10 Point contacts under irradiation

For a long time, point contacts have been used in microwave engineering as detectors, rectifiers, mixers, and harmonic signal sources. Such contacts are usually formed by means of thin wire pressed against a smooth semiconductor or metal surface. The wire represents a kind of antenna to couple in the radiation. A high-frequency current penetrates in the bulk of contact through this wire. An outline of the experimental configuration for the light or high-frequency irradiation of point contact is shown in Fig. 10.1. As the contact

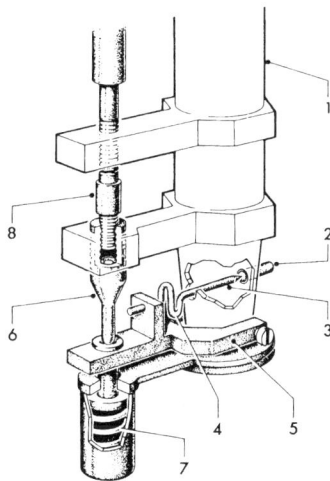

Fig. 10.1. Schema of experimental setup for irradiation of contact: 1 – light cone, 2 – sample, 3 – pointed wire, 4 – spring loading, 5 – cantilever, 6 – movable plunger, 7 – spring, 8 – differential screw. After van der Heijden et al. (1984).

resistance is much smaller than the surrounding environment impedance, the high-frequency radiation produces a high-frequency current in the leads. Inasmuch as the $I - V$ characteristic of a point contact deviates from the Ohms law, in other words, it is strongly nonlinear both in the thermal and the ballistic regime, the contacts independent on the current regimes can be used as a tool for rectifying, mixing of electromagnetic signals, and so on from the microwave up to optical frequencies. Consequently, a microscopic physics of processes in the contacts under the influence of irradiation evokes considerable interest.

10.1 Relaxation processes

Let us take a brief look at characteristic times or frequencies of the transport processes in constriction to compare them with an external action. The electron relaxation time caused by the electron–phonon scattering τ_{e-ph} is easy to estimate by dividing a mean free path by the Fermi velocity of electrons, which gives a value within 10^{-13}–10^{-14} s for the Debye phonons. The reverse phonon–electron processes are slower, roughly speaking, as a sound (phonon) velocity s is smaller compared with the Fermi one. They are characterized by the time $\tau_{ph-el} = 10^{-9}$–10^{-10} s. More precisely, as was shown by Kulik (1985), the phonon–electron relaxation frequency is

$$f_{ph-el} = 1/\tau_{ph-el} \simeq \omega_D \lambda (s/v_F); \qquad (10.1)$$

here, λ is the EPI constant and ω_D is the Debye frequency of phonons. It is worth noting that with a frequency increase to the level comparable with the frequency of electron–phonon relaxation, an energy quantum of radiation is comparable with the phonon energies. Therefore, the corpuscular theory of the absorption and the emission of phonons under the radiation action should be considered. This results in a broadening of features in the point-contact spectra and the appearance of satellites at higher frequencies [Omelyanchouk and Tuluzov (1983)].

For contacts in the thermal limit (see Section 3.5), temperature relaxation time depends on their diameter and the specific characteristics of materials. Corresponding to this case, the relaxation frequency is

$$f_T = 1/\tau_T \simeq \lambda_T/(4C_h \rho_m d^2); \qquad (10.2)$$

here, λ_T is the thermal conductivity, ρ_m is the specific density of material, and C_h is the specific heat. f_T is of the order of 10^9 Hz for the $1\,\Omega$ contact and essentially depends [see (10.2)] on their diameter. At the frequencies above those of the thermal relaxation, the contact temperature no longer traces the modulating high-frequency current. As a result, the possibilities of the "temperature modulation spectroscopy" mentioned in Section 3.5 are limited at higher frequencies. Summarizing, the study of the contact conductivity at high frequencies makes it possible to clear up the nature of various nonstationary processes in metals; namely, it allows us to determine the phonon–electron relaxation time, to elucidate the quantum effect in detection for ballistic contacts, and to investigate the thermal relaxation in small-size conductors in the thermal regime.

10.2 Rectification in point contacts

High-frequency ac current $i\cos\omega t$ induced in a contact under irradiation has an extra dc voltage contribution [see (4.1)] that is proportional to the second derivative. Chopping a radiation signal by the low-frequency ω_R, an ac voltage amplitude

$$V_R = \frac{i^2}{4}\frac{d^2V}{dI^2}\Big|_{I_0} = \frac{i^2}{4}R_d\frac{dR_d}{dV}\Big|_{I_0} \tag{10.3}$$

can be measured, where I_0 is the applied dc current and $R_d = dV/dI$. This method enables us to measure d^2V/dI^2 in another way, overcoming detecting difficulties at high frequency. As was shown by van der Heijden et al. (1980), the rectifying signal of the Cu contact at 525 GHz reveals a phonon structure similar to that as measured in a traditional way (Fig. 10.2). As mentioned

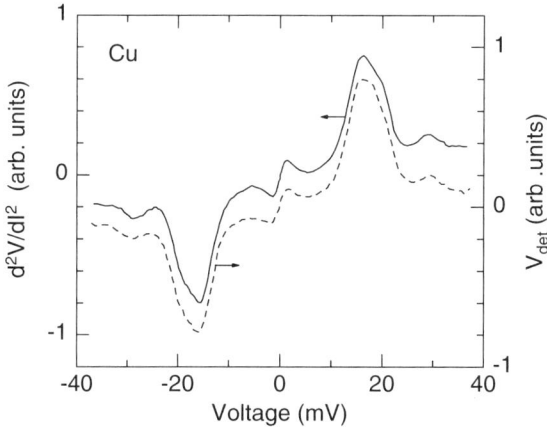

Fig. 10.2. d^2V/dI^2 (solid curve) and laser-induced signal at $f = 525$ GHz (dashed curve) of the Cu - Cu contact. The contact resistance is 5 Ω and T=1.5 K. Data taken from van der Heijden et al. (1980).

by the authors, the dominant detection mechanism can be attributed to the electron–phonon interaction responsible for a nonlinear $I - V$ characteristic. In this case, a necessary condition is that the characteristic frequency $1/\tau_{e-ph}$ should be higher than the timescale imposed by the radiation. In other words, a voltage over a contact must follow the induced ac current. Recall that by the above-mentioned experiments, $1/\tau_{e-ph} = 10^{13} - 10^{14}$ s^{-1} is indeed larger than the alternating signal.

In the opposite case of a thermal limit, the characteristic relaxation time is 10^{-9} s. Balkashin et al. (1982) measured the detection signal of the Ni contact for various modulation frequencies. Although the observed signal was similar

to d^2V/dI^2 below 10^9 Hz, at higher frequencies, the spectrum shape began
to alter (Fig. 10.3). As the heat relaxation in a contact is slower compared

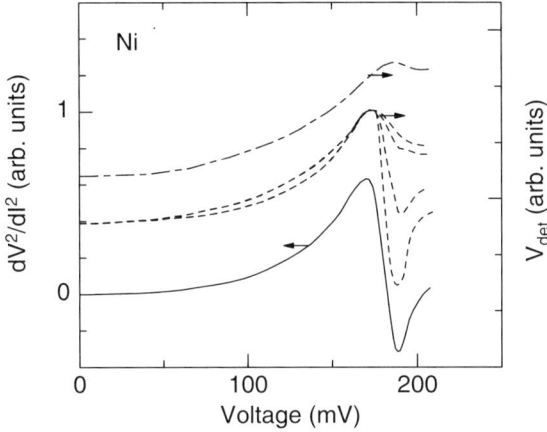

Fig. 10.3. d^2V/dI^2 (bottom curve) and rectifying signal (dashed curves) of 3.3 Ω
Ni-Ni contact at room temperature and at different irradiation frequencies: 0.3, 1,
12, and 16.6 GHz from the bottom dashed curve to the top one. The upper dashed-
dot theoretical curve is calculated by (10.4). The curves are offset vertically for
clarity. Data taken from Balkashin et al. (1982).

with the 10-GHz alternating signal, ac current modulation does not alter
an $I - V$ characteristic resulting in vanishing of a rectifying signal and its
deviation from d^2V/dI^2. On the other hand, the radiation dissipation in the
constriction can rise its temperature. This leads to a detectable change in the
voltage across a constriction [Lysykh et al. (1989)]:

$$V_\mathrm{d} = \frac{\sqrt{2}i^2}{8} R_\mathrm{s} \frac{\mathrm{d}R_\mathrm{s}}{\mathrm{d}V}, \tag{10.4}$$

where $R_\mathrm{s} = V/I$ is the static resistance. As seen from Fig. 10.3, there is
a resemblance between the curve measured at highest frequency and the
calculated one according to (10.4).

A significant broadening of spectra occurs at the high irradiation frequencies when $\hbar\omega$ is comparable with the phonon energy. This broadening
for point-contact spectra, as follows from the Omelyanchouk and Tuluzov
(1983) theory, is caused by a quantum detection effect. A time averaged
$I - V$ curve is given by the expression

$$I_\mathrm{dc} = \sum_{n=-\infty}^{\infty} J_n^2 \left(\frac{ev_1}{\hbar\omega}\right) I_0 \left(V_0 + n\frac{\hbar\omega}{e}\right), \tag{10.5}$$

where $I(V)$ is an unperturbed $I - V$ characteristic, v_1 is the high-frequency alternating voltage, and J_n is the n-order Bessel function. It was shown experimentally by van der Heijden et al. (1984) that at low temperatures under radiation, the response of the normal metal point contact deviates from the classic detection at frequency 2523 GHz, which approaches the phonon frequency (Fig. 10.4). Moreover, they fitted the experimental curve by (10.5)

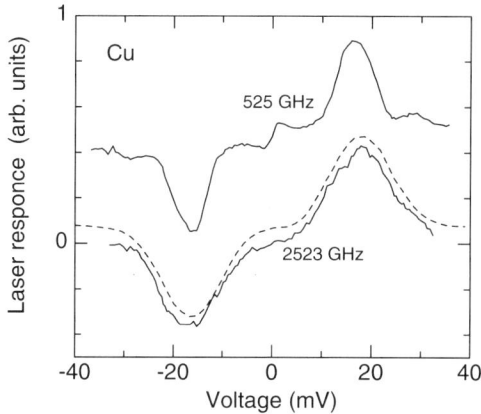

Fig. 10.4. Laser-induced signal of the 42 Ω Cu-Cu contact at 1.2 K and frequencies 525 and 2523 GHz. The dashed curve is a calculation by (10.5) using the measured $\mathrm{d}^2V/\mathrm{d}I^2$ without radiation. The curves are offset vertically for clarity. Data taken from van der Heijden et al. (1984).

and obtained a good correspondence between both curves. It is pertinent to note that at a low power level, (10.5) implies that the spectrum structure is smeared over a voltage range of the order of $\hbar\omega/e$.

10.3 Study of phonon relaxation

At helium temperatures, a voltage-biased point contact is a source of nonequilibrium phonons created by energized electrons. The accumulation of these phonons in the constriction region caused by a short elastic mean free path of the phonons appears to be the main reason for an additional contribution to the point contact spectra known as background (see Section 3.3). Kulik (1985) predicted a frequency dependence of the phonon component in the EPI spectrum at frequencies above $1/\tau_{\mathrm{ph-el}}$. The magnitude of the effect is determined by the relation between the inelastic relaxation rate of nonequilibrium phonons and the frequency of the external radiation incident on the contact. It was shown that the background component of the spectra decreases with the increase of the external frequency f as

$$\eta \sim (1 + (f/f_{\text{ph}-\text{el}}))^{-1}. \qquad (10.6)$$

A suppressing of the point-contact spectrum background with increasing an altering frequency from 1–3 kHz to 80 GHz was observed by Yanson et al. (1985) (Fig. 10.5) in qualitative agreement with (10.6). During the

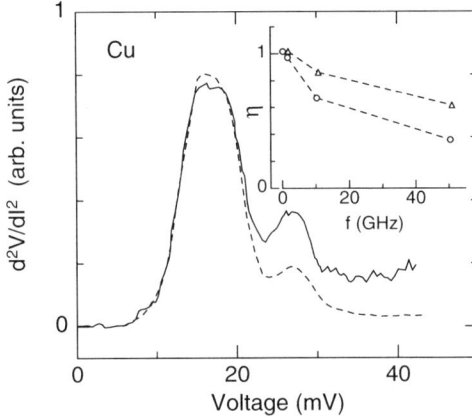

Fig. 10.5. d^2V/dI^2 of the 6.3 Ω Cu-Cu contact obtained at the sound frequency (solid curve) and at frequency 80 GHz (dashed curve). Inset shows a frequency dependence of the background suppression factor for the other 3.6 Ω contact at two bias voltages: 40 mV (triangles) and 60 mV (circles). The dashed lines connect symbols for clarity. Data taken from Yanson et al. (1985).

further study, Balkashin et al. (1987, 1991) determined a nonequilibrium phonon–electron relaxation frequency equal to 0.8, 5, and 30 GHz for Au, Cu, and Be, correspondingly. The frequency rise is in line with an increase of the Debye energy from Au to Be (Fig. 10.6) in qualitative agreement with (10.1).

Balkashin et al. (1987) found the decreasing of the background suppression factor for the contact diameter below 13 nm. They consider that the nonuniform relaxation of phonons also occurs because of their random walking from the contact center to the thermostat with a rate $f_r = sl_r/3d^2$ (s is the sound velocity). If the frequency of this process is higher than the external frequency, some of phonons will escape from the contact, leading to background suppression factor decrease. If f_r is equal to the external frequency at the diameter value at which deviation from the constant value of the background suppression factor takes place, an elastic mean free path of nonequilibrium phonons l_r will be of order 1 nm.

At the same time, the authors mentioned a serious noncoincidence with a theory, by finding a saturation of the background frequency dependence, while in accordance with (10.6), it should monotonically decline to zero with

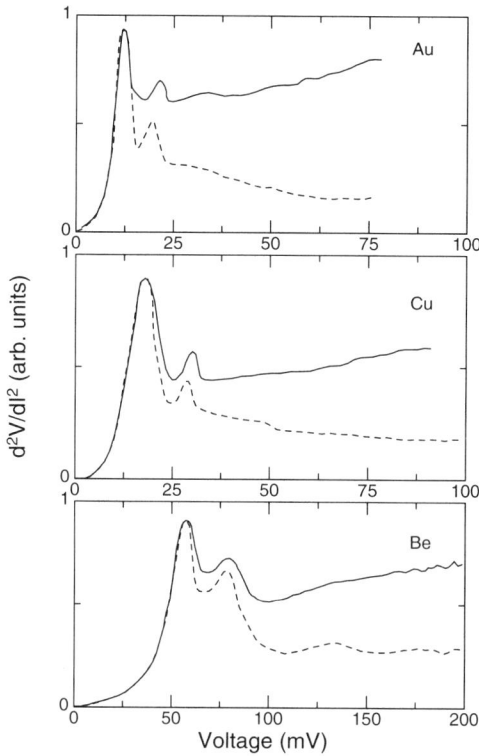

Fig. 10.6. Point-contact spectra of Au, Cu, and Be measured at audio frequency (solid curves) and microwave frequency (4.8 GHz for Au and 80 GHz for Cu and Be) at helium temperatures. The curves are scaled vertically to coincide at the maximum. Data taken from Balkashin (1992).

increasing the frequency. This effect was explained by an additional contribution to the signal associated with a difference of the $I - V$ curves under and without irradiation during the signal chopping. At high frequency, when the $I - V$ characteristic of the contact does not depend on the frequency, induced by the microwave current nonequilibrium phonons make a contribution to the signal detecting at the chopping frequency, proportional to the second derivative. This contribution is associated with the internal contact heating by radiation (bolometric effect), which is discussed below.

10.4 Thermal effects

Balkashin and Kulik (1990) studied the response of the ballistic spear–copper thin-film point contacts to optical laser radiation with $\lambda = 0.63\,\mu$m at helium temperatures. The contacts were irradiated both from the spear and the opposite film side. In both cases, the dependence of response versus a bias voltage was similar to the EPI spectrum of copper (Fig. 10.7). Note that

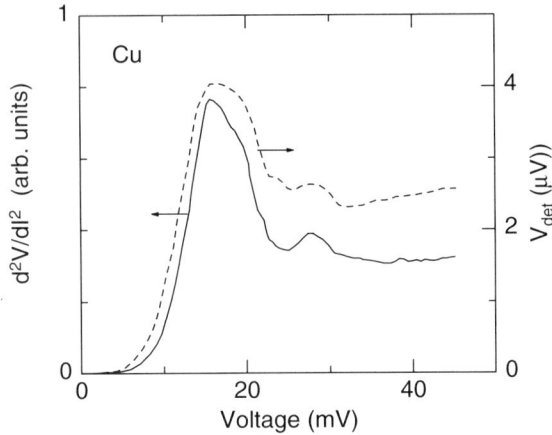

Fig. 10.7. Response of a Cu contact in the case of laser irradiation with $\lambda = 0.63\,\mu$m (dashed curve) and its $\mathrm{d}^2V/\mathrm{d}I^2$ (solid curve) at the helium temperature. Data taken from Balkashin and Kulik (1990).

the energy quantum was about 2 eV in this experiment, i. e., much higher than that used by van der Heijden et al. (1984), where the laser-induced broadening of the spectra was observed compared with the spectra at the lower frequency (Fig. 10.4). The latter is not the case in the experiments of Balkashin and Kulik (1990). They proposed the thermal origin (source) of the effect. Energized electrons excited by photons with the energy 2 eV relax in the bulk material generating nonequilibrium phonons. Thermalization of these phonons leads to an increase of the temperature both in the bulk material and in the contact. The signal amplitude was found to depend critically on the heat transfer conditions in support of the proposed model.

In the thermal regime, point contacts have a characteristic temperature relaxation time of about 10^{-9} s, at which a rectifying signal changes its shape qualitatively, as seen in Fig. 10.3. Detailed measurements of the high-frequency response of metallic contacts with a small electron mean free path were carried out by Balkashin et al. (1992). They came to the following conclusions: (1) a measured signal as high as 4.3×10^{12}Hz is caused by heating of point contact by the high-frequency transport current, (2) the nature of the signal is associated with heating of surrounding medium at higher frequencies, and (3) the thermal contribution to the response signal occurs for ballistic contact above frequency 2.8×10^{13}Hz as well. The latter is the same "bolometric" effect mentioned at the end of the previous Section.

References

Balkashin O. P. (1992) Sov. J. Low Temp. Phys. **18** 470.

Balkashin O. P. and Kulik I. I. (1990) Sov. J. Low Temp. Phys. **16** 166.

Balkashin O. P., Kulik I. I. and Moskalets M. V. (1992) Sov. J. Low Temp. Phys. **18** 192.

Balkashin O. P., Yanson I. K. and Pilipenko Yu. A. (1987) Sov. J. Low Temp. Phys. **13** 222.

Balkashin O. P., Yanson I. K. and Pilipenko Yu. A. (1991) Sov. J. Low Temp. Phys. **17** 114.

Balkashin O. P., Yanson I. K., Solov'jev V. S. and Krasnogorov A. Yu. (1982) Sov. Tech. Phys. **27** 522.

Kulik I. O. (1985) JETP Lett. **41** 370.

Lysykh A. A., Duif A. M., Jansen A. G. M. and Wyder P. (1989) Phys. Rev. **B 39** 12560.

Omelyanchouk A. N. and Tuluzov I. G. (1983) Sov. J. Low Temp. Phys. **9** 142.

van der Heijden R. W., Jansen A. G. M., Stoelinga J. H. M., Swartjes H. M. and Wyder P. (1980) Appl. Phys. Lett. **32**(2) 245.

van der Heijden R. W., Jansen A. G. M., Stoelinga J. H. M., Swartjes H. M. and Wyder P. (1984) J. Appl. Phys. **55**(4) 1003.

van der Heijden R. W., Swartjes H. M. and Wyder P. (1984) Phys. Rev. B **30** 3513.

Yanson I. K., Balkashin O. P. and Pilipenko Yu. A. (1985) JETP Lett. **41** 373.

11 PCS of semimetals, semiconductors, and dielectrics

The distinctive feature of semimetals and semiconductors is, first of all, the low density of charge carriers compared with ordinary metals. Thus, in typical semimetals as arsenic, antimony, and bismuth, the carrier density n decreases from about 2×10^{20} (As) to 3×10^{17} (Bi) per cm^3, whereas in metals, n is 10^{22}–10^{23} cm^3. These values can be still lower by several orders of magnitude in semiconductors depending on the doping level. Accordingly, the Fermi energy decreases from several eV in metals to tens of meV and lower in semimetals (Fig. 11.1). Because of the band structure peculiarities, an effective mass of charge carriers both in semimetals and semiconductors can be one order of magnitude lower as compared with the metals. The small effective mass leads to the decrease of the Larmor radius to a value comparable with the contact size already in an easily attainable fields of about a few Tesla. The effect of a magnetic field on the EPI spectra in As and Sb was described in Section 8.3. Additionally, the de Broglie wavelength at the Fermi energy $\lambda_B \propto n^{-1/3}$ increases in semimetals up to tens of nanometers (Fig. 11.1), i.e., it may become comparable with the contact dimension. This leads to the influence of the quantum interference effects on the contact conductivity. Again, the low carrier density increases a screening length $r_s \propto n^{-1/6}$ (Fig. 11.1) in the case of semiconductor-metal contact, resulting in Schottky barrier formation (see Fig. 2.5) with a low concentration of carriers at the surface. All mentioned features lead both to the new interesting phenomena in point contacts as well as to difficulties in their interpretation, in particular how to separate bulk properties from the surface influence.

11.1 Localization effects

An anomalous decrease of differential resistance has been observed for some antimony contacts as bias reaches characteristic phonon energies. That is, the d^2V/dI^2 spectrum becomes negative (Fig. 11.2), but the extremes caused by the electron–phonon interaction are retained [Yanson et al. (1985)]. Such contacts were distinguished by a small elastic mean free path of electrons of the same order as the de Broglie wavelength. The result was explained, accounting for the phase coherence along the electron trajectories leading to the weak localization of the current carriers. EPI in this case destroys phase

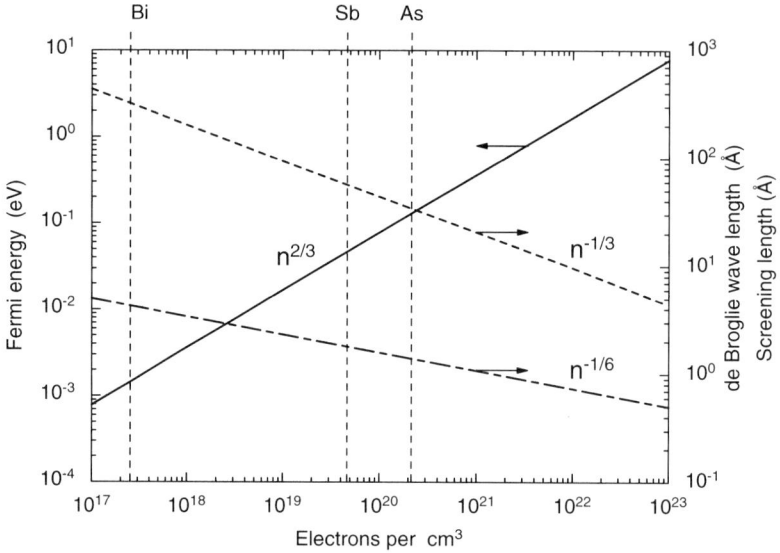

Fig. 11.1. Dependence of the Fermi energy (solid line), de Broglie wavelength (dashed line), and screening length of the Thomas–Fermi model (dot-dashed line) versus density of carriers in the free electron approximation. The vertical dashed lines mark electron density for semimetals Bi, Sb, and As.

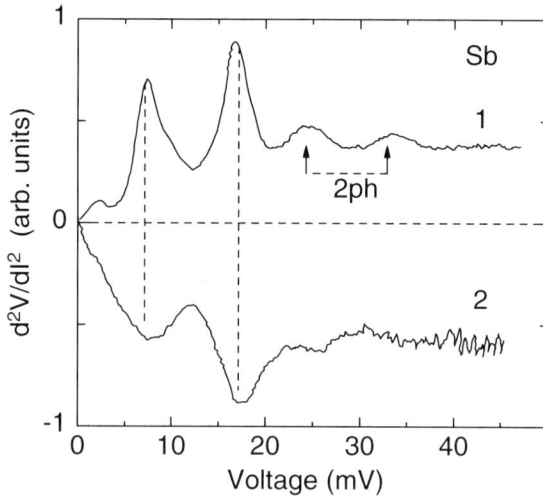

Fig. 11.2. d^2V/dI^2 spectrum of the pure antimony contact (curve 1, $R_0 = 6.3\,\Omega$) along with spectrum of the dirty one (curve 2, $R_0 = 6.1\Omega$) at 4.2 K. A region of the double-phonon peaks is marked by arrows. Data taken from Yanson et al. (1985).

coherence and increases conductivity, contrary to the usual resistance increase in ballistic contacts caused by EPI. Theoretical analysis of the conductivity of dirty contacts between clean conductors with an elastic scattering length comparable with the de Broglie wavelength [given by Itskovich et al. (1987)] confirms mentioned experimental observation for Sb. Analogous results were obtained by Gribov et al. (1991) by the investigation of another semimetal As. Additionally, for As, the negative magnetoresistance was measured for contacts with negative spectra. This is in favor of the localization nature of the effect, because it is well established that the magnetic field destroys the weak-localization regime.

Yanson et al. (1992) measured the transition from the inverse spectrum to the classic EPI spectrum on the same Sb contact during its successive short-circuiting from 24 Ω to about 1 Ω (Fig. 11.3). They supposed that char-

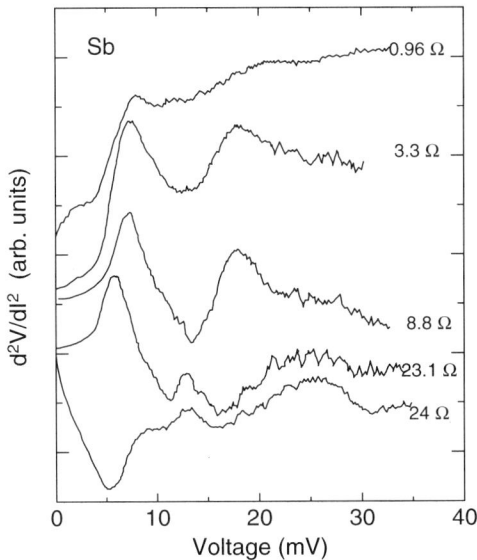

Fig. 11.3. d^2V/dI^2 spectra of a Sb-Sb contact at helium temperature after successive short-circuit of its resistance from 24 Ω (bottom curve) to 0.95 Ω (top curve). The curves are offset vertically for clarity. Data taken from Yanson et al. (1992).

acteristic mean free path of the 24 Ω contact with $d \sim 50$ nm was about 20 nm because of the excessive contamination of the surface. The latter value is comparable with $\lambda_B \sim 5$ nm. The increase of the contact diameter leads to the increase of the effective elastic mean free path, because deeper pure remote bulk regions are included in the constriction. This results in depressing of the weak localization regime and transformation of the curve into the typical EPI spectrum. At further decrease of the contact resistance, the spectrum features smear and the background grows, indicating a transition to the thermal regime.

Most of the point-contact spectra of the other Bi semimetal obtained by Yanson et al. (1992) have the shape of a negative zero-bias anomaly (similar to the spectra of graphite in the following Fig. 11.5) corresponding to the maximum of the differential resistance at $V = 0$. In some cases, the peculiarities close to the energy of intervalley acoustic and intravalley optical phonons were observed in the negative part of the d^2V/dI^2 dependence. The authors interpreted the spectrum as an inverse spectrum of bismuth. Later d^2V/dI^2 characteristics clearly reflected EPI in Bi were measured by Arutyunov et al. (1996) by special prepared microbridges with controlled geometry (diameter 0.1–3 μm, length 0.5–20 μm) and various degrees of perfection. The EPI spectrum contains two maxima at 4 and 11 mV corresponding to the intervalley acoustic and intravalley optic phonons (Fig. 11.4, curve 1). The material quality aggravation in a microbridge leads to the appearance of a low-energy positive anomaly and to the broadening of the intravalley peak.

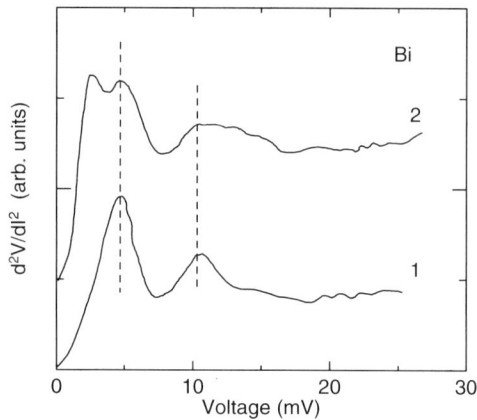

Fig. 11.4. Point-contact spectrum for two Bi microbridges at 1.6 K with diameter 0.3 μm and length 3 μm and different metal quality reflecting by residual resistance ratio: 1 – $R_{300}/R_{4.2}$=7.5, 2 – $R_{300}/R_{4.2}$=1.75. Data taken from Arutyunov et al. (1996).

The point-contact spectra of the last semimetal graphite were measured by Sato et al. (1988). For some contacts, the negative d^2V/dI^2 demonstrates features at about 20 and 60 mV, the position of which is close to the expected maxima of phonon density of states (Fig. 11.5). As shown in the inset, this structure becomes blurred near 20 K. For contacts with relatively small resistance, the position of the valleys corresponds to the characteristic phonon energies (Fig. 11.5, curve 2). Probably, they reflect EPI in graphite in the localization regime.

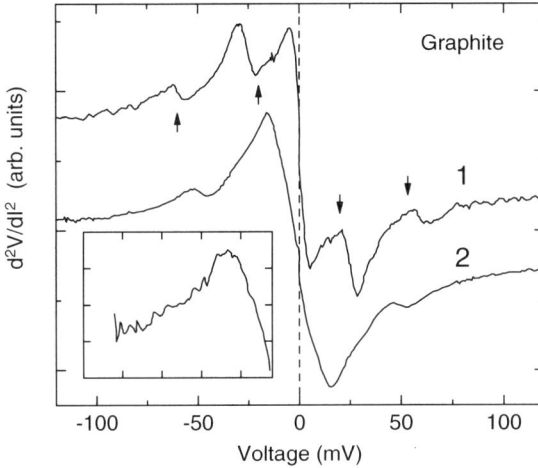

Fig. 11.5. Point-contact spectrum for two contacts between highly oriented pyrolitic graphite and Mo needle with the resistance 108 Ω (curve 1) and 56 Ω (curve 2) at 1.5 K. Arrows mark position of the peaks in the phonon density of states determined from the neutron inelastic scattering. The inset shows a part of the spectra (1) at the negative bias at about 20 K. Data taken from Sato et al. (1988).

11.2 Thermal transport

The extensive investigation of point contacts in semiconductors was carried out by the stuttgarter group starting by the work of Gerlach-Meyer and Queisser (1983). The general experimental arrangement and electric circuit used are depicted in Fig. 11.6. The contacts were established by touching two wedge-shaped samples crosswise against each other by a mechanical feedthrough in combination with the piezoelectric drive. The wedges were etched to remove surface defects and an oxide layer and were mounted in the UHV chamber. An ion sputter gun was used for the surface cleaning by ion bombardment, and the Auger spectrometer was applied to control the oxide layer. Whereas one of the samples was coupled to a heat sink, the other one was thermally isolated and equipped with a small heater for the case of the thermal resistance and thermopower measurements. The thermal resistance, thermoelectric power, and electrical resistance were registered simultaneously.

The first effect observed in semiconductor point contacts with n-doped silicon was asymmetrically produced Joule heating in the downstream of the

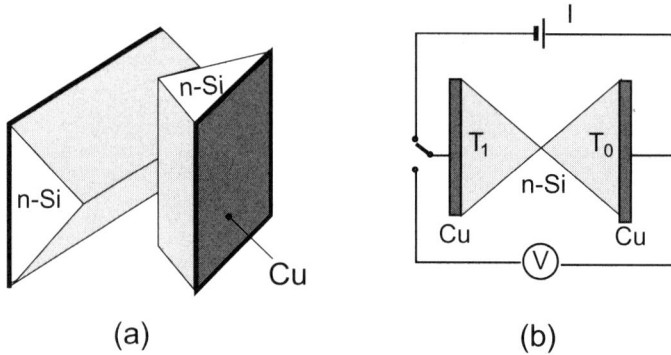

Fig. 11.6. (a) Experimental schema: two wedges of n-type Si are mounted on the Cu blocks. (b) Electrical circuit: One sample is kept at the bath temperature T_0, and the temperature of another one may rise to T_1 as a result of current flow. After the current flow (about 10 min), a switch is thrown to measure the difference in the temperature via Seebeck thermovoltage. Adapted from Trzcinski et al. (1987).

current [Gerlach-Meyer and Queisser (1983)]. This asymmetric heat generation leads to temperature imbalance between two legs of contact, which was detected by Seebeck voltage measurements between two wedges after passage of the current for approximately 10 min. The increase of the contact resistance enhances the asymmetry (Fig. 11.7), which arises, in the author's opinion, because of the Knudsen regime of ballistic electron transport, when geometrical dimension of the contact is reduced compared with the electron mean free path. In the pertinent simplified model, the heat generation is assumed to be proportional to the resistance ($Q = I^2 R$) and the asymmetry Y is defined as a relation between the ballistic resistance and total (Wexler) contact resistance:

$$Y \sim R_{\mathrm{Sh}}/R_{\mathrm{W}} \approx K/(K+1), \tag{11.1}$$

where $K = l/a$ is the Knudsen factor and $a = d/2$ is the contact radius. To compare this formula with the experimental data (Fig. 11.7), the authors calculated the contact dimension using (3.18). The mean free path was estimated using the Drude formula. The further experiments by Trzcinski et al. (1987) on n- and p-type Si and n-type GaAs showed that this size effect is a general phenomenon for restricted geometry independently on the semiconductor material, type of charge carriers, and doping. However, as mentioned by Weber et al. (1989) in the case of semiconductors, the calculation of the contact radius from electrical resistance yields values smaller than the actual geometrical size, especially with resistance increasing. They measured the thermal and electrical resistance of the n-type Si contacts ($n = 1.7 \times 10^{19}$ cm^{-3}) simultaneously to estimate the contact size. By analogy with (3.18), the phonon part of the thermal resistance can be written as

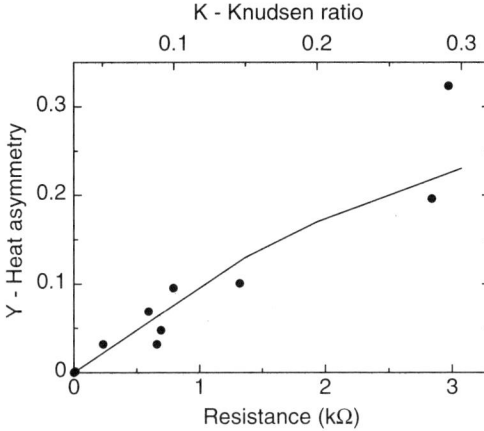

Fig. 11.7. Heat asymmetry (symbols) as a function of contact resistance or corresponding Knudsen ratio. The solid curve represents (11.1), assuming the electron mean free path of 2 nm. Data taken from Gerlach-Meyer and Queisser (1983).

$$W_{\mathrm{ph}} = \frac{16K\rho_{\mathrm{p}}}{3\pi d} + \frac{\rho_{\mathrm{p}}}{d}, \qquad (11.2)$$

where $K = l_{\mathrm{ph}}/d$, $\rho_{\mathrm{p}}^{-1} = k$ is the lattice thermal conductivity and l_{ph} is the mean free path of phonons. The latter was estimated according to the kinetic formula for the heat conduction $k = Csl_{\mathrm{ph}}/3$, where C is the specific heat and s is the sound velocity. The contact size determined by the above formula was roughly equal to the value derived from the Wexler formula at $d \sim 1\,\mu$m, whereas the discrepancy reached as much as 30 times for the smallest ($d \sim 0.01\mu$m) contact [Weber et al. (1989)]. Measurements with a weakly doped silicon, as mentioned by the authors, gave an even larger difference. The presence of a tunnel resistance at the interface was supposed as a main reason.

During further investigation, Weber et al. (1991) studied the character of the potential barrier arising at the surface in the constriction as a function of the contact size, oxide thickness, temperature, and doping level. Although the potential barrier seemed to be exclusively caused by oxidation for heavily doped Si ($n = 1.7 \times 10^{19} \mathrm{cm}^{-3}$), properties of weakly doped material ($n = 2.8 \times 10^{16} \mathrm{cm}^{-3}$) were considerably affected by surface charge. It turned out also that the transmission of the tunnel barrier strongly depends on the contact size, because of compression of the oxide under mechanical pressure. After cleaning the surface from the oxide, the contact resistance behaves like that for a clean contact (Fig. 11.8).

Another phenomenon studied in semiconductor point contacts by the same group is the suppression of thermopower with decreasing of the contact size [Trzcinski et al. (1986)]. This experiment for the semiconductors

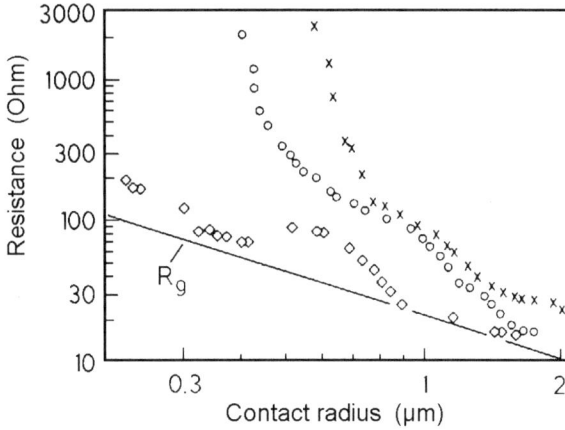

Fig. 11.8. Electrical resistance versus radius for silicon ($n = 1.7 \times 10^{19} \mathrm{cm}^{-3}$) point contacts with various oxide thickness d_{ox}, measured at room temperature ($\diamond$ - $d_{\mathrm{ox}}{=}0$, $\circ$ - $d_{\mathrm{ox}}{=}6\,\text{Å}$, $\times$ - $d_{\mathrm{ox}}{=}20\,\text{Å}$). Solid line represents behavior of the geometrical resistance according to (3.18). After Weber et al. (1991).

is equivalent to the experiment done by Shklyarevskii et al. on metals (see Section 7.2). No electrical current was applied, and only one electrode was heated. Drop of thermoelectric voltage takes place with increasing the contact resistance (Fig. 11.9). This sudden drop was ascribed by the quenching of the phonon drag when the phonon mean free path becomes comparable with the contact dimension. However, these results were critically discussed after experiments by Weber et al. (1992). They found reduction of thermopower in contact of heavily doped silicon, where phonon drag effect should be negligibly small (Fig. 11.10). Nevertheless, the reduction of the thermopower was almost the same both for the clean and oxidized contacts still testifying the phonon nature of the effect. Then Weber et al. (1992) used moderate doped silicon where the phonon drag effect is very strong. The sample geometry was varied to minimize the mechanical pressure. The measured thermopower versus temperature for two contacts with different size is shown in Fig. 11.11. It is seen that thermopower of a small contact is much smaller than the bulk one and corresponds to the diffusion part of thermopower. For large contact deviation from the bulk thermopower occurs only by lowering temperature below 200 K, that is, by increasing a mean free path of phonons. At 200 K, the phonon mean free path has been estimated to be about 5 μm, which is one-third of the contact radius. It proves the phonon drag suppression model. In the case of the heavily doped samples, deformation of the band structure caused by the high mechanical pressure has been considered as a reason for thermopower suppression. It is well known that the thermopower as well as the electrical resistivity of silicon are strongly reduced by mechanical pressure.

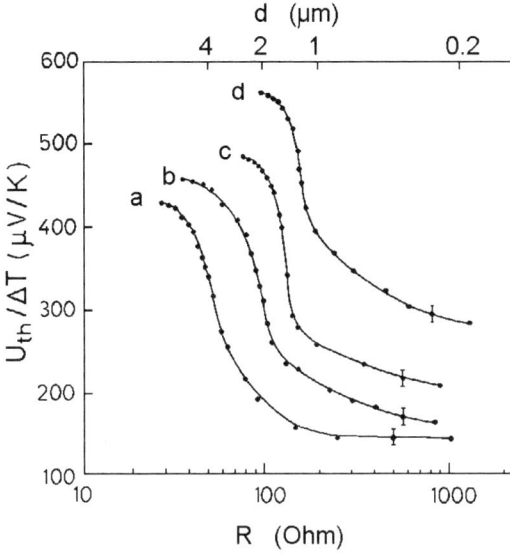

Fig. 11.9. Thermoelectric voltage divided by temperature difference $\Delta T = T_2 - T_1$ versus contact resistance of doped silicon ($n = 8 \times 10^{17} \mathrm{cm}^{-3}$) point contacts. Curves are measured for fixed $T_1 = 300$ K and $T_2 = 342$ K (a), $T_2 = 388$ K (b), $T_2 = 406$ K (c), and $T_2 = 450$ K (d). After Trzcinski et al. (1986).

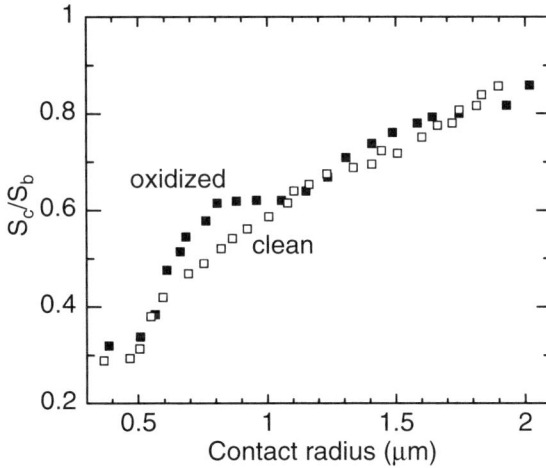

Fig. 11.10. Normalized to the bulk value thermopower S_c/S_b of clean and oxidized silicon ($n = 1.7 \times 10^{19} \mathrm{cm}^{-3}$) point contacts versus their radius, measured at the room temperature. Because of high impurity concentration (doping), the phonon drag effects should be negligible. Data taken from Weber et al. (1992).

The analogous quenching of the phonon-drag thermopower was observed in the moderately doped n- and p-type Ge by Le Hang Nguyen et al. (1996).

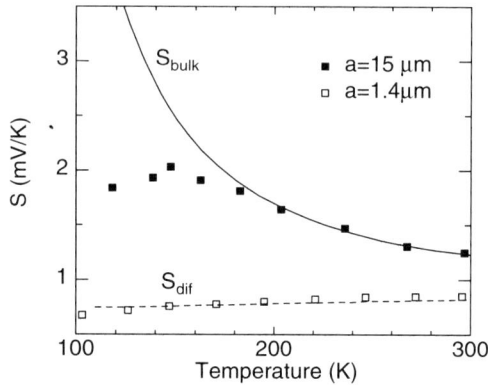

Fig. 11.11. Temperature dependence of the thermopower for two point contacts (symbols) with different sizes on the base of doped silicon ($n = 2.6 \times 10^{16} \mathrm{cm}^{-3}$). The solid curve is the bulk thermopower S_b, and the dashed curve is the calculated diffusion part S_{dif} of the bulk thermopower. Data taken from Weber et al. (1992).

11.3 Electrical transport

The first attempt to measure ballistic contacts with semiconductors was undertaken by Pepper (1980). The author prepared a structure between Al film and degenerated Si. The film was separated from n^+ diffusion in Si by a 500–1000 Å thick thermally grown SiO_2. Filamentary contacts between Al and n-Si ($n=1$–3×10^{20} cm^{-3}) with the resistance 5–10 kΩ were formed by application of high voltage. The author argued that neither tunnel barrier nor depleted region are present in the contact, which implies use of the heavy n-doping and junction with very small barrier heights. At the same time, he also mentioned that observation of the current increase at $eV > \epsilon_F \simeq 200\,\mathrm{mV}$ faster than V, indicates that space charge effects are important in the depleted Si around the junction region. Turning to the $\mathrm{d}^2I/\mathrm{d}V^2$ spectra in Fig. 11.12, one sees a series of features both below and above the maximal phonon energy. A number of minima and maxima were identified by Pepper (1980) with some phonon energies or with their combinations, as well as even with vibrations of doping impurities and plasmons. However, there are some observations have to be explained: (1) Both minima and maxima in the spectra are taken as the consequence of EPI or impurity scattering, (2) the spectra are strongly asymmetric with respect to the bias polarity [for another

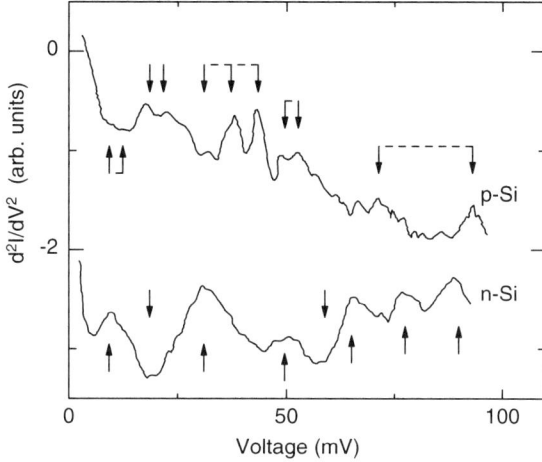

Fig. 11.12. $\mathrm{d}^2 I/\mathrm{d}V^2$ for electron injection in phosphor (upper curve) and As-doped Si (bottom one) at 4.2 K. Both curves have negative values. The principal phonons participating in the intervalley and intravalley scattering along with their combination (features above 70 mV) as well as vibration of P and As atoms in the range of 30–45 mV are indicated by arrows. Data taken from Pepper (1980).

polarity, only Al phonons have been considered by Pepper (1980b)], and (3) the principal features of the "phonon nature" persist at the temperature as high as 77 K, whereas the spectra smearing according to (4.3) in this case should be about 40 mV, which is of the order of the characteristic phonon energy.

Theoretical investigation of semiconducting point contacts began with papers by Vengurlekar and Inkson (1983), Shekhter (1983), and Kulik and Shekhter (1983). The former ones developed the theory to explain Pepper's experiments. Supposing some artificial model of contact representing a spherical metal dip penetrated into a sample with the barrier region between metal and semiconductor, Vengurlekar and Inkson (1983) showed that inelastic electron–phonon relaxation can produce dips in $\mathrm{d}^2 I/\mathrm{d}V^2$. However, the nature of peaks or role of donor species was left out of consideration.

Kulik and Shekhter (1983) mentioned that the voltage bias in the case of the semiconductor can be easily made comparable with the characteristic Fermi or thermal energy or even higher than those. This leads to the substantially nonlinear $I - V$ curves caused by the energy dependence of the charge-carriers relaxation rate, and indicative features can appear already in the differential resistance. The authors considered a situation in which the relaxation of electrons by optical phonons dominates and both the elastic and inelastic mean free paths of electron–optical phonon interaction are smaller than the contact length (they used the contact model in the shape of a long channel). In this case, a charge transfer along the channel is realized as a diffusive drift of electrons in the electric field interrupted at certain points by the emission of optical phonons. Besides, they ignored reabsorption of nonequilibrium phonons, supposing that the latter have momenta of the order of the Debye momenta and therefore cannot be reabsorbed by electrons; the momentum of which is small, corresponding to the low Fermi energy. Under the mentioned condition, the dI/dV curve represents a series of a sharp peaks at the energy $n\hbar\omega_0$, where ω_0 is the optical phonon frequency and $n=1$, 2,. . . .

Pong-Fei Lu et al. (1985) studied electrical transport through the microchannels in oxide between the 3-μm-thick n-type $In_{0.53}Ga_{0.47}As$ single-crystal epitaxial layer and a thick In film. They observed an oscillatory conductance at both bias polarities with a period close to the LO-phonon energy in InGaAs (Fig. 11.13). It is seen from the figure that the oscillations decrease in amplitude with the temperature increasing and vanish in perpendicular to the contact (channel) axis magnetic field. All of the facts give a strong support for the model proposed by Kulik and Shekhter (1983) of LO-phonon emission in the one-dimensional transport of electrons through the semiconducting channels.

Itskovich et al. (1984) considered semiconducting point contacts both between two semiconductors and a semiconductor and a metal in the case of weak screening $r_s \gg d$ and large electron inelastic relaxation length. The contact was taken as an orifice in a thin insulating layer [Fig. 3.1(a)]. The accumulation region of effective thickness r_s was assumed to be formed near both sides of a hole if the band-bending potential U is positive, and the depletion layer formation at $U < 0$ results in the appearance of the Schottky barrier [Fig. 2.5(a)] with an exponentially small current. Under this condition ($U > 0$), a voltage drop in the contact is of the order of $V(d/r_s)$, which is small compared with the total voltage. Correspondingly, an electron distribution function in the accumulation region is similar to that in the center of the metallic contact (Fig. 3.2). The calculated $I - V$ characteristic is very nonlinear so that differential conductivity goes to zero at critical voltage $V_c = \pm 1.74\,U$ for the symmetrical semiconductor contact [Fig. 11.14(a)]. In the case of the metal - semiconductor contact, the $I - V$ characteristic is highly asymmetric: The current increases quickly at the positive potential on the metal side, and at electron injection from the semiconductor side into the metal at $V_c = U$, the charge density in the semiconductor is zero, resulting in zero differential conductivity. The correction of the $I - V$ curve caused by inelastic processes is of the order of d/l_{in}, and maxima in d^2V/dI^2 spectra corresponding to the EPI function should be revealed. The latter is dependent both on the energy of phonons emitted and on the electron energy. It is interesting that the contribution of both materials to the spectra should be of the same order of magnitude for metal-semiconductor contacts. It should

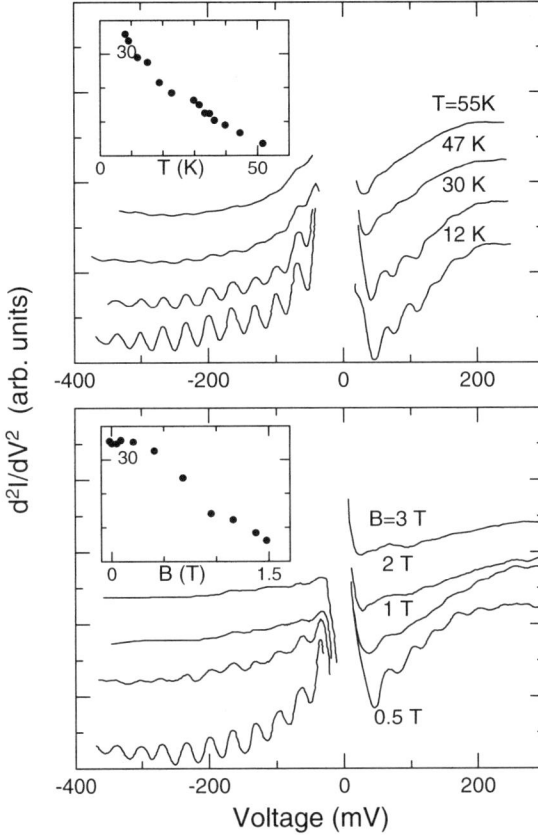

Fig. 11.13. Upper panel: d^2I/dV^2 versus voltage of the In-In$_{0.53}$Ga$_{0.47}$As contact at different temperatures. Inset shows the oscillation amplitude at $V = -200$ mV as a function of temperature. Bottom panel: d^2I/dV^2 versus voltage of the same contact at different magnetic fields parallel to the surface or perpendicular to the contact axis. Inset shows the oscillation amplitude at $V = -200$ mV as a function of magnetic field. Data taken from Pong-Fei Lu et al. (1985).

be noted that in the case of strong screening, these results are qualitatively the same as received by Vengurlekar and Inkson (1983).

Itskovich and Shekhter (1984) analyzed further a model of a dirty contact ($l_{el} \ll L$) in the shape of a revolution hyperboloid (Fig. 3.1), the length L of which is larger compared with its diameter. In addition to the elastic scattering, the Schottky barrier with small transparency D was considered ($D \ll l_{el}/L$), and screening length was supposed to be small $r_s \ll d$. In this case, whole voltage drops at a barrier and the electrical field in constriction can be neglected. Under condition $l_{in} \gg d$, the relaxation of hot electrons is caused by the successive emission of phonons. As a result, the point-contact

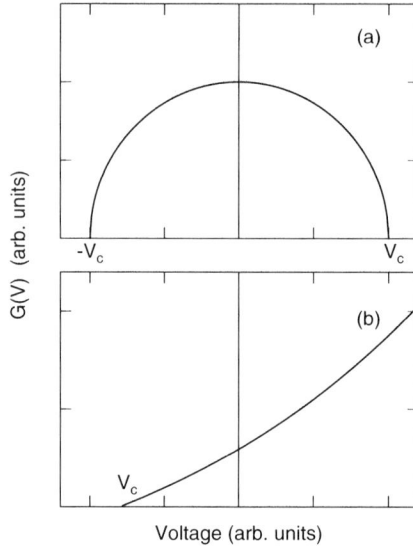

Fig. 11.14. Normalized conductivity versus voltage for (a) symmetric semiconductor–semiconductor junction, and (b) for metal–semiconductor junction (positive V corresponds to the positive potential on metal side). After Itskovich et al. (1984).

spectrum contains peaks at multiplied phonon frequencies and at the sum of those frequencies. However, for a pure contact, such a multiphonon contribution should vanish.

From the above-mentioned theories, it follows that depending on the relation between the number of parameters (an elastic and inelastic mean free paths, a screening length, a shape and size of a constriction, presence of a barrier, etc.), the nonlinearity of the $I-V$ characteristic in the semiconductor point contacts can be determined by different mechanisms. As it is difficult to know these parameters in the real point contacts *a priori*, the puzzling experimental spectra as presented in Fig. 11.12 are observed. It should be also noted that usually presented at the semiconductor interface, surface states (not accounted in the above theories) complicate additionally the problem. This is a possible reason why the number of theoretical papers dealing with electrical properties of semiconductor point contacts is so far prevalent over the experimental ones.

Traditional mechanical point contacts between heavy doped n- and p-type of Ge have been investigated by Naidyuk et al. (1987). The contact resistance was 3–10 kΩ, and typical dV/dI represented a monotonous falling curve with resistance decrease by a factor of two at about 200 mV. Correspondingly, d^2V/d^2I has a negative sign with a minimum at 15–30 mV and saturation above 100 mV. Additionally, most of the d^2V/dI^2 spectra contained a number of features – peaks and dips similar to those shown in Fig. 11.12. The authors concentrated their attention on the spectra of several contacts with relatively low resistance ($< 100\Omega$) (Fig. 11.15). They demonstrate a few pronounced

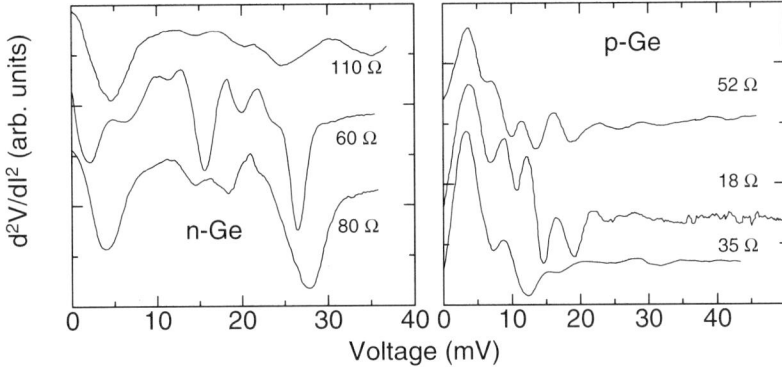

Fig. 11.15. $\mathrm{d}^2V/\mathrm{d}I^2$ versus voltage of the n- and p-doped Ge ($n = 1.2 \times 10^{19}$ cm^{-3}, $p = 10^{18}$ cm^{-3}) contacts at $4.2\,\mathrm{K}$ with different resistance marked by each curve. The curves are offset vertically for clarity. Data taken from Naidyuk et al. (1987).

minima in the region of phonon frequencies, and no noticeable features above the Debye energy (about $40\,\mathrm{meV}$) were resolved. The position of low- (2–$4\,\mathrm{mV}$) and high- (25–$28\,\mathrm{mV}$) energy dips corresponds well to the intra- and intervalley scattering for Ge. The authors explained the negative sign of the spectra in the frame of localization effect like in the case of semimetals. Note that density of carriers in this case is of the same value as for semimetals. Turning to the less doped p-type Ge, the point-contact spectra exhibit mainly the low-energy maxima. They are positioned approximately with 4–6-mV periodicity and were explained[1] by multiphonon scattering.

11.4 Contacts between dielectrics

In the case of dielectrics, only phonon transport through the contact can take place. The heat flux in dielectric contacts in the collisionless, ballistic

---

[1] $\mathrm{d}^2V/\mathrm{d}I^2$ curves in Fig. 11.15 for p-doped Ge correspond to a zero-bias minimum in $\mathrm{d}V/\mathrm{d}I$ at $|V| < 10\,\mathrm{mV}$. According to the unpublished results obtained by the same authors, this minimum vanishes in a magnetic field of about $3\,\mathrm{T}$. Moreover, temperature increasing above $6\,\mathrm{K}$ depresses the minimum as well. Therefore, superconductivity of Ge (according to Buckel and Witting (1965), the high-pressure phase of Ge has $T_c \simeq 5.4\,\mathrm{K}$) in the contact region caused by the local high pressure is the more probable reason for the observed structure for p-Ge below 10–15 mV.

for phonons limit was calculated by Bogachek and Shkorbatov (1985). The dimension of contact was assumed to be macroscopical with respect to the characteristic phonon wavelength λ_{ph} and small compared with the phonon–phonon (l_{ph-ph}) and phonon–impurity (l_{ph-im}) relaxation length. At low temperatures with respect to the Debye one, only low-frequency phonons are excited with linear dispersion law resulting in the heat-flux temperature dependence:

$$Q(T) = \frac{\pi^2}{120} \frac{A}{\hbar^3} D(T_1^4 - T_2^4). \tag{11.3}$$

The temperature dependence $Q \sim T^4$ can be served as a criterion of the ballistic phonon transport and was proved by measurements of Stefanyi et al. (1990, 1992). However, a complex form of $Q(T) \sim F(T) T^4$ with $F(T)$ function having a maximum at about 1 K was found by Stefanyi et al. (1992) at very low temperatures. Shkorbatov et al. (1996) assumed that this occurs because of diffraction of phonons, owing to their effective wavelength increase as $\lambda_{ph} = aT_D/T$ (a is the lattice constant) with temperature decreasing. Therefore, the temperature at which a diffraction effect originates $T_d \sim T_D(a/d)$ is about 0.1–1 K. According to the calculations, the maximum in $F(T)$ is more expressive and shifted to the lower temperature with increasing of L/d ratio, where L is the contact length (Fig. 11.16).

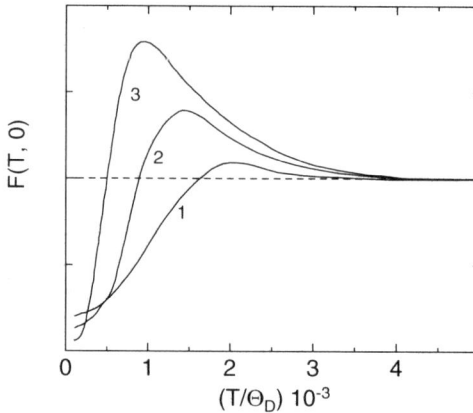

Fig. 11.16. Reduced phonon energy flux function $F(T)$ for different contact length to the diameter ratio: $L/d=1.5$ (curve 1), 2.5 (curve 2), and 5 (curve 3). Shkorbatov et al. (1996).

As seen from the contents of this chapter, point-contact measurements provide a possibility to investigate many phenomena in the semimetals, semiconductors, and dielectrics. However, in most cases, the property of interface and method of contact preparation play a crucial role here. Therefore further progress is connected with preparing artificial structures with well-defined geometry and especially with controlled material properties in the constriction (or at interface), which should correspond to the bulk one.

References

Arutyunov K. Yu., Gitsu D. V., Kondrya E. P., Nikolaeva A. A. and Rybaltchenko L. F. (1996) Physica B **218** 35.

Bogachek E. N. and Shkorbatov A. G. (1985) Sov. J. Low Temp. Phys. **11** 353.

Buckel W. and Witting J. (1965) Phys. Lett. **17** 187.

Gerlach-Meyer U. and Queisser H. J. (1983) Phys. Rev. Lett. **51** 1904.

Gribov N. N., Samuely P., Kokkedee J. A., Jansen A. G. M., Wyder P. and Yanson I. K. (1991) Phys. Rev. Lett. **66** 786.

Itskovich I. F., Kulik I. O. and Shekhter R. I. (1984) Solid State Commun. **50** 421.

Itskovich I. F. and Shekhter R. I. (1984) Sov. J. Low Temp. Phys. **10** 229.

Itskovich I. F., Kulik I. O. and Shekhter R. I. (1987) Sov. J. Low Temp. Phys. **13** 659.

Kulik I. O. and Shekhter R. I. (1983) Phys. Lett. A **98** 132.

Le Hang Nguyen, Riegel H., Asen-Palmer M. and Gmelin E. (1996) Physica B **218** 248.

Naidyuk Yu. G., Koshkin I. V. and Lysykh A. A. (1987) Sov. J. Low Temp. Phys. **13** 57.

Pepper M. (1980) J. Phys. F **13** L709, L717, L721.

Pong-Fei Lu, Tsui D. C. and Cox H. M. (1985) Phys. Rev. Lett. **54** 1563.

Sato H., Sakamoto I., Yonemitsu K. and Hishiyama Y. (1988) J. Phys. Soc. Japan **57** 2456.

Shekhter R. I. (1983) Sov. Phys. and Techn of Semicond. **17** 1463.

Shkorbatov A. G., Feher A. and Stefanyi P. (1996) Physica B **218** 242.

Stefanyi P., Feher A. and Orendacova A. (1990) Phys. Lett. A **143** 259.

Stefanyi P., Feher A. and Shkorbatov A. G. (1992) Sov. J. Low Temp. Phys. **17** 107.

Trzcinski R., Gmelin E. and Queisser H. J. (1986) Phys. Rev. Lett. **56** 1086.

Trzcinski R., Gmelin E. and Queisser H. J. (1987) Phys. Rev. B **35** 6373.

Vengurlekar A. S. and Inkson J. C. (1983) Solid State Commun. **45** 17.

Weber L., Gmelin E. and Queisser H. J. (1989) Phys. Rev. B **40** 1244.

Weber L., Lehr M. and Gmelin E. (1991) Phys. Rev. B **43** 4317.

Weber L., Lehr M. and Gmelin E. (1992) Phys. Rev. B **46** 9511.

Yanson I. K., Gribov N. N. and Shklyarevskii O. I. (1985) JETP Lett. **42** 195.

Yanson I. K., Shklyarevskii O. I. and Gribov N. N. (1992) J. Low Temp. Phys. **88** 135.

12 PCS of superconductors

Starting from the Meissner's (1958) work, superconducting point contacts attracted much attention during the subsequent years both as a powerful instrument for investigation of the superconducting state and as a possibility of their utilization in application. This interest increased incredibly, especially after the breaking work of Josephson (1962) dealing with the weakly coupled superconductors. In the 1960s and 1970s many studies were done concerning the Josephson effects, Andreev reflection, subharmonic gap structure, critical and excess current, and so on using point contacts. In this chapter we restrict overview of research in this field starting from the late 1970s, when, in general, the quantitative description of the $I - V$ characteristics of S-c-S and S-c-N contacts was reached mainly after the theoretical studies by Kulik and Omelyanchouk (1977, 1978), Artemenko et al. (1979), and Blonder et al. (1982) (see the references in Chapter 3). By that time also, search for the phonon features in the superconducting contacts by PCS began [Khotkevich and Yanson (1981), Yanson et al. (1983)]. In this chapter, we present investigations separated by the range of energies probed, namely, studies of the superconducting energy gap structure in the $I - V$ characteristics at low energy and the effect of EPI and nonequilibrium phonons on the conductivity of superconducting contacts at higher energies.

12.1 Phonon structure in superconducting contacts

As was shown by Khlus and Omelyanchouk (1983) and Khlus (1983) (see also Section 3.7), the contribution to the current through the clean S-c-S or S-c-N contact would be expressed at $eV \gg \Delta$ as a sum of four terms:

$$I(V) = V/R_0 + I_N^{(1)}(eV) + I_{exc}^{(0)} + I_{exc}^{(1)}(eV), \tag{12.1}$$

where $I_N^{(1)}$ is the negative increment to the current of the order of d/l_{in} as in the normal state because of EPI, $I_{exc}^{(0)}$ is the excess current [see (3.34) and (3.36)] constant at $eV \gg \Delta$, and $I_{exc}^{(1)}$ is the energy-dependent correction to the excess current of the order of $I_{exc}^{(0)}(d/l_{in})$, which represents a small contribution to the $I - V$ curve proportional to Δ/eV for bias $eV \gg \Delta$. On

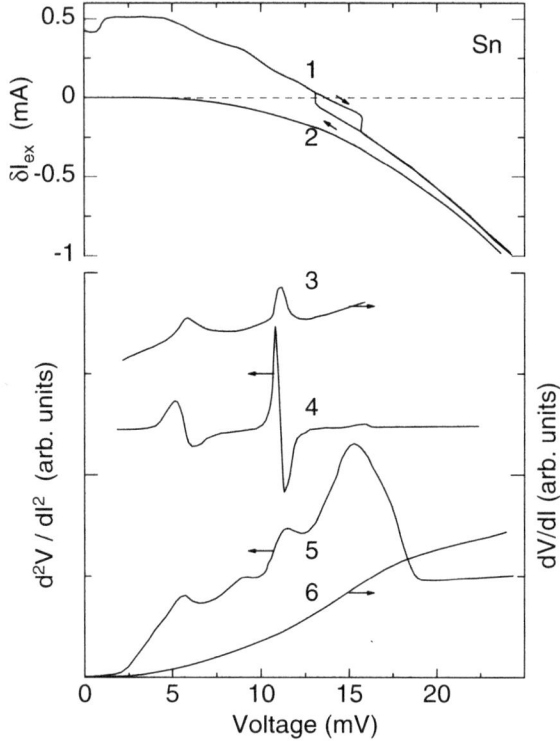

Fig. 12.1. Deviation of current from the ohmic behavior (dashed horizontal line) in normal (curve 2) and superconducting (curve 1) states for a Sn point contact with $R_0=2.55\,\Omega$. Curves 3 and 4 represent dV/dI and d^2V/dI^2 dependencies for the same contact in the superconducting state, respectively, whereas curves 5 and 6 are corresponding dependencies for the normal state. The measuring temperature was 1.6 K, and the magnetic field of 90 mT was used to drive contact into the normal state. Note that the modulation current amplitude was $\sqrt{10}$ times smaller in the superconducting state. The derivatives are offset vertically for clarity. Data taken from Khotkevich and Yanson (1981).

this account, only slight modification of EPI spectrum in superconducting state is expected, although the phonon peaks might appear already in the first derivative (3.39) at definite condition as mentioned by the above theory.

The first attempt to study EPI in the superconducting contacts was undertaken for tin by Khotkevich and Yanson (1981). It was found that the differential resistance curves possess maxima (see Fig. 12.1, curve 3) in the region of characteristic phonon frequencies, although no correlation between the peak position and maxima of the phonon spectrum was observed. Moreover, because of the large nonlinearity of the $I - V$ curve in the superconducting state, shallow EPI features were difficult to resolve. The maxima in the differential resistance in Fig. 12.1 reflect a decrease of the excess current found to be no more constant at $eV > 4\,\mathrm{mV}$. The latter decreases with the bias voltage exhibiting even hysteresis at about $15\,\mathrm{mV}$ by reversing the current sweep through the contact (Fig. 12.1, curve 1). This was pointed out on the thermal nature of the mentioned nonlinearity, supposedly because the heat transfer from the contact to the bulk in the superconducting state is less efficient compared with the normal state. The important result of the study of S-c-S contacts by Khotkevich and Yanson (1981) is establishing correlation between a magnitude of the excess current and the absolute intensity of the point-contact EPI function in the normal state, both of which depend on the elastic mean free path of electrons or, in other words, on the contamination degree of the contact region. For a clean contact with a maximal intensity of an EPI spectrum, product $I_{\mathrm{exc}}R_{\mathrm{N}}/\Delta$ was found to be about 2.3. This value is close to the theoretical prediction $8/3$ given by (3.36).

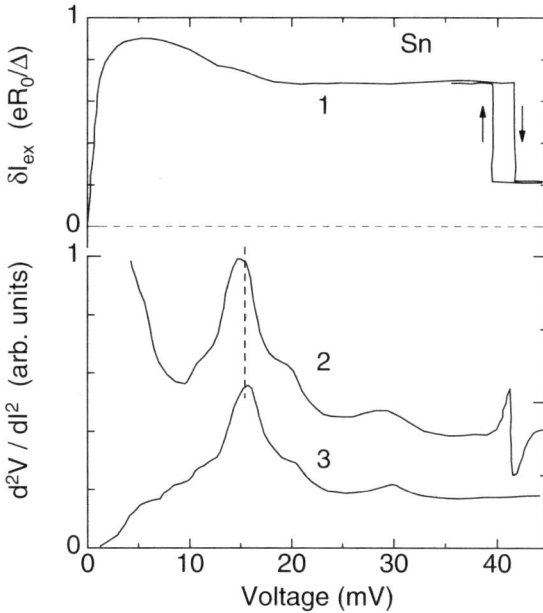

Fig. 12.2. Excess current (curve 1) along with $\mathrm{d}^2V/\mathrm{d}I^2$ in the superconducting (curve 2) and the normal (curve 3) state at $1.6\,\mathrm{K}$ for a Sn-Cu heterocontact with $R_0 = 8.8\,\Omega$. The magnetic field was used to drive contact into the normal state for curve 3. Data taken from Yanson et al. (1984a).

As further studies by Yanson et al. (1984a) showed, the heating effects appear to be overcome by using Sn-Cu heterocontacts, with a copper having higher thermal conductivity than that of the superconducting tin. The authors found that d^2V/dI^2 curves in the normal and superconducting states were practically identical in the region of the phonon spectrum (Fig. 12.2). When Sn becomes superconducting, one can see a small shift of the main maximum to the lower energy by the magnitude of the order of $\Delta(Sn) = 0.6\,\text{meV}$, followed by a slight increase in the width and intensity of the spectrum. The excess current remains almost constant up to about 40 mV, where transition to the nonequilibrium state occurs with suppression of the superconducting order parameter. In general, this confirms the results of the above-mentioned theory developed by Khlus and Omelyanchouk. The further experimental data of EPI study in the pure superconducting point contacts with Pb, Sn, and In are given by Kamarchuk et al. (1986) and Kamarchuk and Khotkevich (1987).

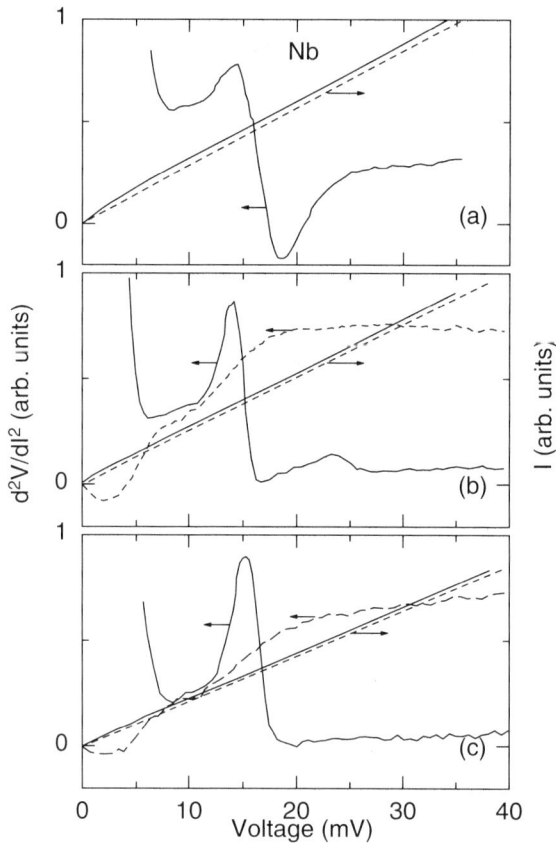

Fig. 12.3. $I - V$ characteristic and their d^2V/dI^2 dependencies (solid curves) for the dirty Nb contacts at $T = 4.2$–4.5 K: (a) $R_0 = 194\ \Omega$, (b) $R_0 = 17.5\ \Omega$, and (c) $R_0 = 29.2\ \Omega$. Dashed curves represent d^2V/dI^2 in the normal state at $T = 10$ K for contacts in (b) and (c). Dashed straight lines show normal state $I - V$ characteristics. Data taken from Yanson et al. (1983).

More unexpected results exhibited measurements of the dirty point contacts in the superconducting state [Yanson et al. (1983, 1984b)]. It was found that the spectra reveal maxima either in the first or second derivative of the $I-V$ curve, located at characteristic phonon energies. However, the maxima disappeared after the transition to the normal state (Fig. 12.3). The authors emphasized the nonthermal nature of the phenomenon contrary to Khotkevich and Yanson (1981) because the excess current remains almost constant in the whole energy range studied. Estimated energy relaxation path $\sqrt{l_{in}l_{el}}$ for the contacts turned out to be comparable with the contact diameter, whereas superconducting coherence length ξ was found to be larger than the diameter. Therefore, in the normal state at $\sqrt{l_{in}l_{el}} \leq d$, there is no separation of electrons into two groups with difference in energy by eV hindering the spectroscopy of the quasiparticle excitations. However, in the superconducting state, in the case $\xi > d$, on both sides of the contact, the chemical potentials of the Cooper pairs are constant and differ by eV. Perhaps inelastic single electron transition between the pair levels may be responsible for the observed phonon singularities in dirty contacts. Another possibility for occurrence of mentioned peculiarities will be considered for strong coupling superconductors in the end of this paragraph.

During the further study of Ta–Cu and Ta–Au heterocontacts, Yanson et al. (1987) was able to trace a smooth transition from the spectrum similar to the ordinary EPI spectrum of Ta to the anomalous spectrum described by Yanson et al. (1984) (Fig. 12.4). Although spectra for superconducting contacts being in the ballistic limit have essentially the same shape corresponding to the EPI spectrum [Fig. 12.4(a)], they become progressively more different by transition to the diffusive limit [Fig. 12.4(b)] transforming to the series of N-shape spikes [Fig. 12.4(c)] for nonballistic ($d > \sqrt{l_{in}l_{el}}$) contact. The position of these features corresponds to the energy of phonons with a small group velocity, which are slowly escaping from the contact region and may locally reduce the superconducting gap. In the normal state, there are no spectral features of any sort in d^2V/dI^2 of the dirty contacts.

Yanson et al. (1987) proposed the energetic diagram (Fig. 12.5) to interpret the phonon spectroscopy in the nonballistic S-c-N contact in the superconducting state. Although the condition $d \gg \sqrt{l_{in}l_{el}}$ does not hold for the normal excitations in superconductor for dirty contacts, the condensate of electrons reaches the interface, holding the electrochemical potential constant over the distance from the interface of the order of the inelastic electron–phonon relaxation length. Respectively, the phonons with the maximal energy eV can be created near the interface (see Fig. 12.5 for details). The phonons with small escape velocity $\partial\omega/\partial q = 0$ are accumulated in the contact. The reabsorption of the nonequilibrium phonons in the superconductor leads to a decrease of Δ near the interface and correspondingly to decrease of the excess current manifesting itself in the derivatives of the $I-V$ curves. Obviously, the mechanism proposed here may also operate in the S-c-S contact

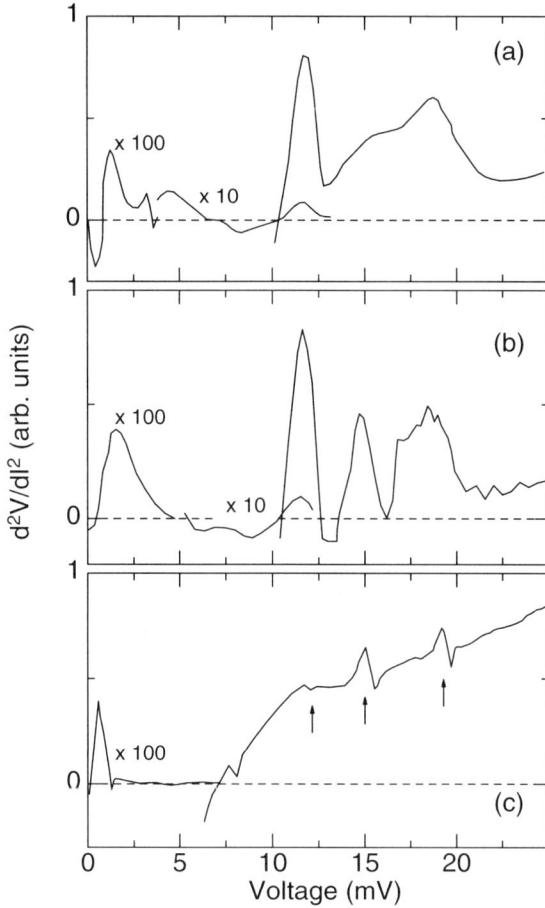

Fig. 12.4. d^2V/dI^2 characteristics of Ta–Cu (a) and Ta–Au heterocontacts (b) and (c) at $T \simeq 1.5$ K. The contact resistance is (a) 26.5 Ω, (b) 19 Ω, and (c) 0.76 Ω. Parts of curves labeled ($\times 10$) and ($\times 100$) correspond to the reduction of the modulating voltage by a factor of $\sqrt{10}$ and 10, respectively. Contribution of the noble metals in the spectra is difficult to distinguish likely because of their higher Fermi velocity [see (3.27)]. This can bear evidence that charge carriers in Ta have lower Fermi velocity. The arrows in (c) indicate the position of the main maxima in the EPI spectra of Ta (see Fig. 5.14). Data taken from Yanson et al. (1987).

with a dimension greater than both the energy relaxation length of electrons and the electrodynamic coherence length $\xi = (\xi_0^{-1} + l_{el}^{-1})^{-1}$.

  Summarizing, it is shown that the reabsorption of the nonequilibrium phonons in the normal state gives rise to the background in the EPI spectra, whereas for the dirty superconducting contacts, some additional peculiarities corresponding to the phonons with the small group velocity ($\partial\omega/\partial q = 0$) appear in the spectra. The same, apparently, is the nature of the sharp features

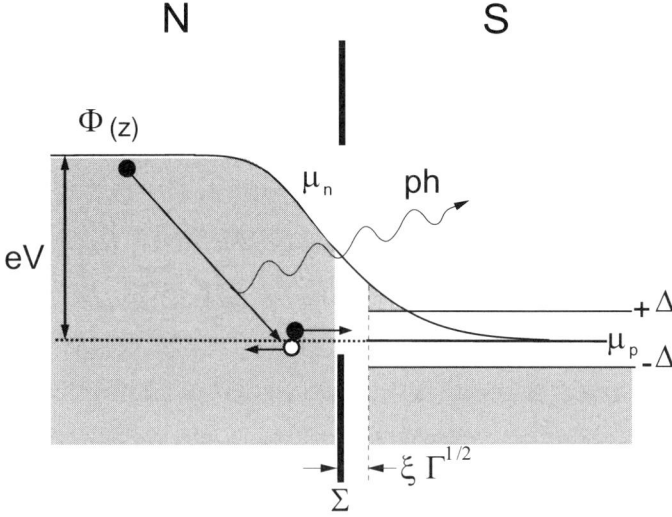

Fig. 12.5. Energy diagram of an S-c-N contact in an opaque screen Σ. Here μ_n is the electrochemical potential of the quasiparticles, μ_p is the electrochemical potential of the pairs, ξ is the coherence length, and $\Gamma \simeq 0.16(\hbar/\tau_{ph}k_B T_c)$ is the depairing factor. The tilted arrow marks the inelastic transition of the electron (closed circle) with the phonon emission to a free state (hole, open circle), which originates because of Andreev reflection of the bottom electron (closed circle). Adapted from Yanson et al. (1987).

observed in $\mathrm{d}^2V/\mathrm{d}I^2$ of S-c-S contacts with NbSe$_2$ [Bobrov et al. (1985)], Nb$_3$Sn [Yanson et al. (1985)], as well as in the case of S-c-N contacts Tc-Ag [Zakharov and Tsetlin (1985)].

Let us turn to strong coupling superconductors represented by the archetypical Pb. In this case, the theory by Omelyanchouk et al. (1988) predicts another mechanism of nonlinearity in S-c-N contacts. Namely, there is an elastic contribution of the order of $(\Delta/\hbar\omega_D)^2$ for $\Delta \ll \hbar\omega_D$ in the $I - V$ curve because of frequency dependence of the energy gap function $\Delta(\hbar\omega)$. At $T = 0$, the total difference conductance is

$$\frac{\mathrm{d}I}{\mathrm{d}V}(eV) = \frac{1}{R_0}\left(1 + \left|\frac{\Delta(\hbar\omega)}{\hbar\omega + \sqrt{(\hbar\omega)^2 - \Delta^2(\hbar\omega)}}\right|^2\right)_{\hbar\omega = eV}. \qquad (12.2)$$

As $I_{\mathrm{exc}} \sim \Delta/R$ and the total current $I \sim \hbar\omega_D/R$, the elastic correction to the net current does not depend on the contact diameter and is of the order of $I(\Delta/\hbar\omega_D)$, which can be greater than the inelastic contribution $\sim I(d/l_{\mathrm{in}})$. The experimental curves presented in Fig. 12.6 together with the calculation according to (12.2) confirm in general the presence of the discussed elastic component. In this case, the Eliashberg EPI function can be extracted from the point-contact characteristics in the same way as in the tunneling

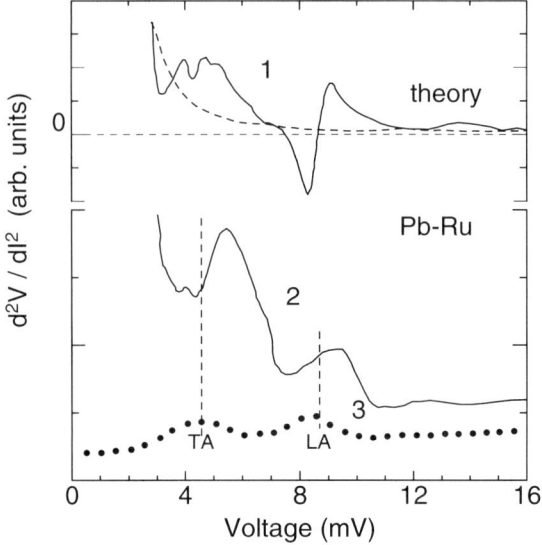

Fig. 12.6. Calculated d^2V/dI^2 (curve 1) using (12.2) along with measured dependence (curve 2) for a Pb-Ru heterocontact. Dashed theoretical curve is calculated for the case $\Delta(\hbar\omega) = \Delta_0 = $ const. Dotted curve 3 shows the spectrum in the normal state with the same ordinate scale as curve 2. The phonons feature in (2) are shifted to the higher energy compared with curve (3) roughly by $\Delta(\text{Pb})\sim 1.3\,\text{mV}$. Ru has a small phonon density of state below $10\,\text{mV}$ (see Fig. 5.17); therefore, its contribution to the displayed spectrum is negligible. Data taken from Khotkevich et al. (1990).

spectroscopy (see Section 2.4), provided that the inelastic processes make negligible contribution to the spectra, which is especially valid in dirty superconductors.

12.2 Spectroscopy of the energy gap

12.2.1 Simple metals

The first experiments to prove the BTK theory (see Section 3.7.1) carried out by Blonder and Tinkham (1983) showed that the study of $I-V$ characteristic of Nb-Cu contacts can provide information about the superconducting gap. dV/dI of S-c-N contact at $T \ll T_c$ allows us to estimate the gap value from the position of the minimum directly (see Fig. 2.4). To receive more reliable data, especially at $T \to T_c$, a fit by virtue of (3.30) should be done. S-c-N contacts with three simple metals Nb, Sn, and Zn were studied extensively during the past both to check the BTK theory and to receive macroscopic characteristics of the superconducting state, e.g., the gap value and its temperature behavior. The metals mentioned have an order of magnitude difference in the critical temperature between Nb and Zn. At the same time, whereas Zn represents an extreme type-I superconductor with superconducting coherence length $\xi_0 \simeq 2000$ nm about two orders of magnitude larger than the London penetration depth λ_L, Nb is on the threshold to type-II superconductor with both lengths of the order of 40 nm.

As seen from calculations in the Fig. 3.15(b), a minimum develops in the dV/dI curves of S-c-N contact around zero bias by decreasing temperature below T_c with the width of the order of 2Δ. The first effect is decreasing of the zero-bias contact resistance R_0 up to two times at $Z = 0$ by lowering temperature [see Fig. 3.15(a)] or in the other words in the case of barrier-less interface. The behavior of $R_0(V = 0, T)$ is sensitive to the mentioned Z-parameter, and the R_N/R_0 temperature dependence is shown in Fig. 12.7. The increase of the contact conductance R_0^{-1} transforms to its decrease at

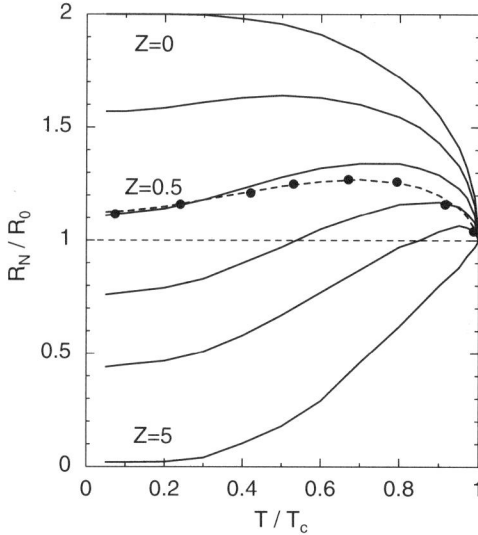

Fig. 12.7. Inverse zero-bias resistance R_0 as a function of reduced temperature for various $Z = 0$, 0.3, 0.5, 0.7, 1 and 5 (from the top curve to the bottom one) calculated according to (3.30). The dashed curve represents the case $Z=0.5$ and $\Gamma/\Delta_0=0.1$ (see text for details). The closed circles are experimental data for Zn-Ag contact from Fig. 12.8.

$Z > 0$ and temperature lowering. For the further rise of the reflection at the interface, the transition to the tunnel regime occurs with $R_0^{-1} \to 0$. The experimental results for Nb-Ag contacts by Reinertson et al. (1990, 1992) be-tween $T_c/2$ and T_c behave qualitatively as calculated R_N/R_0 curves with $Z = 0.3$–0.6. The data for Zn-Ag contact, as shown in Fig. 12.7, perfectly follow the theoretical curve in the whole temperature range if one takes into account small broadening of the density of states described by the Γ parameter [see (3.32)]. Of course, the dependence of R_N/R_0 below T_c is not sufficient to test the BTK-theory. It is a generally accepted calculation of $I-V$ curves derivatives by (3.30). Figure 12.8 shows the behavior of $\mathrm{d}V/\mathrm{d}I$ below T_c for Zn-Ag contact together with the calculated curves. Including only a small broadening parameter $\Gamma \simeq 0.05\,\Delta_0$, which is kept constant for the whole temperature range, the calculated theoretical curves are almost indistinguishable from the experimental ones. $\Delta(T)$ temperature dependence for Zn obtained by the

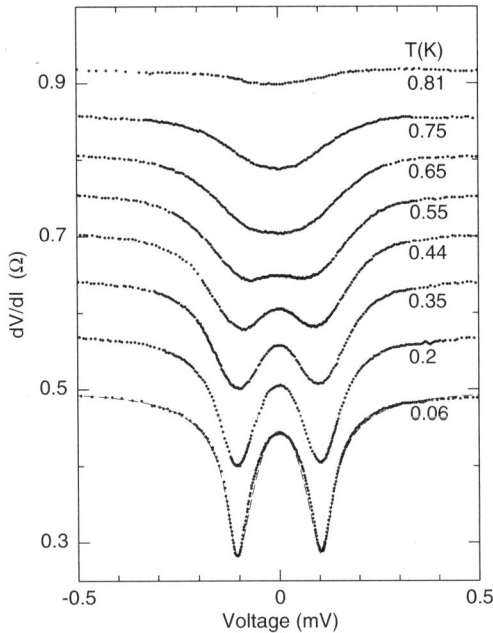

Fig. 12.8. $\mathrm{d}V/\mathrm{d}I$ curves versus V (dots) for a Zn-Ag contact with $R_N = 0.5\,\Omega$ and $T_c = 0.82$ K for different temperatures. Fit parameters for the theoretical curve at $T = 0.06$ K (dashed line) are $\Delta = 110\ \mu eV$, $\Gamma = 6\mu eV$, and $Z = 0.5$. The fit curves are almost indistinguishable from the data also for the higher temperatures and therefore are not shown. The fit is done without other adjustable parameters. Data taken from Naidyuk et al. (1996a).

fitting procedure for a number of contacts is presented in Fig. 12.9. A part of contacts [Fig. 12.9(b)] gives $\Delta(T)$ in perfect accordance with the BCS behavior along with $2\Delta_0/k_B T_c \simeq 3.5$ close to theoretical prediction 3.53. $\Delta(T)$ for other contacts [Fig. 12.9(a)] with $2\Delta_0/k_B T_c \simeq 3$ is somewhat below the BCS curve, whereas for contacts with $2\Delta_0/k_B T_c \simeq 3.8$, $\Delta(T)$ is above the BCS behavior [Fig. 12.9(c)]. The example of Zn-Ag contacts presented testifies that the BTK theory gives a perfect description of the experimental situation and can be used for determination of the superconducting gap in the whole temperature range as well as other parameters (e. g., like Z and Γ) from the dV/dI curves of the S-c-N constriction.

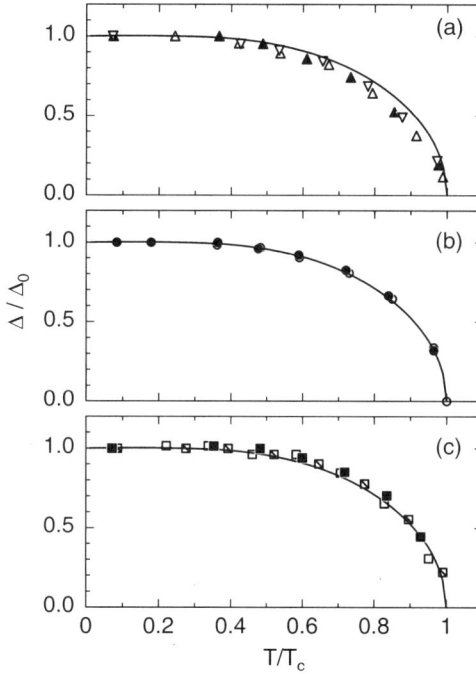

Fig. 12.9. Reduced $\Delta(T)$ for different Zn-Ag contacts with $2\Delta_0/k_BT_c = 3.0 \pm 0.1$ (a), 3.45 ± 0.05 (b), and 3.75 ± 0.05 (c). Different symbols indicate different contacts. Data taken from Naidyuk et al. (1996a).

Let's discuss the nature of the barrier parameter Z. In most cases the Z value for Zn-Ag contacts was found to be between 0.4 and 0.5. Normal reflection of the quasiparticles at the interface caused by mismatch of the Fermi velocities v_F yields $Z \simeq 0.13$ according to (3.31) with textbook data $v_F^{Ag} = 1.39 \times 10^6$ m· s^{-1} and $v_F^{Zn} = 1.82 \times 10^6$ m· s^{-1}. It is interesting to note that the same Z value is often found both for the high-T_c materials (see Figs. 13.12 and 13.13) and for the heavy-fermion superconductors (see Section 14.3.2). This indicates that the difference in the Fermi velocities plays a negligible role in all of these cases, and other physical reasons determine Z.

According to calculation of Mazin et al. (2001), dV/dI of a diffusive S-c-N contact with no interface barrier ($Z = 0$) has the similar shape as that of a ballistic S-c-N contact with $Z = 0.55$. They concluded that it is very difficult to discern the effect of a finite Z in dV/dI for a ballistic contact from the effect of diffusive transport. In view of this fact mentioned above the Z value of about 0.5 for different materials is very likely connected with a diffusive transport in the contact. Of course, it is not excluded that some natural barriers (e. g., oxides) are present at the interface influenced Z value as well.

A practical method to evaluate the superconducting gap provides the study of S-c-S contacts where processes of the multiple Andreev reflection (MAR) dominate (see Section 3.7.2). This leads to the peculiarities of the dV/dI curves at energies $2\Delta/n$ $(n = 1, 2, 3 \ldots)$, which are getting more pronounced with an increase of the Z parameter. Experimentally subharmonic gap structure was observed in numerous papers in the 1960s (see, e. g., Barnes (1969) and the references therein) by using point-contact tunneling. The authors concurred that metallic shorts were present when subharmonic gap structure was appeared. In the early studies of superconductors by PCS, Khotkevich and Yanson (1981) specified subharmonic gap structure with n=1,2 for Sn point contacts. Later, mechanically controllable break junction measurements in Nb manifested a rich and pronounced structure up to $n = 4$ (Fig. 12.10). In general, this dV/dI curve corresponds to the calculations presented in Fig. 3.18 satisfactorily. Smith et al. (1996) measured the temperature dependence of the subharmonic gap structure up to $n = 4$ for Nb point contacts, which was found to obey the standard $\Delta(T)$ BCS dependence.

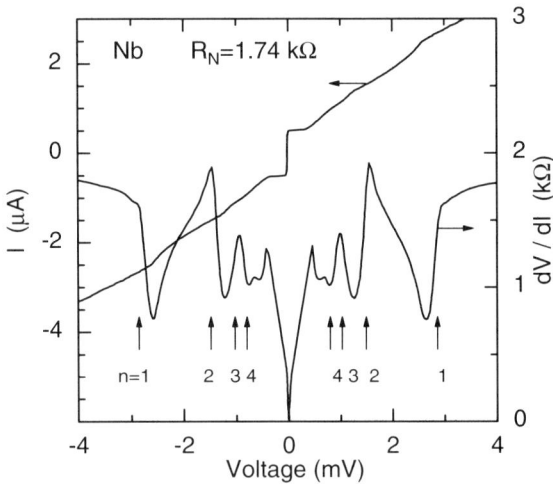

Fig. 12.10. $I(V)$ and $dV/dI(V)$ curves of a Nb mechanically controllable break junction in the superconducting state. The vertical arrows indicate the position of $2\Delta/n$ with $\Delta = 1.47$ meV. Data taken from Muller et al. (1992).

12.2.2 Magnetic field behavior

The study of the S-c-N contacts with ordinary superconductors in a magnetic field attracted considerably less attention. To our best knowledge, until now, there has been no comprehensive BTK-like theory of Andreev reflection in a magnetic field. Likewise, detailed measurements of S-c-N contacts for the conventional superconductors in a magnetic field are rare. One of the first experiments in this field was done by Asen and Keck (1992, 1994), during which an increase of the critical magnetic field above its bulk value for the Ta-Ag contact was observed. Moreover, the double-minimum structure in dV/dI broadens in a magnetic field so that a distance between the minima even increases with a field followed by a decrease at the higher fields. During the further investigations of Sn, In, and Nb, Naidyuk et al. (1996) demonstrated that only modified BTK theory, which includes smearing described by the Γ parameter of the electronic DOS, can depict the dV/dI curves, especially their magnetic field dependence (Fig. 12.11). The authors reveal both temperature

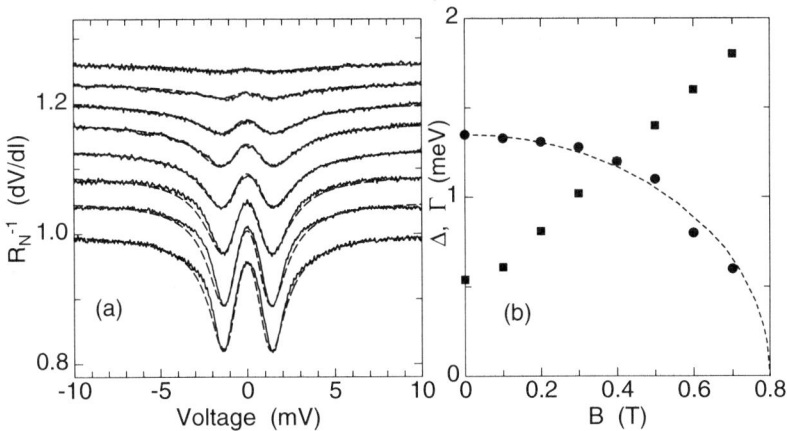

Fig. 12.11. (a) dV/dI curves versus V for the Nb-Ag contact with $R_N = 4.9\,\Omega$ at $T=1.5\,K$ in different magnetic fields B (T): 0, 0.1, 0.2, 0.3, 0.4, 0.5, 0.6, and 0.7 (from the bottom curve to the top one). The dashed curves represent fit by (3.30) and (3.32) with Δ and Γ shown in (b). The curves are offset vertically for clarity. (b) Magnetic field dependence of Δ (circles) and Γ (squares) for the contact in (a). The dashed line shows behavior of the pair potential of a thin film of a type-I superconductor $\sqrt{1 - (B/B_c)^2}$ [Tinkham (1980)]. Data taken from Naidyuk et al. (1996).

and magnetic field dependence of the Δ and Γ from the fit. For Nb, at zero field, $\Delta(T)$ was found that corresponds to the BCS curve and Γ is temperature independent. The decrease of Δ in Nb in the magnetic field was close to pair potential behavior of a thin film of a type-I superconductor and Γ increases almost linearly with a field (Fig. 12.11).

The detailed study of the magnetic field dependence of the Andreev reflection structure in the S-c-N contacts was carried out by Naidyuk et al. (1996a) for superconducting Zn. Two distinct types of dV/dI curves behavior were observed in a magnetic field (Fig. 12.12). Some of the contacts showed abrupt

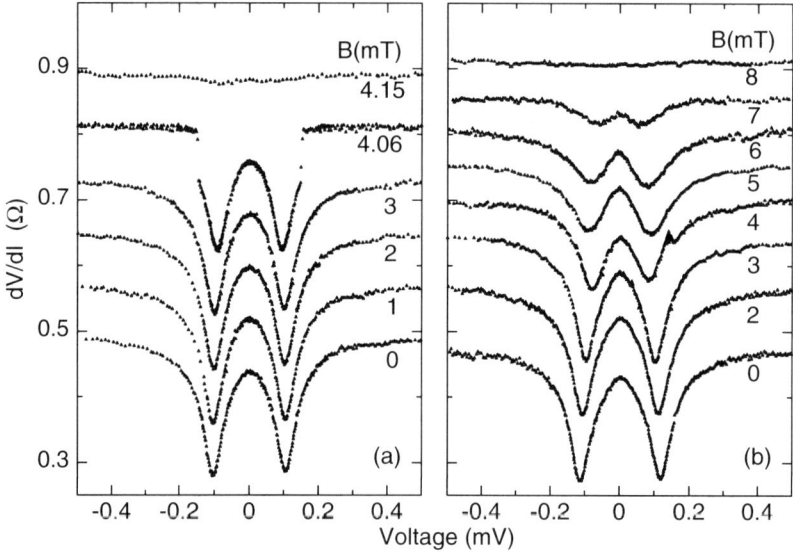

Fig. 12.12. Magnetic field dependence of the dV/dI curves versus V for two Zn-Ag contacts both with $R_N = 0.5\,\Omega$ and $T_c = 0.825 \pm 0.005\,\mathrm{K}$. The curves are offset vertically for clarity. Data taken from Naidyuk et al. (1996a).

disappearing of the double minimum in the dV/dI curves at a magnetic field close to the critical one [Fig. 12.12(a)]. Unlike, for the other contacts, the corresponding structure in the dV/dI curves decreases continuously with a field increase vanishing at the field value higher than $B_c^{\mathrm{bulk}} \simeq 6\,\mathrm{mT}$ [Fig. 12.12(b)]. The former type of behavior corresponds to the first-order transition of the type-I superconductor to the normal state in the magnetic field, whereas the latter looks like the second-order one. Correspondingly, $\Delta(B)$ drops to zero at about bulk critical magnetic field in the former case and decreases continuously in the latter case (Fig. 12.13). These two types of behavior are connected with a different structure of the metal in the constriction. For a clean type-I superconductor, the transition into the normal state at critical magnetic field B_c^{bulk} is of the first order. In the case of the Zn-Ag contact [Fig. 12.12(a)], a critical field is a little bit smaller than B_c^{bulk}, presumably on account of the demagnetization effect originating from the contact geometry. The disorder in

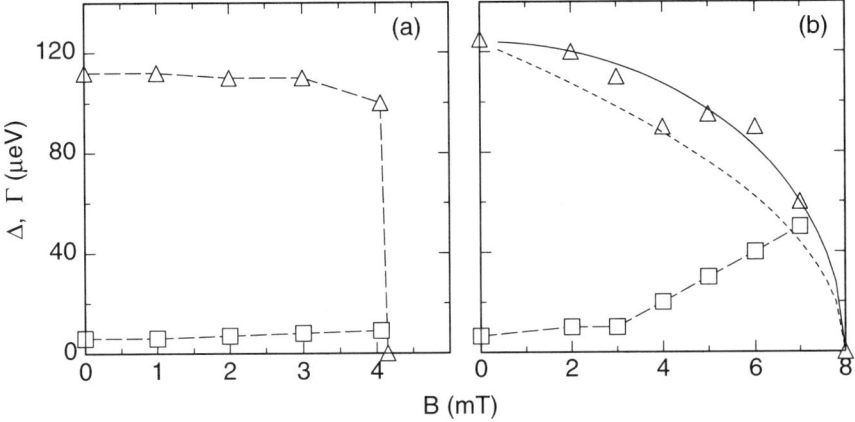

Fig. 12.13. Magnetic field dependence of $\Delta(B)$ (triangles) and $\Gamma(B)$ (squares) at $T \simeq 0.06\,\mathrm{K}$ for two Zn-Ag contacts from Fig. 12.12. Solid and dashed curves in (b) are $\Delta \propto \sqrt{1-(B/B_c)^2}$ [as for a thin films, Tinkham (1980)] and $\Delta \propto \sqrt{1-B/B_c}$ [as for a type-II superconductor, Maki (1969)], correspondingly. Long-dashed lines connect the symbols for clarity. Data taken from Naidyuk et al. (1996a).

the constriction gives rise in shortening of the electron mean free path. When l becomes smaller than ξ_o, so much so $\xi \simeq (\xi_o l)^{1/2} < \sqrt{2}\lambda \simeq \sqrt{2}\lambda_\mathrm{L}(\xi_o/l)^{1/2}$ (where λ_L is the London penetration depth), this leads to establishing a type-II superconducting state. This evokes an increase of the critical field $B_{c2} \simeq \Phi_o/\xi^2$ and a second-order transition to the normal state.

Naidyuk et al. (1996a) discussed probable reasons of the type-II behavior of a superconductor in the contact. They mentioned that the first-order transition was observed only for contacts with $2\Delta_0/k_\mathrm{B}T_\mathrm{c} \leq 3.5$, whereas contacts with a second-order transition showed both smaller and larger values than 3.5 of the reduced gap. This observation suggests different possibilities for the reduction of the mean free path. First, the strong structural disorder in the contact may result in softening of the phonon spectrum of the metal leading to enhanced electron–phonon coupling corresponding to the strong coupling superconductors, which have enlarged reduced gap value compared with the BCS one. Second, the impurity contamination of the constriction does not necessarily change the phonon spectrum and electron–phonon interaction; hence, it can result only in a reduction of the mean free path without altering the reduced gap. On the other hand, Zn is anisotropic metal with a large anisotropy in EPI spectra (Fig. 5.8). The random orientation of crystal in the constriction might cause the deviation of the reduced gap from the BCS value.

Turning to the discussion of Γ, it should be noted that Γ at $B = 0$ is often attributed to the inelastic scattering, i.e., $\Gamma = \hbar/\tau_{in}$, with the inelastic scattering rate τ_{in}^{-1} [Dynes et al. (1978)]. However, e.g., in the case of

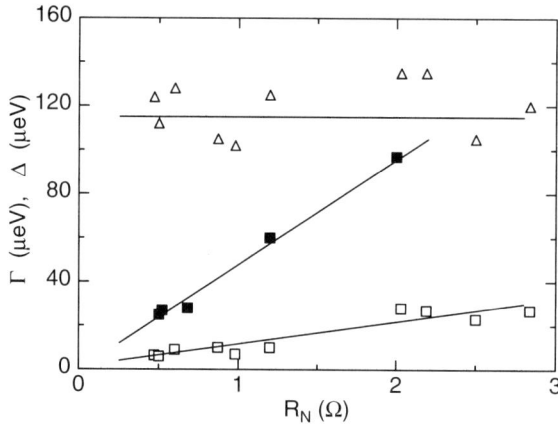

Fig. 12.14. Dependence of the superconducting gap (open triangles) and the Γ parameter (open squares) on the point-contact resistance R_N at $T < 0.1\,\mathrm{K}$ for Zn-Ag contacts together with Γ for Zn-UPt$_3$ contacts (closed squares). Solid lines are to guide the eye. Data taken from Naidyuk et al. (1996b).

Zn-Ag contacts, it is difficult to associate Γ with "usual" inelastic scattering processes such as electron–phonon and electron–electron interaction because these processes are negligible at low temperatures ($<1\,\mathrm{K}$) for simple metals. Figure 12.14 shows that Γ increases roughly linearly with the contact resistance R_N, and Δ shows a large amount of scattering between $105\,\mu\mathrm{eV}$ and $135\,\mu\mathrm{eV}$; however, the scattering is independent of R_N. There is no evidence for decreasing Δ_0 with Γ increase, which might be expected if Γ is related directly to the pair breaking. Furthermore, Naidyuk et al. (1996a) did not find any correlation between the zero-field Γ and the type of superconductor-normal-metal transition in the magnetic field. Hence, the origin of a finite Γ at $B = 0$ should be clarified.

Γ increases significantly in a magnetic field [Fig. 12.13(b)] for contacts with the type-II transition. Indeed, pair-breaking processes give rise to smearing the electronic DOS at the gap edges. It is also well-known that the pair-breaking parameter α increases in the magnetic field and $\alpha \sim B$ for the type-II superconductor [Maki (1969)]. Moreover, Strässler and Wyder (1967) showed that the DOS smearing for a thin film in a parallel magnetic field depends on the mean free path l. For a large l, this smeared DOS is similar to the DOS modified by Γ. Hence, pair-breaking processes might qualitatively explain the increase of Γ with B.

The zero-field Γ and its increase with a point-contact resistance is a few times higher for Zn-UPt$_3$ contacts than for Zn-Ag ones (Fig. 12.14). Furthermore, Γ is considerably higher for Nb contacts (Fig.12.11) and especially for high-T_c materials (see, e. g., Figs. 13.12 and 13.13) suggesting that S-N interface conditions play an important role here. The short coherence length in the latter materials can be one of the reasons as well.

12.2.3 Compounds

Superconducting intermetallic $AuIn_2$ with $T_c = 0.22$ K has been investigated by Gloos and Martin (1996). Although $AuIn_2$ is a compound, it has a large mean free path of about $5\,\mu m$ at low temperatures as well as of the momentum and Fermi velocity comparable with that in Cu. The superconducting coherence length $\xi_0 = \hbar v_F / \pi \Delta_0 \simeq 11\mu m$ is also extremely large, testifying that $AuIn_2$ is a type-I superconductor. The dV/dI curves of $AuIn_2$-Cu contacts with the resistance 0.1–1 Ω exhibit a pronounced double-minimum structure, which can be well described by the BTK model giving $2\Delta_0/k_B T_c = 3.5$. The calculations yield Z parameters of about 0.5, whereas the Fermi velocity mismatch should lead to $Z = 0.05$ only, that points to a natural barrier at the interface or to diffusive regime.

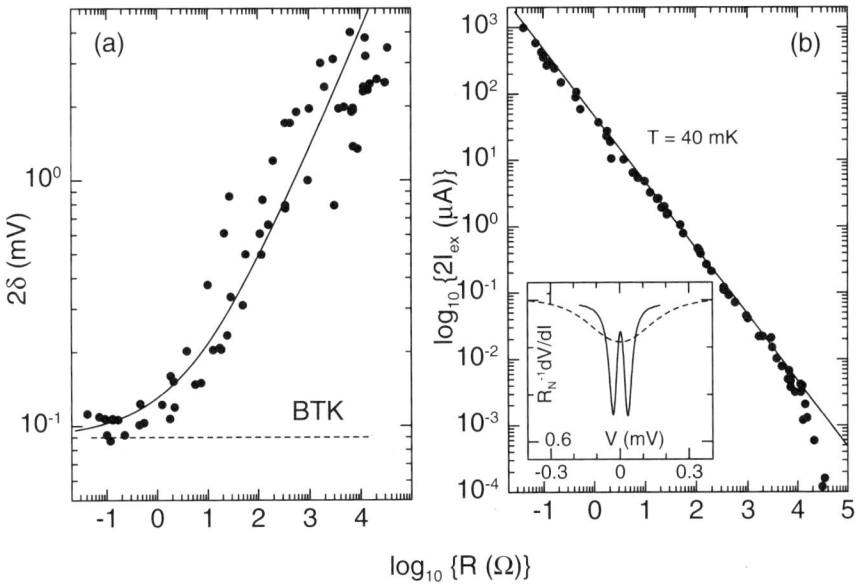

Fig. 12.15. (a) Width of the superconducting minimum in dV/dI at the half heights and (b) the excess current I_{exc} versus resistance for $AuIn_2$-Cu contacts at $T = 0.04$ K. The solid lines represent (a) $2\delta = 0.09\,\text{meV} + 0.04\,\text{meV}\,\sqrt{R(\Omega)}$ and (b) the BTK prediction for the excess current considering $2\Delta_0 = 0.064\,\text{meV}$ and $Z = 0.5$. Horizontal dashed line in (a) is the width of an ideal BTK-type spectrum. Inset in (b) shows the dV/dI dependence at $T = 0.04$ K for a contact with $R_N = 0.13\,\Omega$ (solid curve) and 50 Ω (dashed curve). Data taken from Gloos and Martin (1996).

Gloos and Martin (1996) showed that the relative change of the zero-bias resistance by transition into the superconducting state decreases with increasing a contact resistance and the superconducting minimum broadens (Fig. 12.15(b), inset). However, a shallow minima with a width increased up to 5 mV, which is almost 100 times $2\Delta_0$, was still resolved even at a resistance as high as $10\,\mathrm{k}\Omega$. It is interesting that the excess current still can be described by the BTK behavior $I_{\mathrm{exc}} \propto R_{\mathrm{N}}^{-1}$ in the whole resistance range proving that the superconducting anomalies originate from Andreev reflection. This current falls out rapidly only above $13\,\mathrm{k}\Omega$ by transition to the tunnel regime [Fig. 12.15(b)]. The I_{exc} data testify that lateral confinement in the point contact does not suppress Andreev reflection. When the resistance increases, the spectra broadening is attributed to the energy dispersion of ballistic electrons caused by scattering at defects or at the strain field in the contact.

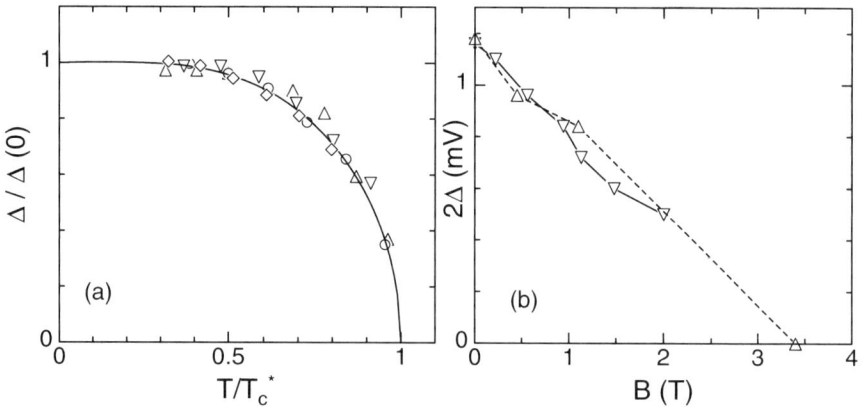

Fig. 12.16. (a) Temperature dependence of reduced Δ for a few CeRu$_2$-Cu contacts along with the BCS curve (solid line). (b) Magnetic field dependence of 2Δ for two CeRu$_2$-Cu contacts. The different symbols for both figures correspond to the different contacts. The lines connect the symbols for clarity. Data taken from Naidyuk et al. (1998).

Intermetallic compound CeRu$_2$ represents a type-II superconductor. Naidyuk et al. (1998) measured temperature and magnetic field dependence of the $\mathrm{d}V/\mathrm{d}I$ curves of CeRu$_2$-Cu contacts. Interestingly, the pronounced double-minimum structure vanishes at $T_{\mathrm{c}}^* \simeq 5.2\,\mathrm{K}$, noticeably below the bulk critical temperature $T_{\mathrm{c}}^{\mathrm{bulk}} = 6.2\,\mathrm{K}$. Above T_{c}^*, $\mathrm{d}V/\mathrm{d}I$ curves still exhibit a broad and shallow superconducting minimum, but it was impossible to fit by the BTK model these curves reasonably for any Δ. For different contacts, T_{c}^* was between 4.4 and 5.25 K, supposedly owing to the local superconducting properties in the contact region. $\mathrm{d}V/\mathrm{d}I(V)$ characteristics below T_{c}^* were found to correspond well to the BTK model, what allows us to establish the temperature and the magnetic field dependence of the superconducting gap Δ in this region (Fig. 12.16). It is seen that $\Delta(T)$ dependence fits well the BCS behavior with $2\Delta_0/k_B T_c^* = 3.1 \pm 0.1$, whereas the gap decreases approximately linear in a magnetic field. It differs from that which is expected for a type-II superconductor. The extrapolated gap vanishes at field $B_c^* \simeq$ 3.5 T well below the upper critical field $B_{c2} \simeq 6$ T; however, it is close to the irreversibility field separating the solid and liquid vortex lattice. The coincidence of B^* with irreversibility field [Huxley et al. (1993)] points to a change of the vortex dynamics as a possible reason for gap vanishing.

In the subsequent paper, Moskalenko et al. (2001) presented the comparison of the superconducting gap for the $CeRu_2$ poly- and single crystals. The common procedure for determination of $2\Delta_0/k_B T_c^*$ leads to the value of 3.2 for $CeRu_2$ single crystal in spite of higher $T_c^* \simeq 5.9$ K. This was naturally explained considering the difference in the quality of samples. The contacts made on the more perfect $CeRu_2$ single crystal exhibit better superconducting properties than those with the polycrystal. Further, the authors supposed the presence of paramagnetic impurities (likely extra Ce ions), which determined the quality of the samples or contacts. They proposed an original procedure for Δ correction, taking into account a pair-breaking effect of impurities. As a result, the gap for a hypothetic "clean" $CeRu_2$ sample presses for the value $2\Delta_0/k_B T_c \approx 3.75$, both using data for single and polycrystal, indicating that $CeRu_2$ is a "moderate" strong-coupling superconductor.

Solanki-Moser et al. (1987b) investigated superconducting ThIrSi and α-$ThSi_2$ compounds crystallized in the tetragonal structure with the critical temperature 6.5 and 3.16 K, respectively. Although the differential resistance of heterocontact with Mo reveals a double-minimum structure for the first compound, the same characteristics for α-$ThSi_2$ exhibit only a narrow single zero-bias minimum. After the BTK fit of curves at lowest temperature, they received $2\Delta_0/k_B T_c = 2.9$ for ThIrSi and about 1 for α-$ThSi_2$. It was concluded that the BCS theory is evidently not an appropriate description for these silicides.

Resuming this paragraph, we present Table 12.2.3 with superconducting characteristics of metals and compounds investigated by S-c-N and S-c-S contacts. Most of the data correspond well to the BCS theory and the values from textbooks. An enhanced value of the gap for Sn and In appears to be connected with a disorder induced in these soft materials by contact creation.

12.3 Spectroscopy of the spin polarization

Andreev reflection on the N–S interface (see Section 3.7.1) involves formation of a Cooper pair from spin-up and spin-down electrons situated in the normal (N) metal. If we suppose that electrons in N metal have only spin-up or that they are 100% polarized, then, intuitively, it is clear that Andreev reflection

Table 12.1. Superconducting gap of simple metals and compounds determined from the $\mathrm{d}V/\mathrm{d}I\,(V)$ curves of the S-c-N point contacts using both BTK and modified BTK fit or extracted from the subharmonic gap structure (SGS) for the S-c-S contacts. Here T_c is the bulk critical temperature. We referred only to the data that represent a clear BTK-like double-minimum feature or distinct SGS in the $\mathrm{d}V/\mathrm{d}I\,(V)$ curves.

Sample	N/S	$2\Delta_0$, meV	$2\Delta_0/k_BT_c$	$\Delta(T)$	Method	Reference
In	Cu	1.2	4.1	-	BTK	Naidyuk et al. (1996)
Nb	Cu	2.95	3.6	-	BTK	Blonder et al. (1983)
	Nb	2.95	3.7	-	SGS	Muller et al. (1992)
	Ag	2.7	3.35	BCS	m-BTK	Naidyuk et al. (1996)
	Nb	3.05	3.85	BCS	SGS	Smith et al. (1996)
Pb	Co, Ni, Cu	2.7	4.35	-	m-BTK	Upadhyay et al. (1998)
Sn	Sn	1.1	3.4	-	SGS	Khotkevich et al. (1981)
	Ag	1.3	4.1	-	BTK	Naidyuk et al. (1996)
	Mo	1.25	3.9	-	BTK	Solanki-Moser (1987a)
Zn	Ag	0.238±0.024	3.35±0.35*	BCS	m-BTK	Naidyuk et al. (1996)
	UPt_3	0.23±0.01	3.3±0.1*	BCS	m-BTK	Naidyuk et al. (1996)
$AuIn_2$	Cu	0.064	3.5*	BCS	BTK	Gloos and Martin (1996)
$CeRu_2^{**}$	Cu	1.35±0.05	3.1±0.1*	BCS	BTK	Naidyuk et al. (1998)
	Mo	0.9	1.7	-	BTK	Solanki-Moser (1987a)
ThIrSi	Mo	1.65	2.9	-	BTK	Solanki-Moser et al. (1987b)

* Calculation $2\Delta_0/k_BT_c$, the temperature T_c at which the BTK-like superconducting minimum in $\mathrm{d}V/\mathrm{d}I$ vanished was used.
** For the more perfect $CeRu_2$ single crystal, Moskalenko et al. (2001) received $\Delta \simeq 1.6$ meV and $2\Delta_0/k_BT_c=3.2$. Additionally, accounting pair-breaking effects $2\Delta_0/k_BT_c=3.75$ was estimated (see Section 12.2.3 for details).

will be suppressed because of lack of spin-down electrons. Thus, in the first approximation, the presence of spin polarization leads to reducing of the Andreev reflection and correspondingly to the diminution of N-c-S contact conductivity at $eV < \Delta$. The measured value of spin polarization P_n depends

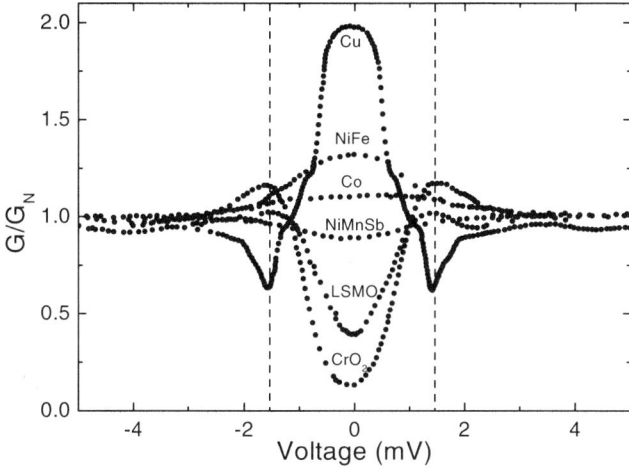

Fig. 12.17. Reduced differential conductance $G = dI/dV(V)$ for contact between superconducting Nb and different magnetic (spin polarized) metals $Ni_{0.8}Fe_{0.2}$, Co, NiMnSb, $La_{0.7}Sr_{0.3}MnO_3$, CrO_2, including nonmagnetic Cu. The zero-bias conductivity decreases with increasing of spin polarization factor P_{pc}. The vertical dashed lines mark position of the superconducting gap in Nb. Data taken from Soulen et al. (1998).

on the experimental technique and can be defined by the generalized formula [Nadgorny et al. (2001)]:

$$P_n = \frac{\langle N(\epsilon)_\uparrow v_{F\uparrow}^n \rangle - \langle N(\epsilon)_\downarrow v_{F\downarrow}^n \rangle}{\langle N(\epsilon)_\uparrow v_{F\uparrow}^n \rangle + \langle N(\epsilon)_\downarrow v_{F\downarrow}^n \rangle}, \qquad (12.3)$$

where $N(\epsilon)_\uparrow v_{F\uparrow}^n$ and $N(\epsilon)_\downarrow v_{F\downarrow}^n$ are the majority and minority spin density of states and the Fermi velocities, respectively. The spin polarization for charge transport in point contacts P_{pc} corresponds to $n = 1$ for ballistic and $n = 2$ for diffusive regime, respectively.

In the absence of the interfacial scattering ($Z = 0$), the zero-bias conductivity $G_0 = dI/dV(V = 0)$ as shown by Soulen et al. (1998) depends on P_{pc} by simple relation:

$$G_0/G_N = 2(1 - P_{pc}), \qquad (12.4)$$

where G_N is the normal state conductivity. Equation (12.4), eventually, is evident at least for two opposite cases, $P_{pc} = 0$ and 1, and it was also anticipated by de Jong and Beenakker (1995).

Soulen et al. (1998) measured $G(V)/G_N$ (see Fig. 12.17) for point contacts between superconducting Nb and a variety of metals: $Ni_{0.8}Fe_{0.2}$, Ni, Co, Fe, NiMnSb, $La_{0.7}Sr_{0.3}MnO_3$, and CrO_2, where spin polarizations vary from 0.35 to 0.9. The authors supposed that Z appears to be minimal for the presented contacts, and using (12.4) allows a direct estimation of P_{pc}. The curves in

Fig. 12.17 for Cu ($P_{pc} = 0$) with doubled conductance at zero bias, and that for half metallic CrO_2 ($P_{pc} \to 1$) with vanished zero-bias conductance, even visually support this claim.

Ji et al. (2001) analyzed the complete conductance-voltage curve of a Pb-CrO_2 point contact using the modified BTK model to extract the polarization reliably. They determined the spin polarization of single-crystal CrO_2 films to be close to 1, namely, 0.96 ± 0.01 and showed that a surface barrier layer $Z \neq 0$ strongly reduces the apparent spin polarization of CrO_2, e.g., down to 0.7 at $Z = 1$.

Nadgorny et al. (2001) observed a significant increase from 0.6 to 0.9 in the current spin polarization in colossal magnetoresistive manganite $La_{0.7}Sr_{0.3}MnO_3$ with the rise of residual resistivity. This counterintuitive trend was explained by a transition from ballistic to diffusive transport in the contact.

Upadhyay et al. (1998) studied Pb contacts with ferromagnetic Ni and Co. They received polarization 32% and 37% for Ni and Co, respectively. Strijkers et al. (2001) measured Nb-Me contacts also with Me=Ni, Co, and Fe. They showed that polarization for Fe, Co, and Ni depends substantially on the quality of the metal–superconductor contact. The value of the intrinsic spin polarization of the current can be obtained by extrapolation to $Z = 0$, resulting in 0.37 ± 0.01, 0.45 ± 0.02, and 0.43 ± 0.03 for Ni, Co, and Fe, respectively.

There are a number of theoretical investigations of ferromagnet–superconductor (F–S) junction conductivity. Žutić and Valls (2000) also took into account Fermi wave-vector mismatch between the F and S regions. They pointed out that because of the Fermi wave-vector mismatch, spin polarization can even enhance Andreev reflection and give rise to a zero-bias conductance peak for an s-wave superconductor. Žutić and Valls (2000) concluded that to determine spin polarization, the appropriate Fermi wave-vector mismatch should be taken into account properly. Dimoulas (2000) predicted enhancement of the Andreev reflection probability for quasiparticles in the minority spin band only. This enhancement takes place below the superconducting gap near the gap edge.

12.4 Large contacts

Under a large contact, we consider the constriction with a dimension much greater than both elastic and inelastic mean free paths of electrons as well as the superconducting coherence length. Thus, such contacts are beyond the above-mentioned theories. The typical $I - V$ curves for the large S-c-N contacts on the base of amorphous Zr_2Ni film are shown in Fig. 12.18(a). The characteristic feature is here a jump of a voltage at reaching some critical current I_c, which is more expressed by lowering the contact resistance. This jump is accompanied by a hysteresis at reversing the current through the constriction for low ohmic ($< 1\,\Omega$) contacts. With resistance increasing, the

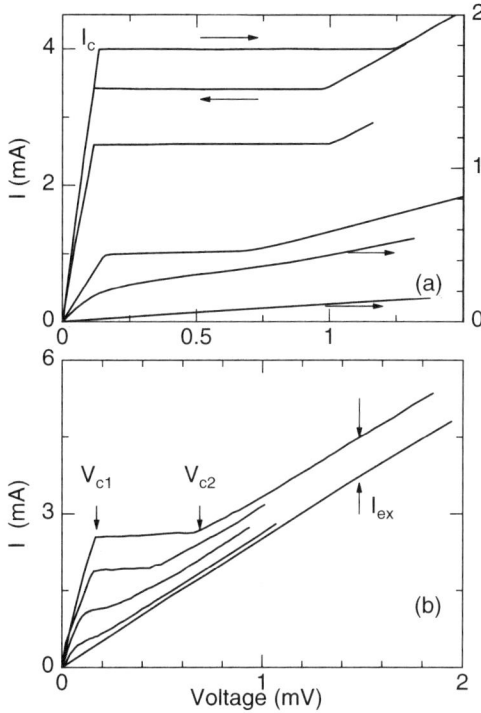

Fig. 12.18. (a) $I - V$ characteristics of Zr_2Ni-Cu contacts at 1.5 K. The resistance of the contacts in the normal state R_N is 0.4, 0.405, 1.02, 2.88, and 9.55 Ω (from the top curve to the bottom one). For the upper curve, a hysteresis loop is seen. (b) Temperature dependence of $I - V$ curves for contact with $R_N = 0.42$ Ω at $T = 1.55$, 1.88, 2.3, 2.5, and 2.6 K (from the top curve to the bottom one). Data taken from Naidyuk and Kvitnitskaya (1991).

voltage jump gradually transforms into the reversible inflection point. The same transformation occurs also with temperature increasing [Fig. 12.18(b)]. I_c was found to be proportional to the contact diameter for contacts between amorphous superconductor Zr_2Ni ($T_c \simeq 2.5$ K) and normal metal Cu or Ag. It was assumed that, by analogy with the Silsbee rule for a superconducting wire, the point contact turns into a resistive state when a magnetic field induced by the transport current exceeds the critical field B_c or the lower critical field B_{c1} for a type-II superconductor. Using expression $B = \mu_0 I / \pi d$ for a magnetic field produced by current at the surface of the wire with diameter d and Maxwell equation $R_N = \rho / 2d$ for resistance of heterocontact in normal state (here we discard the resistivity of the normal-metal counter-electrode, which is negligible compared with the resistivity of an amorphous compound), the critical current is given by

$$I_c = \frac{\pi d B}{\mu_0} = \frac{\pi \rho B_{c1}}{2 \mu_0 R_N}. \tag{12.5}$$

By investigation of another amorphous superconductor $(Mo_{0.55}Ru_{0.45})_{0.2}P_{0.8}$, Häussler et al. (1996) showed that (12.5) fits quantitatively experimental data without any adjustable parameter (Fig. 12.19).

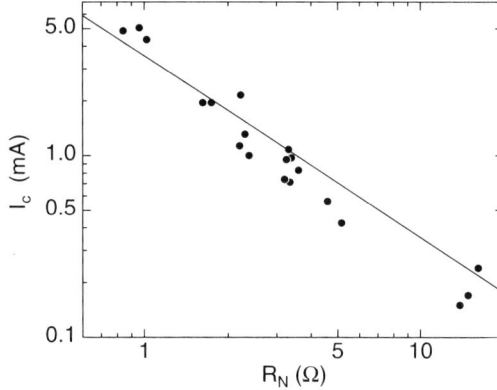

Fig. 12.19. Dependence of the critical current I_c versus resistance for the $(Mo_{0.55}Ru_{0.45})_{0.2}P_{0.8}$–Ag contacts. The solid line is according to (12.5). Data taken from Häussler et al. (1996).

Naidyuk and Kvitnitskaya (1991) explained the voltage jump by accounting the motion of vortices, which are induced by the transport current. The toroidal vortices, appeared on the edge of the contact by reaching I_c and experienced Lorentz force, move toward the constriction center and annihilate. As the range of interaction between vortices is approximately equal to London penetration depth λ_L, this motion is restricted when a radius of the contact is less than λ_L. A crossover should occur at $R_N \simeq \rho/4\lambda_L \simeq 1\Omega$ that agrees with the observations. Naidyuk and Kvitnitskaya (1991) also drew attention to the fact that the size of the low-resistance contacts may exceed the magnetic penetration length λ_L, which results in a concentration of the current near a surface in a layer with thickness λ_L.

Häussler et al. (1996) studied dV/dI curves of $(Mo_{0.55}Ru_{0.45})_{0.2}P_{0.8}$-Ag contacts with $R_N > 1\Omega$ for which a voltage jump in the $I-V$ curve transforms to the inflection point corresponding to the maximum in dV/dI. About one-third of the dV/dI curves exhibited also a shallow double-minimum structure around zero bias. The sharp peaks or maxima at the edges of the minimum at about $1\,mV$, which correspond to I_c, are getting broader with increasing temperature or magnetic field (Fig. 12.20). These structures vanish at reaching B_{c2} or T_c in the bulk compound. Additionally, weak peaks are observed at $V \simeq \pm 3$ mV, corresponding to the current density of about 10^6 A/cm^2.

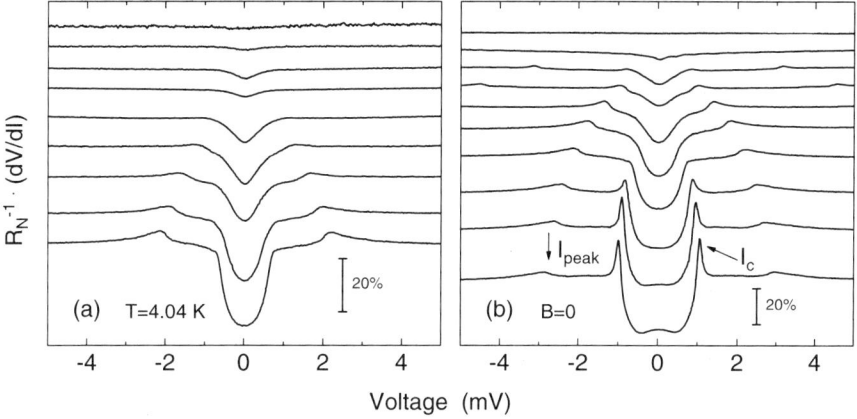

Fig. 12.20. (a) dV/dI curves versus V for the $(Mo_{0.55}Ru_{0.45})_{0.2}P_{0.8}$-Ag contact with $R_N = 2.1\,\Omega$ at $4.04\,K$ in different magnetic fields B (T): 0, 1, 2, 3, 3.5, 4, 4.2, 4.7, and 5.5 (from the bottom curve to the top one), and (b) at zero field for different temperatures T (K): 1.49, 2.28, 3.21, 4.04, 4.59, 5.03, 5.29, 5.40, 5.51, and 5.71 (from bottom to top). The curves are offset vertically for clarity. The arrows show a position of the critical I_c and the peak I_{peak} current. Data taken from Häussler et al. (1996).

The authors associate this structure with exceeding of a pair-breaking current density for a type-II superconductor, estimated according to Tinkham (1980) as

$$j_c = \frac{4B_c^{th}}{3\sqrt{6}\mu_0\lambda_L}. \tag{12.6}$$

Indeed, the temperature dependence of I_{peak} from Fig. 12.20(b) behaves as an order parameter in the BCS theory, whereas the magnetic filed dependence of $I_{peak} \sim \sqrt{1 - B/B_{c2}}$ corresponds to a spatially averaged order parameter of the type-II superconductor in the vortex state predicted by the Abrikosov theory. Similar to that mentioned above, a gap peak was noticed by Xiong Peng et al. (1993) during the investigation of a Nb-Al-thin-film high-transmittance microjunction.

Westbrook and Javan (1999) observed finite-bias dynamic resistance peaks in Ta-W point contacts. They found that the bias voltage at which these peaks occurs is linear in the square root of the contact resistance. This effect was explained by destruction of superconductivity near the contact caused by the magnetic field produced by current flow through the contact. Using the photoresponce measurements on the same contacts the authors concluded that superconductivity is destroyed at the contact when the first peak in dV/dI occurs, thus suppressing the excess current at voltages above the peak.

Another point worth noting is that an external magnetic field less than 0.5 T has no significant effect on $I-V$ curves [Häussler et al. (1996)], although it exceeds B_{c1} about two orders of magnitude. This probably is connected with different topology and pinning of the vortices. Current-induced vortices having a toroidal shape are created directly in the constriction, and magnetic-field induced vortices represent a long tube expanded over a whole sample and therefore have sufficiently higher probability to be pinned.

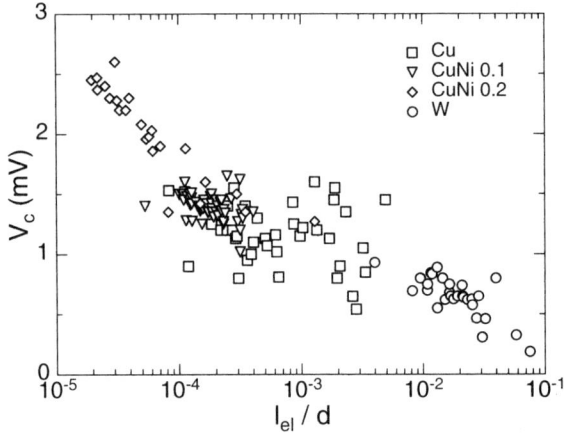

Fig. 12.21. Dependence of the critical voltage V_c versus l_{el}/d ratio for the NbTi-Me contacts with Me = W, Cu, $Cu_{0.9}Ni_{0.1}$, and $Cu_{0.8}Ni_{0.2}$. Data taken from Lysykh et al. (1992).

For low ohmic contacts, a position of the voltage jump V_c corresponding to I_c (see Fig. 12.18) is often close to a value of the superconducting gap, which is why Sullivan and Ross (1967) proposed to use such $I - V$ curves for determination of Δ. Lysykh et al. (1992) thoroughly studied the $I - V$ curves of contacts on the base of the NbTi superconductor to check this observation. They established that V_c critically depends on the purity of the normal counterelectrode used. In the other words, V_c is intimately connected with a regime of a current flow in the constriction. By decreasing the ratio l_{el}/d from 10^{-1} to 10^{-5}, the critical voltage increases from about 0.15 to 2.6 mV (Fig. 12.21); that is, no correlation with the gap value was found[1].

Lysykh et al. (1992) used (3.26) for normal and superconducting states to calculate a diameter and residual resistivity (mean free path) in their contacts. It turned out that the contact resistance in the superconducting state R_s caused by the normal counterelectrode is inversely proportional to the

---

[1] No correlation between V_c and the gap value was found also by Naidyuk and Kvitnitskaya (1991) for the Zr_2Ni-Ag contacts. The V_c scattering was between 0.1 and 1.5 mV.

diameter d for the NbTi-Cu heterocontacts, and $R_s \propto d^{-2}$ for NbTi-W heterocontacts. In this instance, the resistance of the normal metal behaves as Maxwell resistance in the former case and as Sharvin resistance in the latter one. It was explained using the difference in mechanical properties of Cu and W. Namely, by creating point contacts, low-temperature deformation produces more degradation of a material in the constriction in the case of soft Cu compared with the hard W. Further, the critical current has different behavior $I_c \propto d$ for the dirty contact and $I_c \propto d^2$ for the clean one (Fig. 12.22). The latter dependence testifies that the transition at V_c corresponds to the

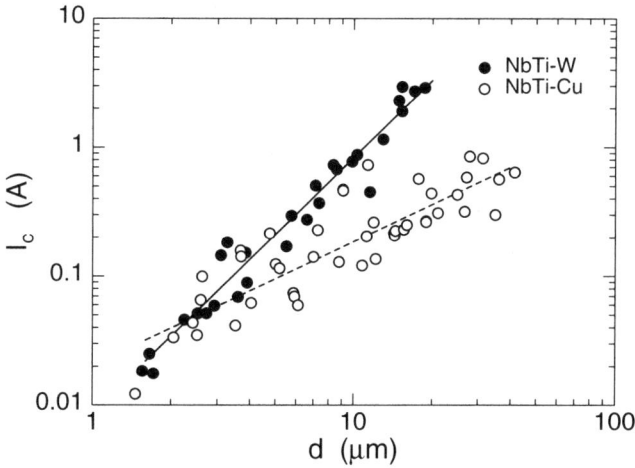

Fig. 12.22. Dependence of the critical current I_c on the contact diameter for the NbTi-Me contacts (Me = W and Cu) at 4.2 K. Solid line represents behavior $I_c \propto d^2$, and dashed line is $I_c \propto d$. Data taken from Lysykh et al. (1992).

critical current density $10.9 \pm 0.7 \times 10^6$ A/cm^2 for all NbTi-W contacts investigated. It means that the voltage jump at V_c in these contacts is caused by reaching the critical current density.

Let us turn to dirty NbTi-Cu contacts. As shown by MacDonald and Leavens (1983), the short elastic mean free path $l_{el} \ll d$ leads to the nonuniform potential distribution so that the current density is not homogeneous through the constriction cross-section and increases abruptly at the contact periphery. Therefore, the current is distributed mainly in the ring with outer diameter d and width L. This naturally explains the linear dependence instead of the quadratic one of I_c on d. Using the obtained value of the critical current density, Lysykh et al. (1992) calculated the width L of the ring, which turned out to be nearly 100 times higher than l_{el} and almost linearly depends on the mean free path.

References

Artemenko S. N., Volkov A. F. and Zaitsev A. V. (1979) Sol. State Communs. **30** 771; JETP Lett **49** 924.

Asen M. and Keck K. (1992) Sov. J. Low Temp. **18** 602.

Asen-Palmer M. and Keck K. (1994) Z. Phys. B **94** 21.

Barnes L. J. (1969) Phys. Rev. **184** 434.

Blonder G. E. and Tinkham M. (1983) Phys. Rev. **27** 112.

Blonder G. E., Tinkham M. and Klapwijk T. M. (1982) Phys. Rev. **25** 4515.

Bobrov N. L., Rybaltchenko L. F., Obolenskii M. A. and Fisun V. V. (1985) Sov. J. Low Temp. Phys. **11** 510.

de Jong M. J. M. and Beenakker C. W. J. (1995) Phys. Rev. Lett. **74** 1657.

Dimoulas A. (2000) Phys. Rev. B **61** 9729.

Dynes R. C., Narayanamurti V. and Garno J. P. (1978) Phys. Rev. Lett. **41** 1509.

Häussler R., Goll G., Naidyuk Yu. G. and von Löhneysen H. (1996) Physica B **218** 197.

Huxley A. D., Paulsen C., Laborde O., Tholence J. L., Sanchez D., Junod A. and Calemczuk R. (1993) J. Phys.: Cond. Matter **5** 7709.

Gloos K. and Martin F. (1996) Z. Phys. B **99** 321.

Ji Y., Strijkers G. J., Yang F. Y., Chien C. L., Byers J. M., Anguelouch A., Gang Xiao and Gupta A. (2001) Phys. Rev. Lett. **86** 5585.

Josephson B. D. (1962) Phys. Letters **1** 251.

Kamarchuk G. V. and Khotkevich A. V. (1987) Sov. J. Low Temp. Phys. **13** 717.

Kamarchuk G. V., Khotkevich A. V. and Yanson I. K. (1986) Sov. Phys.-Solid State **28** 254.

Khlus V. A. (1983) Sov. J. Low Temp. Phys. **9** 510.

Khlus V. A. and Omelyanchouk A. N. (1983) Sov. J. Low Temp. Phys. **9** 189.

Khotkevich A. V. and Yanson I. K. (1981) Sov. J. Low Temp. Phys. **7** 354.

Khotkevich A. V., Khotkevich V. V., Yanson I. K. and Kamarchuk G. V. (1990) Sov. J. Low Temp. Phys. **16** 693.

Kulik I. O. and Omelyanchouk A. N. (1977) Sov. J. Low Temp. Phys. **3** 459.

Kulik I. O. and Omelyanchouk A. N. (1978) Sov. J. Low Temp. Phys. **4** 142.

Lysykh A. A., Zasetsky G .V. and Bagatsky V. M. (1992) Sov. J. Low Temp. Phys. **18** 419.

MacDonald A. H. and Leavens C. R. (1983) J. Phys. F **13** 665.

Maki K. (1969) in *Superconductivity* edited by R. D. Parks (Marcel Dekker, New York) Vol. 2, p.1035.

Mazin I. I., Golubov A. A. and Nadgorny B. (2001) J. Appl. Phys. **89** 7576.

Meissner H. (1958) Phys. Rev. **109** 686.

Moskalenko A. V., Naidyuk Yu. G., Yanson I. K., Hedo M., Inada Y., Onuki Y., Haga Y., and Yamamoto E. (2001) Low Temp. Phys. **27** 613.

Muller C. J., van Ruitenbeek J. M. and de Jongh L. J. (1992) Physica C **191** 485.

Nadgorny B., Mazin I. I., Osofsky M., Soulen R. J., Jr., Broussard P., Stroud R. M., Singh D. J., Harris V. G., Arsenov A. and Mukovskii Ya. (2001) Phys. Rev. B **63** 184433.

Naidyuk Yu. G., Häussler R. and von Löhneysen H. (1996) Physica B **218** 122.

Naidyuk Yu. G. and Kvitnitskaya O. E. (1991) Sov. J. Low Temp. Phys. **17** 439.

Naidyuk Yu. G., Moskalenko A. V., Yanson I. K. and Geibel C. (1998) Low Temp. Phys. **24** 374.

Naidyuk Yu. G., von Löhneysen H. and Yanson I. K. (1996a) Phys. Rev. B **54** 16077.

Naidyuk Yu. G., von Löhneysen H. and Yanson I. K. (1996b) Chech. J. Phys **46** 711.

Omelyanchouk A. N., Beloborod'ko S. I. and Kulik I. O. (1988) Sov. J. Low Temp. Phys. **14** 630.

Reinertson R. C., Dolan P. J., Craig D. L. Jr. and Smith C. W. (1990) Physica B **165&166** 1615.

Reinertson R. C., Smith C. W. and Dolan P. J. Jr. and Smith C. W. (1992) Physica C **200** 377.

Smith C. W., Reinertson R. C. and Dolan J. P. Jr. (1996) Physica B **218** 119.

Solanki-Moser M. (1987a) Ph. D. Thesis, ETH Zürich. (unpublished)

Solanki-Moser M., Buffat B., Wachter P., Wang Xian Zhong, Czeska B., Chevalier B. and Etourneau J. R. (1987b) J. Magn. Magn. Mater. **63 & 64** 677.

Soulen R. J., Byers J. M., Osofsky M. S., Nadgorny B., Ambrose T., Cheng S. F., Broussard P. R., Tanaka C. T., Nowak J., Moodera J. S., Barry A. and Coey J. M. D. (1998) Science **282** 85.

Strässler S. and Wyder P. (1967) Phys. Rev. **158** 315.

Strijkers G. J., Ji Y., Yang F. Y., Chien C. L. and Byers J. M. (2001) Phys. Rev. B **63** 104510.

Sullivan D. B. and Ross C. E. (1967) Phys. Rev. Lett. **18** 212.

Tinkham M. (1980) *Introduction to Superconductivity* (McGraw-Hill, New York) p.125.

Upadhyay S. K., Palanisami A., Louie R. N. and Buhrman R. A. (1998) Phys. Rev. Lett. **81** 3247.

Westbrook P. S. and Javan A. (1999) Phys. Rev. B **59** 14606; *ibid.* 14612.

Xiong Peng, Xiao Gang and Laibowitz R. B. (1993) Phys. Rev. Lett. **71** 1907.

Yanson I. K., Kamarchuk G. V. and Khotkevich A. V. (1984a) Sov. J. Low Temp. Phys. **10** 220.

Yanson I. K., Bobrov N. L., Rybaltchenko L. F. and Fisun V. V. (1983) Sov. J. Low Temp. Phys. **9** 596.

Yanson I. K., Bobrov N. L., Rybaltchenko L. F. and Fisun V. V. (1984b) Solid State Commun.**50** 515.

Yanson I. K., Bobrov N. L., Rybaltchenko L. F. and Fisun V. V. (1985) Sov. Phys.- Solid State **27** 1076.

Yanson I. K., Fisun V. V., Bobrov N. L. and Rybaltchenko L. F. (1987) JETP Lett. **45** 543.

Zakharov A. A. and Tsetlin M. B. (1985) JETP Lett. **41** 11.

Žutić I. and Valls O.T. (2000) Phys. Rev. B **61** 1555.

13 PCS of high-T_c and other uncommon superconductors

The high-T_c superconductors are distinguished by their high anisotropic properties, extreme sensitivity to the stoichiometry and oxygen contents, the short mean free path, and the coherence length. The two latter points make it difficult to apply directly for those materials mentioned in Chapter 3 both the Kulik–Omelyanchouk or the BTK theory for investigation of critical Josephson current or gap structure by S-c-S weak links and the Kulik–Omelyanchouk–Shekhter theory for study of the EPI interaction. The above-mentioned theories are developed for clean (or diffusive) contacts with the size both well below the coherence length and the inelastic mean free path. Moreover, non-s-wave symmetries of the superconducting order parameter, as well as the non-electron–phonon mediating pairing mechanism are discussed by explanation of the experimental results in the high-T_c materials. Therefore, it is not clear *a priori* whether it is possible to extract spectral information about the bulk properties of the high-T_c materials by the point contacts. However, as mentioned in the previous chapter (see also Fig. 12.5), the energy-resolved spectroscopy of quasiparticles may take place also for the dirty S-c-N and S-c-S contacts at some condition. The mechanism of emergence of singularities in the spectra caused by the inelastic interaction between the carriers and phonons or other quasiparticles is not completely clear. It is obviously connected with partial suppression of the superconducting order parameter, e. g., by the nonequilibrium phonons with the small group velocities ($\partial\omega/\partial q = 0$) and the large density of states leading to the threshold in the excess current. The typical behavior of such phonon singularities in the spectra is their robustness to temperature change and the magnetic field rise. Contrary, the position of the superconducting features caused by reaching the critical current density for some "weak-links" or produced by the heating effects strongly depends on these external parameters, especially at temperatures close to T_c.

Let us turn to the issue about "gap" structure formation in the $I-V$ characteristics for a contact with a short coherence length ξ. Because of inequality $\xi \ll d$ typical for the point contacts with the high-T_c materials, which is totally opposite to the theoretical case, the appearance of some peculiarities at $eV \simeq \Delta$ is not obvious. However, experimental observations show that the $I-V$ curves and their derivatives contain plenty of features just in the energy region of the expected gap. The mechanism of appearance of such peculiarities is not clear whether it is Andreev reflection, critical current effect, heating and so on. Nevertheless, thorough study of these features shows their reproducibility, reasonable temperature and magnetic field behavior, adequate reaction on the tuning of superconducting state by stoichiometry, donation, alloying and so on. Therefore, they can reflect bulk superconducting properties and are someway connected with the superconducting order parameter.

There is a crowd of publications concerning the study of high-T_c compounds properties by point contacts. Our aim is not to refer to all of them, but outline the main progress in the field and demonstrate the possibilities of PCS with respect to this new class of superconductors. The chapter ends with the review of the properties of borocarbides, organic superconductors, magnesium diboride, and Sr_2RuO_4. The former two are not real high-T_c materials; however, share with the latter more common features than with the ordinary superconductors. Magnesium diboride is a very new superconductor with the highest critical temperature about $40\,K$ among the binary compounds. Layered superconductor Sr_2RuO_4 with T_c as high as $1.5\,K$ is debated as a system with p-wave symmetry of the order parameter.

13.1 Selection rules and contact models

As in the case of the conventional metal for the high-T_c point contacts, some quality criteria also should be established, determining whether contact is suitable for the further investigation. Among them are [see also Yanson et al. (1989)]:

First, the critical temperature in the contact T_c^* should be close to T_c of the bulk sample. It means that all superconducting peculiarities will disappear above T_c. It is also very desirable to measure temperature dependence of the contact resistance $R_0(T)$ up to T_c and above in the normal state. $R_0(T)$ should correspond to the theoretical model (see Fig. 12.7) below T_c and to the temperature dependence of the bulk resistivity $\rho(T)$ above T_c.

Second, the voltage dependence of the differential resistance $dV/dI(V)$ should behave similar to $\rho(T)$ above T_c at bias voltage larger than the superconducting gap. For example, high-quality high-T_c materials have linear resistivity increase in the normal state by rising temperature. Correspondingly, differential resistance of contact should increase with voltage at $eV > \Delta$.

Third, pronounced (of the order of Δ/eR_N) excess current almost constant at $eV > \Delta$ must be present in the $I - V$ curves. It testifies to the absence of both the potential barrier in the contact and the noticeable heating effects.

Fourth, the common criterion is that the point-contact resistance R_N should be as high as possible. However, the nonlinearity in the contact conductivity usually diminishes with increasing of R_N. Therefore, there is always some optimum resistance depending on the material and on the preparation method. Another important thing is that a diameter for the S-c-N contact calculated by the Sharvin formula from contact resistance in the superconducting state, where the Maxwell term from high-resistivity high-T_c material vanishes, must correspond to the diameter calculated by the Maxwell formula using the contact resistance in the normal state. Often, the discrepancy is large. This points (1) to the deviation of the contact model from the orifice, (2) to the nonequal filling of the point contact by S or N metal, or (3) to the degradation of the superconducting properties in the contact (or in a part of the contact) with respect to the bulk one.

Fifth, by study of the phonon features, their position should be robust against change of temperature or magnetic field; only their intensity and/or width may depend on these parameters. Contrary, stray spikes in dV/dI or $d^2V/dI^2(V)$, caused by superconductivity degradation along the current lines, move toward zero with temperature or magnetic field increase, remaining almost unchanged in shape.

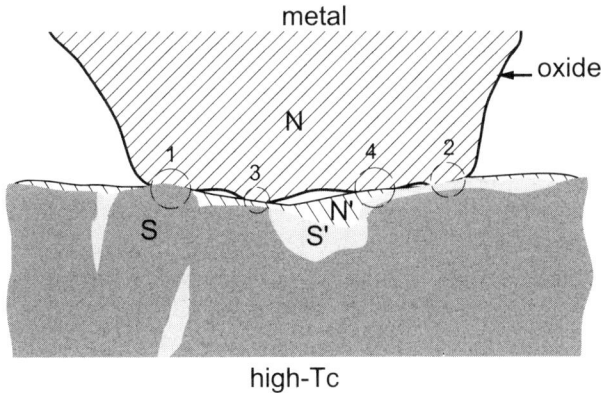

Fig. 13.1. Model of the S-c-N point contact between the high-T_c material and a normal metal: 1 – ideal case of the S-c-N contact, 2 – SS′-c-N contact with superconducting region S′ with lower critical parameters, 3 – SN′-c-N contact with a normal N′ layer, 4 – combination of the latter two models SS′N′-c-N contact.

As in the case of simple metals, the mechanical contact for the high-T_c samples is formed over the large area, whereas the electric contact(s) may only emerge in some place(s) free of dielectric oxide film or semiconducting (degraded) surface layer (see Fig. 13.1), which is typical for the interface with the high-T_c materials. Additionally, superconducting properties of high-T_c are usually degraded close to the surface; therefore, e. g., the following situations can take place in case of high-T_c–normal metal contact: (1) ideal case of S-c-N contact, (2) contact with normal N′ layer (SN′-c-N), (3) contact with superconducting region S′ with lower critical parameters (SS′-c-N), and

(4) combination of the latter two models, namely, SS'N'-c-N contact (all of these cases are shown in Fig. 13.1). Even in the case of a sharp or an ideal S-N boundary, the potential barrier can emerge because of the remarkable difference in the electronic characteristic of the high-T_c and the normal metal. This means that the transparency of the boundary between the metals may be well below unity. If there are no centers destroying Cooper pairs at the interface, the proximity effect "pulling" superconductivity into the normal metal must take place as well. This can lead to an appearance of an additional sharp minimum in dV/dI at $V = 0$, resolved for some spectra presented in this chapter. The depth of this dip increases with the lowering temperature.

It should be mentioned that current spreading in the point contacts with the high-T_c superconductors may be specific. Considering the anisotropic structure of the high-T_c materials with high conducting or superconducting layers (planes) for some directions (usually in the basal plane), the spreading of electrons will be dictated by these ways. If an easy line for the current will be interrupted by some defects, phase slip centers emerge along these lines, at which the electrochemical potential of pairs undergoes a jump. This can lead to the additional manifestation mechanism of the phase singularity in the $I - V$ curves as well as to the nonequilibrium suppression of the excess current at phonon energies, which apparently constitutes the basis of the EPI spectroscopy for these materials. The emergence of the phase slip centers can be proved by external electromagnetic irradiation of contact, leading to the appearance of peak series on dV/dI [Yanson et al. (1989)].

Another point worth noting is that the current distributes itself in the constriction along the route corresponding to the phases with maximal conductivity. Correspondingly, the maximum gap value should usually emerge in the spectra. The question is in what way will the manifestation of gap features occur. The dV/dI curves for contacts with high-T_c are in general different from those expected from the BTK theory for the S-c-N contact. It is not surprising, because as was mentioned above, the theoretical conditions for the electron transport are not fulfilled in the high-T_c materials point contacts. However, the systematic appearance of the double-minimum structure around zero bias in dV/dI with the width of the order of expected gap value may signify that the BTK theory is applicable for this restricted energy region. Therefore, the estimation of the gap value from the position of the mentioned minima is widely used by point-contact measurements with high-T_c.

13.2 La$_{1.8}$Sr$_{0.2}$CuO$_4$

Figure 13.2 shows the $I - V$ curves and their derivatives for La$_{1.8}$Sr$_{0.2}$CuO$_4$-Cu contact at different temperatures and magnetic fields. The characteristics are weakly asymmetric. Here the asymmetry is apparently caused by the dependence of the height of the small potential barrier at the interface on the

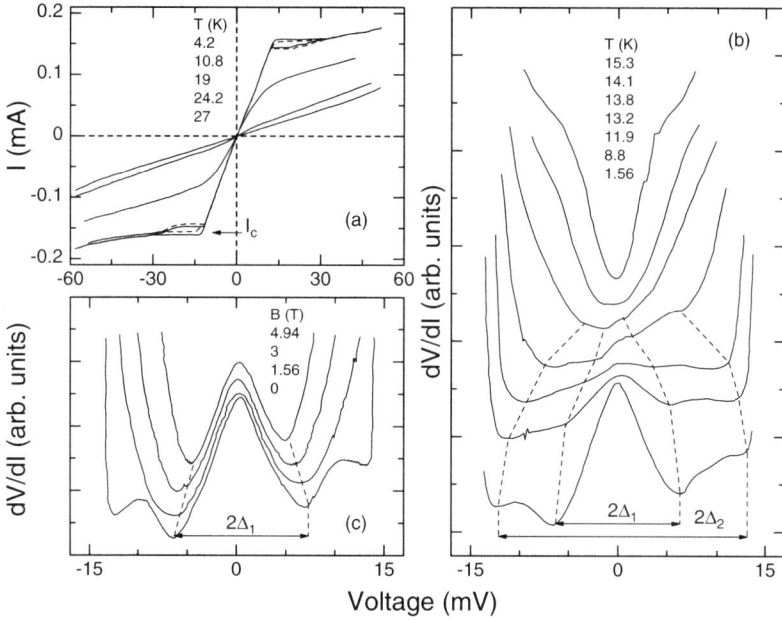

Fig. 13.2. $I - V$ curves (a) and their first derivative (b) of a La$_{1.8}$Sr$_{0.2}$CuO$_4$-Cu contact taken at different temperatures. The normal state resistance is about 1 kΩ. (c) dV/dI dependencies of the same contact in a magnetic field at 1.56 K. Data taken from Yanson et al. (1989).

polarity of the bias voltage by analogy with the tunneling junctions, where asymmetry is very pronounced as a rule. The gap value taken from the position of the minimum in dV/dI is on average 6.5 meV, and the reduced gap $2\Delta_0/k_B T_c^*$ is 5.5. Here T_c^* is the critical temperature at which the superconducting features vanished in the dV/dI curve. It is seen from Fig. 13.2(b) that dV/dI contains the second double-minimum structure at larger bias, which is supposed to be caused by the presence of a second larger gap. The latter may also correspond to the superconducting remote region unperturbed by the contact (see contact model 2 or 4 in Fig. 13.1). The value of the larger gap is approximately twice as large as the first gap with $2\Delta_0/k_B T_c^* \simeq 11$. The temperature dependence of dV/dI presented in Fig. 13.2 gives the critical temperature about 14 K for the first gap, whereas for the second one, $T_c^* \simeq 28$ K, and it is close to the bulk T_c. Figure 13.3 depicts temperature dependence of the mentioned gap structures (viz. minima) along with the excess current. It is seen that all dependencies differ considerably from the prediction of the BCS theory. This difference for the higher temperatures can be partly because the thermal smearing of the electron distribution function is disregarded and a minima position roughly corresponds to the gap. How-

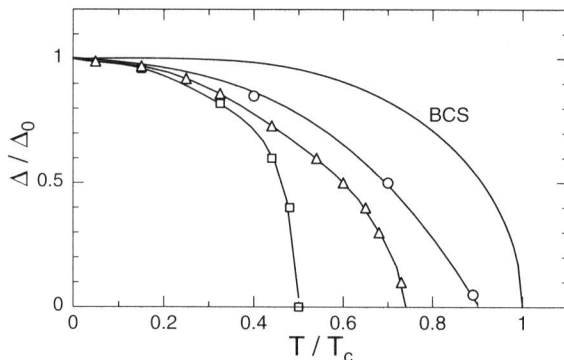

Fig. 13.3. Temperature dependence of the gap minima Δ_1 (squares) and Δ_2 (triangles) as well as excess current (circles) for the contact from Fig. 13.2 along with BCS dependence (solid line). The solid lines connect symbols for clarity. Data taken from Yanson et al. (1989).

ever, the temperature dependence of the gap differs from the BCS curve even in the low-temperature region.

The larger gap minima vanish already in a moderate magnetic field, and smaller ones shift about 30% to the lower voltages [Fig. 13.2(c)]. Such behavior can be naturally explained by the following supposition. Consider a contact model corresponding to the case 2 (N-cN′S) with proximity-induced superconductivity in N′. In this opportunity, the magnetic field does not change Andreev-reflected hole current, when this reflection occurs directly at the interface in the contact center and hole returns to the normal metal in any case. The second boundary with unperturbed material with higher Δ is remote from the interface with the normal metal; therefore, the magnetic field bends the quasiparticles trajectories and hinders backflow of the holes in the normal metal. This leads to a retroreflected hole current decreasing followed by suppression of the Andreev reflection and hence to vanishing of the gap features in the spectra.

Yanson et al. (1987a) discussed voltage jump in the $I - V$ curves from Fig. 13.2(a) accompanied by a small hysteresis. This jump corresponds to some critical current I_c. The presence of the excess current with a value close to I_c evidences that superconductivity in the point contact is not completely destroyed. Apparently, the region of the order parameter slippage is formed, where the quasiparticle transport current above I_c is generated. Hence, I_c corresponds to the maximal supercurrent, the density of which is estimated by $j_c \simeq I_c/d^2 \simeq I_{exc}/d^2 \sim 10^8 \mathrm{A/cm}^2$. This value is of the same order of magnitude as the pair-breaking current density evaluated from the standard formula (12.6) in the theory of superconductivity. Such a large value of the critical current is because by point contact the local properties of the material with the size of the order of the contact diameter are investigated. The contact size was estimated in a few nanometers in the examined case, which is of the order of the superconducting coherence length. Vortices cannot penetrate or move in a such small region that discards energy dissipation.

As was shown in Section 12.1, the phonon features can be seen in dV/dI of "dirty" superconducting point contacts in the form of sharp maxima or N-shape features in d^2V/dI^2. This mechanism, apparently, is responsible for the manifestation of the phonon singularities in the conductivity of contacts with high-T_c materials shown in Fig. 13.4. At low temperatures, the height

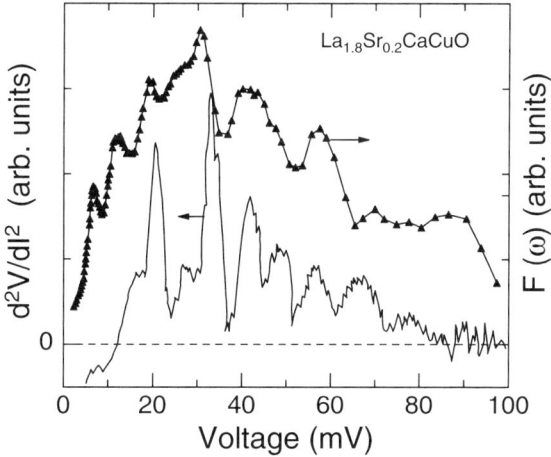

Fig. 13.4. Phonon structure in d^2V/dI^2 for La$_{1.8}$Sr$_{0.2}$CuO$_4$-Cu contact with $R_N \simeq$ 82 Ω from Yanson et al. (1987b). The symbols show the density of the phonon states measured by neutron scattering by Belushkin et al. (1989).

of the spikes in dV/dI of La$_{1.8}$Sr$_{0.2}$CuO$_4$-Cu contacts amounts to the extent of 10% of the normal state resistance, which is nearly one order of magnitude higher than the intensity of the similar spikes for the Ta-Cu contacts (see Fig. 12.4). It was shown that the temperature increase up to 10 K does not remarkably shift the position of the peculiarities above 15 mV (Fig. 13.4), but d^2V/dI^2 curves undergo large changes below this voltage in the so-called gap region. As mentioned by Yanson et al. (1989), the phonon structure can be successfully reproduced for the contact given, but it varies slightly from contact to contact. Nevertheless, some general features of the structure can be reproduced for different contacts. First of all, the most intensive structure is in the energy interval between 20 and 50 mV, where the density of the phonons is maximal according to the neutron measurements (see Fig. 13.4). The peak intensity drops sufficiently at the energy of about 90–100 meV, which is above the maximal phonon frequency. A slight difference in the exact position of the peaks in d^2V/dI^2 for different contacts compared with the maxima in the phonon density of states can be because by point contact, a local characteristic of EPI function for materials having strong anisotropic structure is measured. Finally, $-\mathrm{d}^2 I/\mathrm{d}V^2$ spectra of $La_{1.8}Sr_{0.2}CuO_4$-Al contacts measured by point-contact tunneling [Ekino (1992)] have in general similar structure with sharp peaks positioned close to the maxima in the phonon density of states.

Table 13.1. Superconducting gap in the high-T_c materials determined from the $\mathrm{d}V/\mathrm{d}I\,(V)$ curves of point contacts by double-minimum (DM) position and harmonic and subharmonic gap structure (HGS, SGS), using BTK or modified BTK fit. Two data in the one cell correspond to the presence of two gaps simultaneously (&) or gap distribution (–). Here T_c is the temperature at which the superconducting features vanish in $\mathrm{d}V/\mathrm{d}I$.

Sample	$2\Delta_0$ meV	$2\Delta_0/k_B T_c$	Method	Reference
$La_{1.8}Sr_{0.2}CuO_4$	14&26	6&11	DM	Yanson et al. (1989)
$YBa_2Cu_3O_{7-\delta}$	40	5.2	DM	Goll et al. (1992)
	42±2	5.4	SGS	Zimmermann et al. (1996)
$RBa_2Cu_3O_{7-\delta}$ (R=Y, Yb)	56–66	7.2–8.8	SGS	Aminov et al. (1994)
$SmBa_2Cu_3O_{7-\delta}$	20–60	6–8	HGS	Akimenko et al. (1989)
$Bi_2Sr_2CaCu_2O_{8+x}$	28	8	mBTK	Plecenik et al. (1994)
$(Hg_{0.7}Cr_{0.3})Sr_2CuO_4$	20–30	3.4–5.1	mBTK	Rybaltchenko et al. (1996c)

13.3 1:2:3 compounds

The $\mathrm{d}V/\mathrm{d}I$ curves with well-pronounced double-minimum structure similar to that presented in the previous section were measured for $YBa_2Cu_3O_{7-\delta}$ contacts (with Cu or Ag) by Yanson et al. (1988b) (see Fig. 13.5). Differential resistance $\mathrm{d}V/\mathrm{d}I$ of the contacts often decreases with voltage increasing beyond the gap minimum]Fig. 13.5(b)] representing so-called semiconductor-like behavior. $\mathrm{d}V/\mathrm{d}I$ shown in Fig. 13.5(a) and (b) exhibits an additional zero-bias sharp minimum, which can be suppressed by the magnetic field, microwave irradiation, or temperature increase, whereas the overall gap structure does not practically change. As we mentioned above, this minimum is, apparently, connected, with the proximity effect, i.e., inducing of the superconducting region in the normal metal near the orifice. The spectrum, shown in Fig. 13.5(c), demonstrates a simple flat minimum around $V = 0$ instead of a double-minimum structure. Additionally, this contact has a higher excess current compared with the previous one. According to the BTK theory, such behavior is characteristic for the barrierless interface between two electrodes. The latter is possible when both materials have almost the same Fermi parameters. It points out the possibility of the existence of certain crystallographic directions in $YBa_2Cu_3O_{7-\delta}$ with typical metallic conductivity. Meanwhile, the discussed dV/dI curve has too steep an increase above the minimum to be fitted well by the BTK model with $Z = 0$ [Fig. 3.15(a)]. The maximal gap

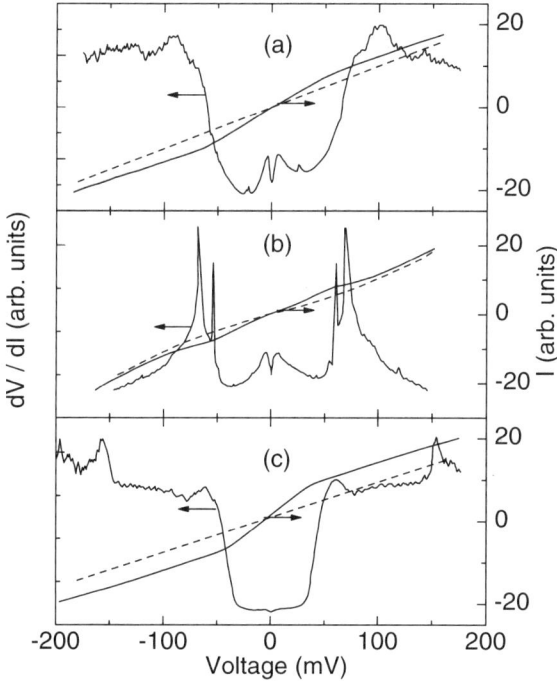

Fig. 13.5. $I-V$ curves and their first derivative of some $YBa_2Cu_3O_{7-\delta}$-Ag contacts measured at $4.2\,K$. The normal state resistance R_N (Ω) is 6.3 (a), 6.9 (b), and 9 (c). Dashed curve in (b) is the $I-V$ characteristic in the normal state at $82\,K$. Dashed straight lines in (a) and (c) are drawn parallel to the corresponding $I-V$ curves at the high voltage to show the presence of the excess current. Data taken from Yanson et al. (1988b).

value estimated from the double-minimum position is about $40\,meV$, which leads to reduced gap $2\Delta_0/k_BT_c^*$ close to 12. The temperature dependence of the gap structure deviates considerably from the BCS behavior as in the previous case of LaSrCuO.

YBa$_2$Cu$_3$O$_{7-\delta}$ single crystal with $T_c \simeq 90$ K was studied by Rybaltchenko et al. (1990). In this case, the metallic-like increase of the differential resistance with the bias voltage was observed as a rule (see Fig. 13.6). A peculiar

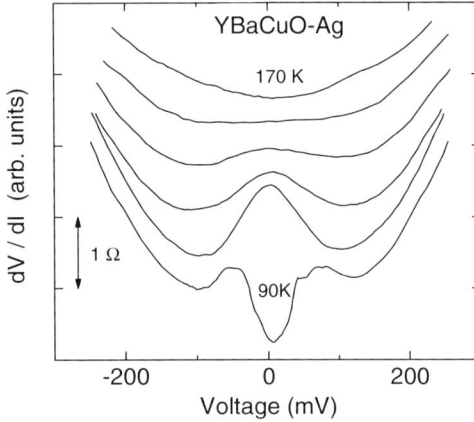

Fig. 13.6. dV/dI dependencies for contact between YBa$_2$Cu$_3$O$_{7-\delta}$ single crystal and Ag measured above T_c at different temperatures T (K): 170, 160, 142, 120, 96, and 90 from the top curve to the bottom one. The curves are offset vertically for clarity. The contact resistance is $4\,\Omega$ at $V = 0$ and 96 K. Data taken from Rybaltchenko et al. (1990).

behavior of dV/dI is seen in the temperature range above T_c, which demonstrates that a broad double-minimum structure is developing. This leads to the speculation about manifestation of a gap in the excitation spectrum of local pairs above T_c assumed in some theoretical models [e. g., Kulik (1988)] or so-called pseudo-gap in the case of underdoped samples, which can be larger compared with the conventional BCS gap (see Deutscher (1999) for the speculations). The excess conductivity of point contacts between 100 and 200 K measured for the same compound was supposed to be caused by fluctuating pairing of electrons above T_c. Above 120 K, the normalized excess conductivity in a sample studied corresponds to the two-dimensional model.

The unusual effect of the excess current suppression in the $I - V$ curves in one polarity and the restoration of its initial value for the opposite polarity was observed in RBa$_2$Cu$_3$O$_{7-\delta}$-Ag (R=Y, Ho) contacts exceeding some critical voltage of about 250 mV (see Fig. 13.7). Such effect has never been observed for contacts with the traditional superconductors. Initial curve 1 in Fig. 13.7 is stable for both voltage polarities below 250 mV. At reaching the bias about $V_c \simeq - 280$ mV, the transition to one of the other curves ($\rightarrow 2 \rightarrow 3 \rightarrow 4 \rightarrow 5$) occurs with lower or even negative excess current depending on the time for which the contact is subjected to this bias voltage.

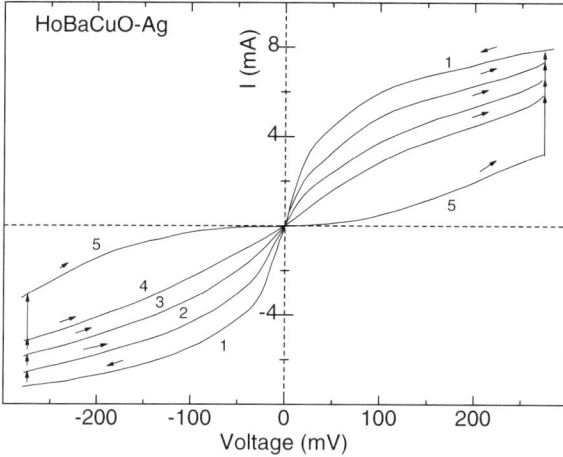

Fig. 13.7. Reversible modification of the $I - V$ curves for $HoBa_2Cu_3O_{7-\delta}$-Ag contact after the attainment of the critical bias voltage of about $|V_c| \simeq 270\,\text{mV}$ (see text for the further explanation). Voltage polarity corresponds to the Ag electrode. Data taken from Rybaltchenko et al. (1991).

The transition rate decrease for the higher voltages, and instant switching may be observed for large biases. The reverse of the bias sign to the opposite voltage $(+280\,\text{mV})$ results in excess current restoration and the $I - V$ curve from type 5 to $4 \rightarrow 3 \rightarrow 2 \rightarrow 1$ (see Fig. 13.7) depending on the time. It should be mentioned that all curves remain stable if the voltage bias does not exceed $\pm\,250\,\text{mV}$. For the different contacts, V_c can vary between 250 and 320 mV. Additionally, the magnetic field up to 4.5 T does not practically affect the critical voltage and the excess current. Measurements of the $I - V$ curves in a wide temperature range as high as 120 K also showed reversible changes; *viz.* the excess current will remain unchanged after returning to the initial temperature if the bias voltage does not exceed the critical voltage. Moreover, the critical temperature determined for the contact with partially suppressed excess current is close to its initial value. The emergence of such metastable states is not accompanied by notable variation of the gap structure in the dV/dI curves. In the author's opinion, the nature of the extraordinary behavior of $I - V$ curve is connected with the oxygen subsystem. The migration of oxygen can take place in the crystal lattice of the high-T_c material near the contact interface stimulated by the strong electric field and high current density. This leads to the local variation of the superconducting properties of material in the point contact region.

Analogous effect was observed by Plecenik et al. (1996) in point contacts between $YbBa_2Cu_3O_{7-\delta}$ thin film and Au. By applying a bias of about 500 mV to the contact during 5–10 s, the differential resistance change from the tunneling-like behavior to the metallic one was observed. If the same

opposite voltage was applied, the reversible restoration of the metallic conductivity would occur. This was explained by modification of the interface
of the high-T$_c$ material caused by oxygen replacement in the unit cell analogously to the latter case.

Akimenko et al. (1991) analyzed a gap-related structure, namely, double
minimum around $V = 0$ or single minimum at $V = 0$ in dV/dI for the high-
T$_c$ compounds, supposing the gap distribution in the contacts. The authors
consider that the origin of such a distribution is a strong gap anisotropy.
They built the histogram (see Fig. 13.8) showing a broad distribution of the

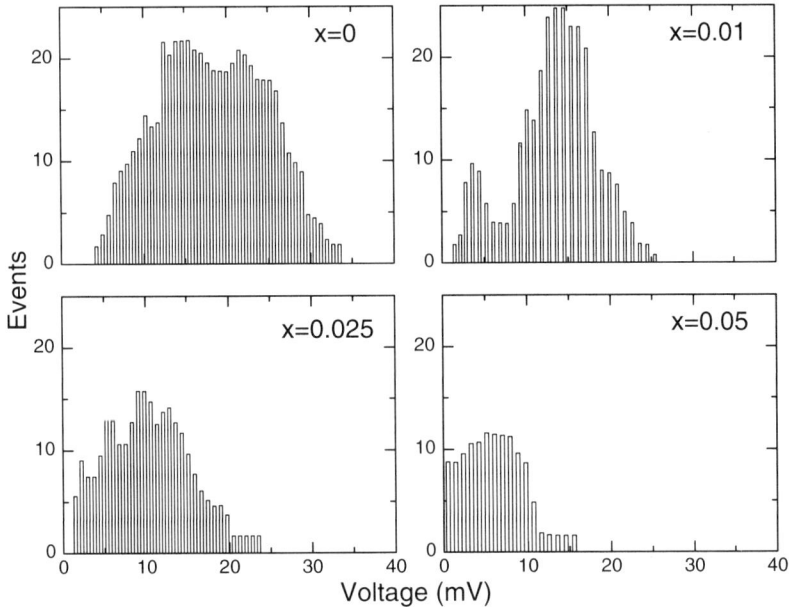

Fig. 13.8. Histogram for the gap-related minimum occurrence in the dV/dI curves
for YBa$_2$(Cu$_{1-x}$Zn$_x$)$_3$O$_{7-\delta}$-Ag contacts. The histograms were built by choos-
ing a half-width of the minimum as a single event. Data taken from Akimenko
et al. (1991).

gaps with well-defined upper and lower limits. With decreasing of the critical
temperature of samples from 92 to 40 K by replacing of Cu by Zn, the gap
distribution also shifts to the smaller voltages. In further experiments with
GdBa$_2$Cu$_3$O$_{7-\delta}$, Akimenko et al. (1992) measured similar double minima in
dV/d$I(V)$ curves; however, they were wide. He compared the results with
calculation by virtue of the BTK model, accounting for a simple rectangular
gap distribution between 7 and 35 meV. Figure 13.9 shows that calculations
describe qualitatively the broad minima in the experimental curves and their
modification by interface transparency decreasing.

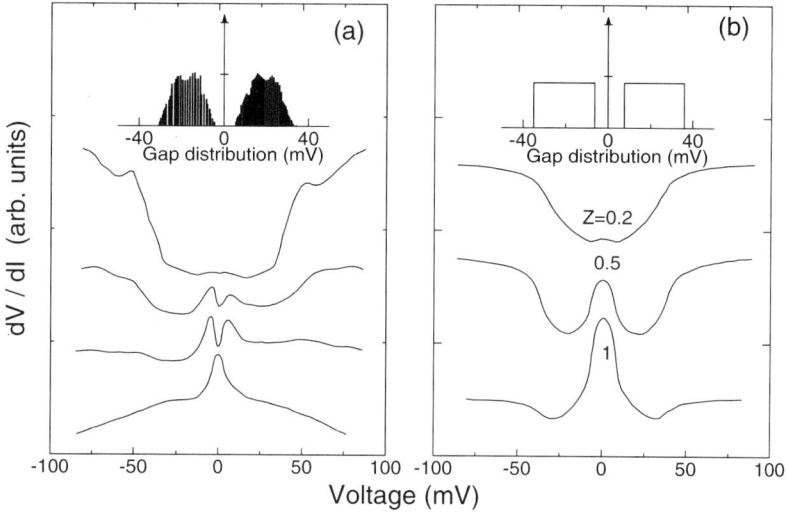

Fig. 13.9. (a) dV/dI curves for a few $GdBa_2Cu_3O_{7-\delta}$-Ag contacts along with a histogram of distribution of the gaps. (b) Calculation according to (3.30) by summarizing over 29 BTK-like spectra with gap values taken equidistant between 7 and 35 meV. Parameter Z is shown for each curve. Data taken from Akimenko et al. (1992).

Nearly BTK-shaped dV/dI curves that exhibit very few additional peculiarities were presented for $YBa_2Cu_3O_{7-\delta}$-Ag contact by Goll et al. (1992) (Fig. 13.10). Derived from this measurement, the temperature dependence of the gap, *viz.* minima position in dV/dI, shows that the latter decreases faster than for the standard BCS superconductors with temperature increasing. Similar temperature dependence of $\Delta(T)$ (see Fig. 13.11, inset) was established by measuring the subharmonic gap structure up to n = 4 for $YBa_2Cu_3O_{7-\delta}$ break-junction by Zimmermann et al. (1996). The samples were broken at helium temperatures. This allowed them to receive good surface quality without oxide contamination and degradation of properties caused by the oxygen diffusion from the surface of high-T_c. The gap value and its temperature dependence was received by comparison with MAR theory for the subharmonic gap structure in the S-N-S contact (see Section 3.7.2). These data are similar to the results shown in Fig. 13.10. Taking into account that using different methods, both mentioned experiments give experimental curves close to that predicted by BTK or MAR theories along with almost the same Δ and similar temperature dependence of Δ; they, apparently, reflect the real behavior of the order parameter in $YBa_2Cu_3O_{7-\delta}$ samples.

Experiments on Ag/YBaCuO bilayers by Hoevers et al. (1988) and Elesin et al. (1993) are worth pointing out. Here *in situ* grown high-T_c film was covered by a thin Ag layer to avoid degradation and oxidation of the interface.

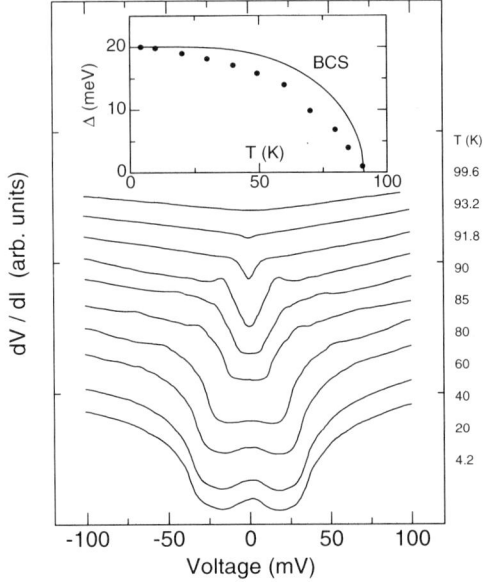

Fig. 13.10. Temperature dependence of the dV/dI curves for a $YBa_2Cu_3O_{7-\delta}$-Ag contact. The curves are offset vertically for clarity. Inset shows temperature dependence of the gap defined as position of the minima in dV/dI. The solid line represents BCS dependence. Data taken from Goll et al. (1992).

Then the point contact between bilayer and sharpened Ag or Au thin wire was formed. The authors observed distinct Andreev reflection features in the dV/dI curves yielding the reduced gap $2\Delta_0/k_B T_c^* \simeq 3.5$. This is sufficiently lower than in the experiments we mentioned above. The reason for low Δ may be a proximity effect at the Ag/YBaCuO interface. Indeed, by measuring Ag-Ag/Pb contacts with Pb covered by Ag film as shown by van Son et al. (1987), the intensity of the Andreev reflection and the double-minimum width in dV/dI decreased with increasing of Ag film thickness.

The break junctions with 90 K-phase single crystalline $RBa_2Cu_3O_{7-\delta}$ (R = Y, Yb), where subharmonic gap structure up to n = 8 was observed, have been investigated by Aminov et al. (1994). These results give the larger value of about 30 meV for the gap with $2\Delta(0)/k_B T_c \simeq 8$.

13.4 Bi and Hg cuprates

Point contacts between thin $Bi_2Sr_2Ca\ Cu_2O_{8+x}$ film with zero-resistance critical temperature 80 K and Au have been investigated by Plecenik et al. (1994). The dV/dI curves in Fig. 13.12 show a pronounced BTK-like double-minimum structure at $|V| < 40$ mV. The modified BTK-fit with broadening parameter

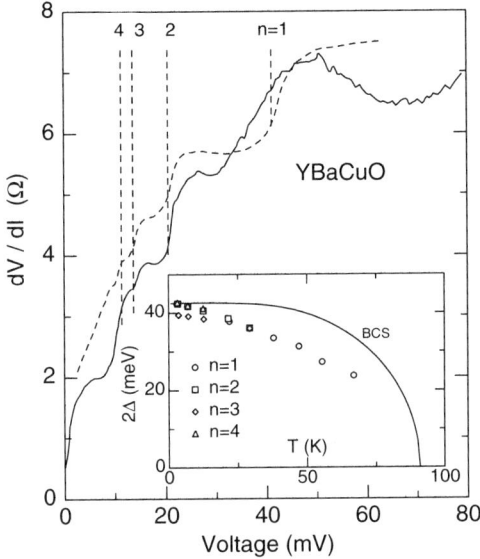

Fig. 13.11. dV/dI (solid curve) for the YBa$_2$Cu$_3$O$_{7-\delta}$ break junction at 4.2 K. The dashed curve is calculated according to the MAR theory with $\Delta = 21$ mV and $Z = 0.3$. Inset shows the temperature dependence of the subharmonic gap structure. The solid line represents BCS dependence. Data taken from Zimmermann et al. (1996).

Γ describes measured dependences well. Derived values of the energy gap decreased from 27 to 18 mV by increasing temperature from 4.2 to 60 K, which is also somewhat faster compared with the BCS theory. The reduced gap is found to be $2\Delta_0/k_B T_c \simeq 8$. The authors suppose that the value of $Z \simeq 0.5$ obtained in most of cases is caused by the mismatch in the Fermi velocity v_F between electrodes. This leads to the estimation of v_F in Bi$_2$Sr$_2$Ca Cu$_2$O$_{8+x}$ about a factor 2.6 lower compared with that in Au. Substitution of $v_F = 5 \times 10^5$m/s and $\Delta = 30$ meV into the well-known equation for the coherence length

$$\xi = \hbar v_F / \pi \Delta \tag{13.1}$$

yields $\xi = 3.5$ nm. However, as we mentioned in the previous chapter, $Z \simeq 0.5$ is found for the S-c-N contacts with different superconductors. This questioned the method of determination of v_F supposing that the reflection at the interface is caused by the difference in the Fermi velocity only. Nevertheless, the mentioned superconducting parameters as stated by Plecenik et al. (1994) are in reasonable agreement with the data of the other authors.

The Hg-based family of the high-T$_c$ materials attracted attention, owing to the highest critical temperatures above 130 K for samples containing three CuO$_2$ layers. However, because of the low chemical stability of the HgBaCuO superconducting phase at ambient pressure, attempts to substi-

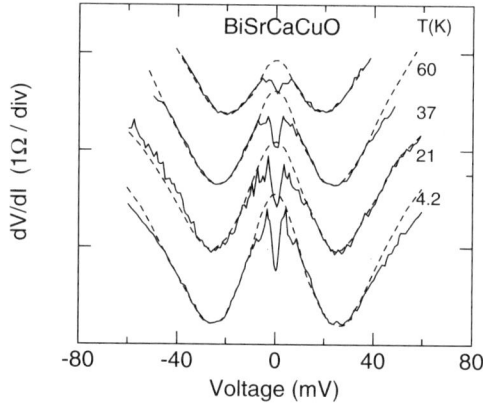

Fig. 13.12. dV/dI curves for a $Bi_2Sr_2CaCu_2O_{8+x}$-Au point contact with $R_0 \simeq 2\,\Omega$ at different temperatures. The dashed curves are calculated according to (3.30) and (3.32). The curves are offset vertically for clarity. For the lowest curve, the fit parameters are $\Delta = 24\,\text{meV}$, $Z = 0.56$, and $\Gamma = 12\,\text{meV}$. Data taken from Plecenik et al. (1994).

tute partially Hg and Ba atoms by less chemically active elements were undertaken. Rybaltchenko et al. (1996c) investigated point contacts based on $(Hg_{0.7}Cr_{0.3})Sr_2CuO_4$ compound with improved chemical stability but lower $T_c = 68\,\text{K}$. The hallmark feature of the $dV/dI(V)$ curves for contacts between $(Hg_{0.7}Cr_{0.3})Sr_2CuO_4$ and Ag is an intensive zero-bias maximum (its value is about 10% of the zero-bias resistance) and symmetrically placed minima, which position shifts from about $\pm 19\,\text{mV}$ to $\pm(10\text{--}13)\,\text{mV}$ by increasing point-contact resistance from 14 to 126 Ω. The fitting of the experimental curves by modified BTK model yields good results for the low-temperature region (Fig. 13.13). However, with temperature increasing, the evident discrepancy appears. It was explained by stronger gap decrease than that expected by the theory and by possible decrease of the tunnel-like barrier height (decrease of Z parameter) with temperature. For most of contacts investigated, the gap value ranged between 10 and 15 meV corresponding to the ratio $2\Delta_0/k_B T_c$ from 3.4 to 5.1.

13.5 Borocarbides

The rare-earth RNi_2B_2C (R=Er, Ho, Dy) borocarbides exhibit moderate critical temperature, however, showing an interesting interplay between superconductivity and magnetic order. For example, Ho-based compounds show reentrant to the normal state transition of magnetization and resistivity in the temperature range just below the antiferromagnetic transition at $T_N = 5\,\text{K}$, which is below the superconducting critical temperature $T_c \simeq 9\,\text{K}$. The

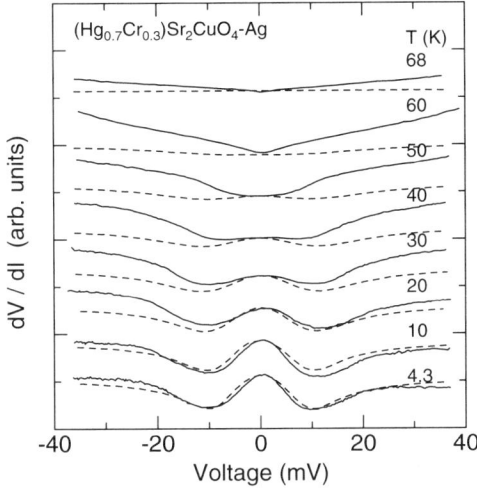

Fig. 13.13. dV/dI curves for a $(Hg_{0.7}Cr_{0.3})Sr_2CuO_4$-Ag point contact at different temperatures. The dashed curves are calculated by (3.30) and (3.32) with $\Delta(T)$ following the BCS curve. The curves are offset vertically for clarity. For the lowest curve, the fit parameters are $\Delta = 10$ meV, $Z = 0.65$, and $\Gamma = 4$ meV. Data taken from Rybaltchenko et al. (1996c).

dV/dI curves of $HoNi_2B_2C$ are shown in Fig. 13.14. It is seen that a pronounced minimum in dV/dI appears at temperature below T_N, whereas only a gentle structure exists between T_N and T_c. By lowering temperature, typical Andreev-reflection double-minima appear. They can be fitted satisfactorily by the BTK model. However, it should be noted that the depth of the minimum is about 10% of the normal resistance, which is sufficiently lower than the expected theoretical decrease of R_N up to 50% at zero bias. Fit for the different contacts yields the gap $\Delta = 1.04 \pm 0.06$ meV with reduced gap $2\Delta_0/k_BT_c$=3.7 by using the critical temperature of about 6.5 K, at which the clear minimum appears.

BTK-like dV/dI curves were also measured for the other both magnetic $ErNi_2B_2C$ and nonmagnetic magnetic YNi_2B_2C compounds. In this case, the temperature at which the structure in dV/dI vanishes coincides with the bulk T_c. Below T_c, a superconducting state with reduced gap $2\Delta_0/k_BT_c$=3.7 develops for both compounds. As it is shown in Fig. 13.15, Δ correspond well to the BCS relation for all three compounds if one takes T_c^* instead of the bulk T_c for Ho-based samples. This means that weak-to-moderate coupling is responsible for the superconductivity of the borocarbides.

The EPI in mentioned systems with R = Ho and Y along with isostructural nonsuperconducting $LaNi_2B_2C$ were studied by Yanson et al. (1997) to clarify the pairing mechanism in these compounds. They found strong interaction of electrons with low-energy "soft-phonon" modes for Y- and Ho-based

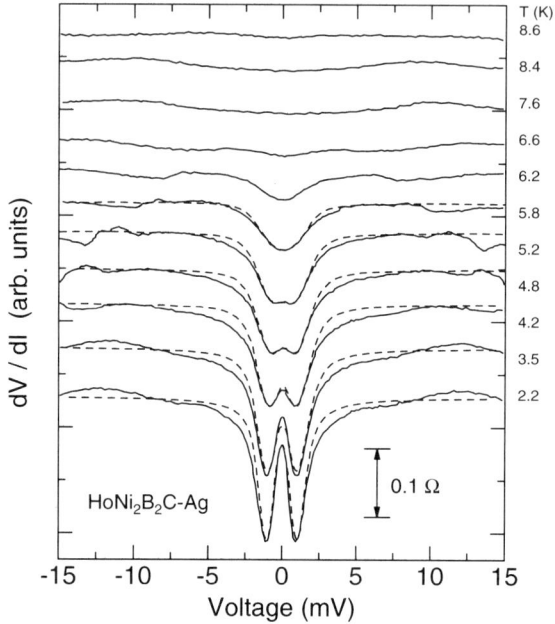

Fig. 13.14. dV/dI curves of a HoNi$_2$B$_2$C-Ag point contact with $R_N = 2.65\,\Omega$ at different temperatures. The normal state background was subtracted from the curves, and they are offset vertically for clarity. The dashed curves with a different vertical scaling factor between 0.16 and 0.29 for each curve are calculated by (3.30). Data taken from Rybaltchenko et al. (1996b).

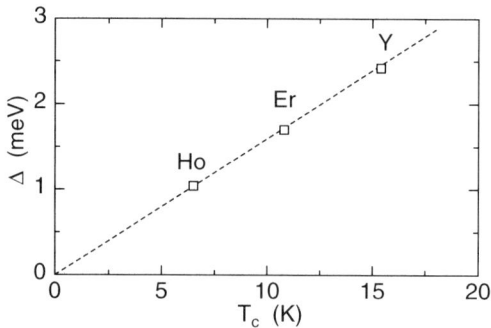

Fig. 13.15. Energy gap Δ as a function of T_c for the RNi$_2$B$_2$C (R = Y, Ho, Er) compounds. In the case of Ho, T_c is taken equal to $T_c^* \simeq 6.5\,\mathrm{K}$. The straight line gives relation $2\Delta/k_B T_c = 3.7$. Data taken from Rybaltchenko et al. (1996a).

superconducting compounds unlike for nonsuperconducting LaNi$_2$B$_2$C. Figure 13.16 shows that this interaction manifests itself in the point-contact spectrum by a strong integral intensity increase of the low-energy part of the spectra in the case of Ho and Y. Surprisingly, the suppression of this low-energy part of the spectra was observed in the order of a few Tesla magnetic field. This can be related to the intimate connection between EPI and magnetism for HoNi$_2$B$_2$C magnetic superconductor, and the influence of the superconducting state on the low-frequency phonon modes for nonmagnetic YNi$_2$B$_2$C [Bullock et al. (1998)].

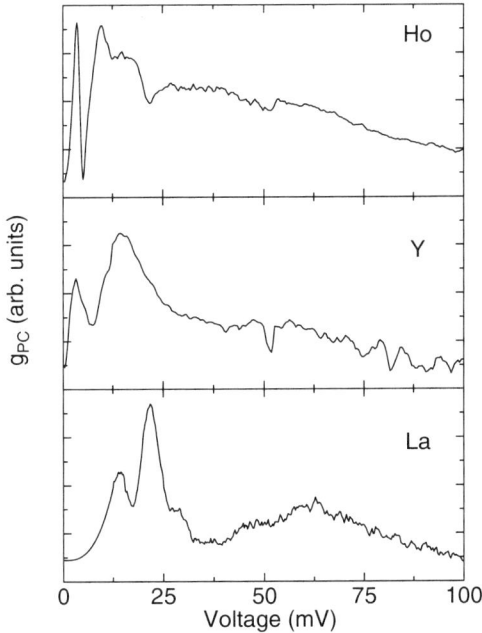

Fig. 13.16. EPI function for RNi$_2$B$_2$C (R = Ho, Y, La) compounds received from d^2V/dI^2 point-contact spectra after subtracting of the background. Data taken from Yanson et al. (1997).

In the following papers, Yanson et al. (2000) investigated both gap structure and electron–quasiparticle interaction in DyNi$_2$B$_2$C. They found gap values satisfying the BCS behavior similar to the above-mentioned borocarbides (both magnetic and nonmagnetic) in spite of the magnetic order with $T_N = 10.5$ K $> T_c \simeq 6$ K. The electron–quasiparticle interaction spectrum in DyNi$_2$B$_2$C was found to exhibit a low-frequency peak around 5 meV becoming discernible at 15 K and growing in intensity, with constant width as the temperature is lowered. Interaction of conduction electrons with coupled crystal-electric-field-phonon excitations was supposed as a conceivable reason for the observed low-energy structure.

13.6 Organic superconductors

A new family of organic superconductors (BEDT-TTF)$_2$X was discovered about 20 years ago. Since that time, the superconducting critical temperature for these compounds increased, with the rate approximately 0.5 K per year to above 10 K because of varying of the anion group X. Organic materials have a layered structure with a very strong anisotropy in conductivity, which differs by a factor of about 100 along the organic layers with respect to the perpendicular direction. Thus, the organic superconductors can be classified as two-dimensional conductors, the property shared with the many high-T_c superconductors. This results in strong anisotropy both for the phonon spectrum and electron–phonon interaction. The spectra measured by point contacts between β-(BEDT-TTF)$_2$I$_3$ and Cu for two main directions are shown in Fig. 13.17. The main peak at about 17 mV in the curve 1 corresponds well

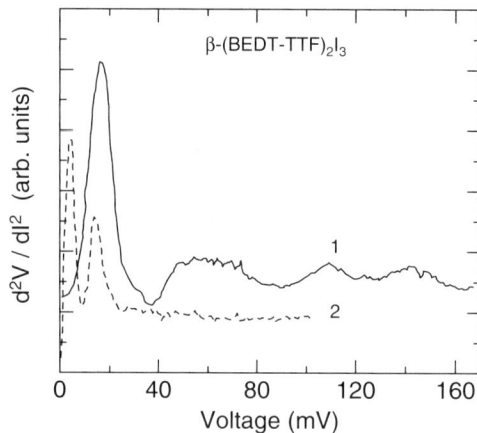

Fig. 13.17. Point-contact spectra of the β-(BEDT-TTF)$_2$I$_3$ - Cu heterocontacts measured at 4.2 K for orientation of contact axis in the plane of layers of organic molecules (curve 1) and for the perpendicular direction (curve 2). Data taken from Kamarchuk et al. (1994).

with the position of the transverse phonons in copper. There are a few less expressed broad maxima at higher voltages between 50 and 150 mV. The energy of these maxima is in the range of intramolecular vibrations as follows from the optical reflection and absorption spectra. The spectrum has only

low-energy peaks at 4 and 15 mV in the perpendicular direction and no in-
dication of Cu phonons. The position of the peaks correlates with Raman
spectrum, which reflects the vibration mode of I_3^- and symmetric stretching
vibrations of iodine atoms. The absence of the Cu phonon peaks in the latter
case is caused by the large difference in the Fermi velocity, which can amount
to three orders of magnitude in this case. In the former case, the presence
of Cu phonons in the spectra is apparently caused by sufficient geometrical asymmetry of contact filled mainly by copper. Similar d^2V/dI^2 spectra
for β-(BEDT-TTF)$_2$I$_3$ were measured in the non-superconducting state with
maxima at 1, 4, and 15 mV by Nowack et al. (1987) a few years before. Subsequent investigations by Nowack (1990) showed that this structure disappears
sharply above 8–9 K, which is close to the critical temperature value of this
material under pressure. Therefore, the authors attributed these peculiarities to be caused by the pressure-induced superconductivity in the contact
regions. Returning to the results obtained by Kamarchuk et al. (1994), measurement of the d^2V/dI^2 spectra are, of course, desirable at higher temperature and in the magnetic field to confirm the phonon or intramolecular
vibrations' nature of the observed peculiarities.

The homocontacts based on α-(BEDT-TTF)$_2$I$_3$ with critical temperature of about 8 K under ambient pressure have been investigated by Weger
et al. (1992). A deep zero-bias minimum was observed for dV/dI with sharp
peaks in the range $V = \pm(8$–$11)$ mV for different contacts, and additional
gentle maximum is seen above 20 mV (Fig. 13.18). The temperature depen-

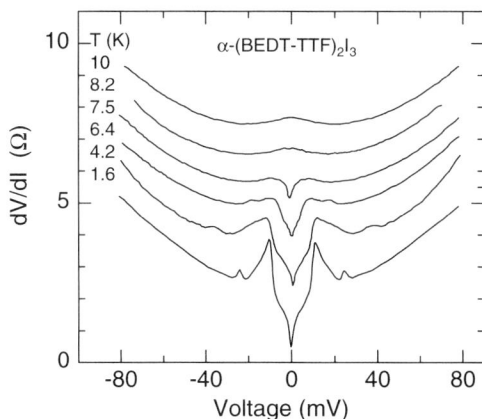

Fig. 13.18. dV/dI curves of α-(BEDT-TTF)$_2$I$_3$ homocontact measured at different
temperatures. The curves are offset vertically for clarity. The differential resistance
is about 5.5 Ω at 80 mV for all curves. Data taken from Weger et al. (1992).

dence of the peak position for some contacts is close to $\Delta(T)$ from the BCS theory. The contacts did not have zero resistance at $V = 0$ so that a region between electrodes must be normal conducting. In this case, there are two N–S boundaries, for both of them Andreev reflection processes can occur. Thus, the authors consider that the peak position corresponds to 2Δ with peak-to-peak separation 4Δ. However, to do so, they received the reduced gap value 3–4 times larger than the standard BCS one. Therefore, an alternative interpretation of peak appearance was proposed. The speculations are based on the presence of the Einstein phonon mode with energy $4\,\text{meV}$ as known from Raman spectra. According to Omelyanchouk et al. (1988), if the interaction of the electrons with the phonon mode is strong, this will lead to the renormalization of the superconductor energy spectrum and developing of peculiarity in $\mathrm{d}V/\mathrm{d}I$ at the energy of phonons plus the gap value. This gives 0–$1.5\,\text{meV}$ for the gap being in line with the BCS value, taking the peak position for the measured curves between 4 and $5.5\,\text{meV}$ and the phonon energy of about $4\,\text{meV}$. Surprisingly, the maxima position is shifted slightly to the higher energy and then smeared out in the magnetic field. As mentioned by the authors, there is still no clear-cut evidence of the spectroscopic nature of these features in $\mathrm{d}V/\mathrm{d}I$.

13.7 MgB$_2$

In the area of superconductivity the entry of the new millennium was distinguished by the fabulous Nagamatsu et al. (2001) report of $40\,\text{K}$ superconductivity in the magnesium diboride. Surprisingly, it was not only the value of the critical temperature in MgB_2, but also the fact that it was reached in a very simple compound and very likely caused by conventional electron–phonon coupling. The latter was proved in the following papers by detailed first-principle calculations of the electron–phonon interaction, giving rise also to retrying to understand the capabilities of the phonon-mediated superconductivity.

It turns out that MgB_2 has two different parts of the Fermi surface as shown by Kortus et al. (2001). One of those represents quasi-two-dimensional cylinders derived from the σ states, which stemmed from the unique case of a covalent band covering the Fermi level. Another part is the three-dimensional tubular network originated from the π states. Correspondingly, Liu et al. (2001) considered two different gaps on the σ and π parts of the Fermi surface in the superconducting state. The larger gap for σ band can be observed in-plane of boron sheets direction, whereas the smaller by a factor 3 gap for π band being nearly isotropic. In the dirty limit, these gaps collapse to one, whose amplitude is in between the two gaps of the clean limit. Support for this picture comes from the numerous experimental observations; among them, the point-contact and tunneling data are the most straightforward. From a number of the early point-contact investigations, Szabo et al. (2001)

were the first who showed the presence of the two-gap structure in the differential conductivity of point contacts, which persists up to the bulk critical temperature.

Combining directional point-contact establishing with the selective removal of the smaller π-band gap by a magnetic field, Gonnelli et al. (2002) proved the existence of two gaps in MgB$_2$ and their anisotropic manifestation in single crystal. They also found that although the large gap follows BCS-like temperature evolution, the small one deviates from BCS behavior at approaching T_c, in agreement with the two-band model.

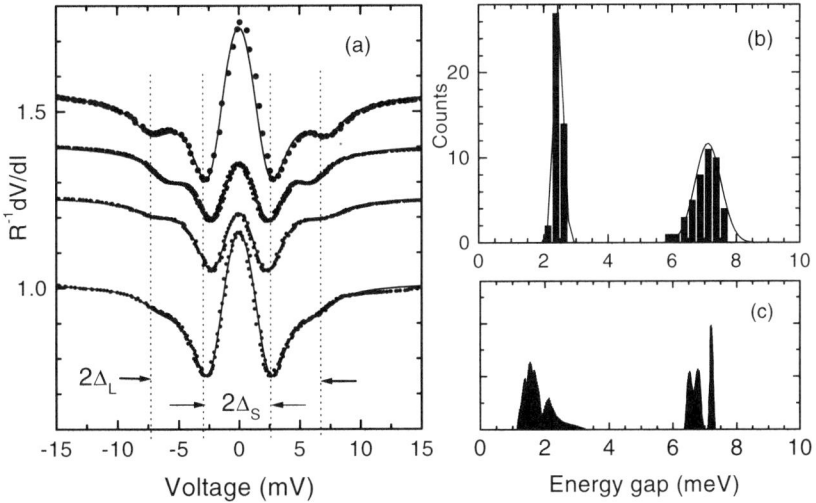

Fig. 13.19. (a) Reduced differential resistance $R_N^{-1} dV/dI$ vs. V with double-gap structure (symbols) measured at $T = 4.2$ K for four MgB$_2$-Ag contacts with resistance between 20 and 47 Ω. Thin lines are theoretical BTK dependencies accounting for both gaps with an appropriate weight factor (between 0.06 and 0.11 for a large gap structure). dV/dI are measured for contacts nominally along the c direction. The curves are vertically offset for clarity. (b) The histograms of superconducting energy gap distribution in c-axis-oriented MgB$_2$ thin film from point-contact study of about 50 contacts. Thin lines show Gaussian fit with maxima at 2.45 and 7 meV. (c) Distribution of gap values over the Fermi surface calculated by Choi et al. (2002). Data taken from Naidyuk et al. (2002).

By this means, MgB$_2$ belongs, or it represents a unique class of multigap superconductors in which two distinct gaps are opened on the different parts (2D and 3D) of the Fermi surface. To illustrate this, some dV/dI are shown in left panel of Fig. 13.19. The curves show a visible two-gap structure, although with shallow features corresponding to a larger gap. This is because the measurements by Naidyuk et al. (2002) were done using c-axis-oriented thin films; that is, the contact axis was aligned mainly in the c direction, for which contribution of the σ band is negligible.

It turns out that the histogram of the gap distribution (presented in the right panel of Fig. 13.19) built on the basis of about 50 spectra has two well-separated and narrow (especially for the small gap) maxima. The observed two distinct maxima in the gap distribution, corresponding to the theoretical calculations of Choi et al. (2002), ruled out surface or multiphase origin of the gap structure and testify about intrinsic superconducting double-gap state in MgB_2.

Thus the specific band structure of MgB_2 with the covalent bands crossing the Fermi level leads to interesting and unusual transport properties of this material both in the normal and superconducting state, which are under spreading study also by utilizing point contacts.

13.8 Sr_2RuO_4

Although Sr_2RuO_4 has the same layered perovskite structure, its electronic states and the nature of superconductivity seems to be different as in the high-T_c compounds. T_c in Sr_2RuO_4 is rapidly suppressed by nonmagnetic impurities, as in most unconventional superconductors. There are also other experimental evidences that the superconducting state of Sr_2RuO_4 below $T_c \simeq 1.5\,K$ is unconventional [Maeno et al. (1994)]. Namely, most of the results are in accordance with symmetry considerations, suggesting that superconductivity in Sr_2RuO_4 is of p-wave (odd-parity) type, analogous to superfluid ^{3}He; albeit views on the gap symmetry of Sr_2RuO_4 are changing in time. Additionally, Sr_2RuO_4 possesses a cylindrical, "quasi-two-dimensional" Fermi surface, and it is well described by the Landau–Fermi liquid model. Accordingly, Sr_2RuO_4 provokes special interest as a unique quasi-two-dimensional spin-triplet superconductor.

The superconducting gap function of Sr_2RuO_4 has been investigated by means of point contacts by Laube et al. (2000). They found two distinctly different types of dV/dI versus V spectra, either with a traditional double-minimum structure or with a single minimum corresponding to the limit of high and low transparency, respectively. The depth of the structures were only 1% to 5% of the background resistance, and the overall amplitude of the spectra was rescaled to compare with the calculations. A zero-temperature gap value of about 2.2 meV higher than the BCS one and temperature dependence, distinct from BCS, for both types of spectra were obtained within a model of p-wave pairing. However, as mentioned by the authors, unambiguous criterion to discriminate between the different spin-triplet pairing states is still lacking, because for three considered symmetries, the theoretically predicted spectra look qualitatively the same.

References

Akimenko A. I., Goll G., von Löhneysen H. and Gudimenko V. A. (1992) Phys. Rev. B **46** 6409.

Akimenko A. I., Goll G., Yanson I. K., von Löhneysen H., Ahrens R., Wolf T. and Wühl H. (1991) Z. Phys. B **85** 5.

Akimenko A. I., Ponomarenko N. M., Gudimenko V. A., Yanson I. K., Samuely P. and Kuš P. (1989) Sov. J. Low Temp. Phys. **15** 686.

Aminov B. A., Wehler D., Müller G., Piel H., Hein M. A., Heinrichs H., Brandt N. B., Chang Sun Hu, Ponomarev Ya. G., Tsokur E. B., Chesnokov S. N., Yusupov K. Ch., Yarygin A. V., Winzer K., Rosner K. and Wolf T. (1994) JETP Lett **60** 429.

Belushkin A. V., Goremychkin E. A., Natkaniec I., Sashin I. L. and Zajaz W. (1989) Physica B **156 & 157** 906.

Bullock M., Zaretsky J., Stassis C., Goldma A., Canfield P., Honda Z., Shirane G. and Shapiro S. M. (1998) Phys. Rev. B **57** 7916.

Choi H. J., Roundy D., Sun H., Cohen M. L., and Louie S. G. (2002) Nature (London) **418**, 758.

Deutscher G. (1999) Nature **397** 411.

Ekino T. (1992) Sov. J. Low Temp. **18** 399.

Elesin V. F., Sinchenko A. A., Ivanov A. A. and Galkin S. G. (1993) Physica C **213** 490.

Goll G., Seemann K., Bräuchle G., von Löhneysen H. Erb A., Müller-Vogt G., Akimenko A. I. and Yanson I. K. (1992) Sov. J. Low Temp. **18** 415.

Gonnelli R. S., Daghero D., Ummarino G. A., Stepanov V. A., Jun J., Kazakov S. M. and Karpinski J. (2002) Phys. Rev. Lett. **89**, 247004.

Hoevers H. F. C., van Bentum P. J. M., van de Leemput L. E. C., van Kempen H., Schellingerhout A. J. G. and van der Marel D. (1988) Physica C **152** 105.

Kamarchuk G. V., Khotkevich A. V., Kolesnichenko Yu. A., Pokhodnya K. I. and Tuluzov I. G. (1994) J. Phys.: Condens. Matter **6** 3559.

Kortus J., Mazin I. I., Belashchenko K. D., Antropov V. P. and Boyer L. L. (2001) Phys. Rev. Lett **86**, 4656.

Kulik I. O. (1988) Sov. J. Low Temp.Phys. **14** 116.

Laube F., Goll G., von Lohneysen H., Fogelstrom M. and Lichtenberg M. (2000) Phys. Rev. Lett. **84**, 1595.

Liu A. Y., Mazin I. I. and Kortus J. (2001) Phys. Rev. Lett. **87**, 087005.

Maeno Y., Hashimoto H., Yoshida K., Nishizaki S., Fujita T., Bednorz J. G., and Lichtenberg F. (1994) Nature (London) **372** 532.

Nagamatsu J., Nakagawa N., Muranaka T., Zenitani Y. and Akimitsu J. (2001) Nature **410**, 63.

Naidyuk Yu. G. Yanson I. K., Tyutrina L. V., Bobrov N. L., Chubov P. N, Kang W. N., Kim H. J., Choi E.-M. and Lee S.-I. (2002) JETP Lett. **75**, 238.

Nowack A. (1990) Ph.-D. Thesis, Köln. (unpublished)

Nowack A., Poppe U., Weger M., Schweitzer D. and Schwenk H. (1987) Z. Phys. B **68** 41.

Omelyanchouk A. N., Beloborod'ko S. I. and Kulik I. O. (1988) Sov. J. Low Temp. Phys. **14** 630.

Plecenik A., Grajcar M., Beňačka Š., Seidel P. and Pfuch A. (1994) Phys. Rev. B **B 49** 10016.

Plecenik A., Grajcar M., Seidel P., Nebel R., Schmauder T., Beňačka Š. and Darula M. (1996) Physica B **218** 209.

Rybaltchenko L. F., Fisun V. V., Bobrov N. L., Yanson I. K., Bondarenko A. V. and Obolenskii M. A. (1991) Sov. J. Low Temp. Phys. **17** 105.

Rybaltchenko L. F., Yanson I. K., Bobrov N. L., Fisun V. V., Obolenskii M. A., Bondarenko A. V., Tret'yakov Yu. D., Kaul A. R. and Graboi I. E. (1990) Sov. J. Low Temp. Phys. **16** 30.

Rybaltchenko L. F., Yanson I. K., Jansen A. G. M., Mandal P., Wyder P., Tomy C. V. and McK Paul D. (1996a) Physica B **218** 189.

Rybaltchenko L. F., Yanson I. K., Jansen A. G. M., Mandal P., Wyder P., Tomy C. V. and McK Paul D. (1996b) Europhys. Lett. **33** 483.

Rybaltchenko L. F., Yanson I. K., Jansen A. G. M., Wyder P., Mandal P. and Mandal J. B. (1996c) Physica B **218** 220.

Szabó P., Samuely P., Kacmarčik J., Klein Th., Marcus J., Furchart D., Miragila S., Marcenat C. and Jansen A. G. M. (2001) Phys. Rev. Lett. **87**, 137005.

van Son P. C., van Kempen H. and Wyder P. (1987) Phys. Rev. Lett. **59** 2226.

Weger M., Nowack A. and Schweitzer D. (1992) Sov. J. Low Temp. Phys. **18** 403.

Yanson I. K., Bobrov N. L., Tomy C. V. and McK Paul D. (2000) Physica C. **334** 33; *ibid.* 152.

Yanson I. K., Fisun V. V., Jansen A. G. M., Wyder P., Canfield P. C., Cho B. K, Tomy C. V. and McK Paul D. (1997) Phys. Rev. Lett. **78** 935.

Yanson I. K., Rybaltchenko L. F., Bobrov N. L. and Fisun V. V. (1987a) Sov. J. Low Temp. Phys. **13** 315.

Yanson I. K., Rybaltchenko L. F., Fisun V. V., Bobrov N. L., Kirzhner V. M., Tret'yakov Yu. D., Kaul A. R. and Graboi I. E. (1988b) Sov. J. Low Temp. Phys. **14** 402.

Yanson I. K., Rybaltchenko L. F., Fisun V. V., Bobrov N. L., Obolenskii M. A., Brandt N. B., Moshchalkov V. V., Tret'yakov Yu. D., Kaul A. R. and Graboi I. E. (1987b) Sov. J. Low Temp. Phys. **13** 315.

Yanson I. K., Rybaltchenko L. F., Fisun V. V., Bobrov N. L., Obolenskii M. A., Tret'yakov Yu. D., Kaul A. R. and Graboi I. E. (1989) Sov. J. Low Temp. Phys. **15** 445.

Zimmermann U., Abens S., Dikin D., Keck K. and Wolf T. (1996) Physica B **218** 205.

14 PCS of heavy-fermion systems

In this chapter, we outline the point-contact experiments with well-known Ce and U heavy-fermion (HF) compounds in the normal and in the superconducting (SC) ground state. The expression "heavy fermions" is used for the compounds for which Sommerfeld coefficient γ in the specific heat of electrons exceeds that for the ordinary metals more than hundred times showing enormous increase of the density of states at the Fermi level or appearance of the heavy quasiparticles. Point-contact investigations could help in better understanding the low-temperature normal, magnetic ordered, and superconducting (SC) ground state of the HF compounds. In the latter case, this study can illuminate their possible unconventional type of superconductivity. As described in Section 3.7, point contacts can be applied as a tool for direct determination of the spatial or directional dependence of the SC order parameter via mechanism of Andreev reflection at the S–N interface or via the multiple Andreev reflection in the S-c-S contacts. The exploring of Josephson effects in the S-c-S weak links can be used to provide additional information about the phase of a complex order parameter.

14.1 Heavy-fermion phenomena

The modern view on the origin and the properties of HF ground state is presented in a comprehensive way, e. g., in reviews by Fisk et al. (1988), Grewe and Steglich (1991), Hess et al. (1993), and references therein. HF compounds contain certain lanthanide (mainly cerium) or actinide (mainly uranium) elements, and they belong to the class of highly correlated electron systems that display dramatically different properties at low temperatures compared with the ordinary metals both in the normal and in the superconducting state. The nature of the low-temperature normal or magnetic as well as SC ground state in HF compounds is under intensive investigation as the first HF superconductor $CeCu_2Si_2$ was discovered by Steglich et al. (1979). A wide variety of ground states in HF systems have been established, $viz.$, the nonmagnetic Fermi-liquid ($CeCu_6$, $CeAl_3$), the antiferromagnetic ($CeAl_2$, U_2Zn_{17}), the superconducting ($CeCu_2Si_2$, UBe_{13}), as well as both types of ordering (UPt_3, URu_2Si_2, UPd_2Al_3, UNi_2Al_3). The crucial role in the formation of the ground state belongs to the partially filled f-electron shell of the cerium or uranium ions, which carry magnetic moments. The large separation of about a few Angstrom between neighboring f atoms compared with the f-shell radius less than $1\,\text{Å}$ hinders direct overlap of the f-electron wave functions; however, the correlation of electrons in the f-shell favors the formation of the local magnetic moment. The localized f-states can hybridize via exchange interaction with the conduction electron states and form a narrow band with the enormous high electronic DOS $N(\epsilon)$ [Fig. 14.1(a)] corresponding to the appearance of quasiparticles with a mass several hundred times higher than free electron mass. This is confirmed by the very large linear coefficient of

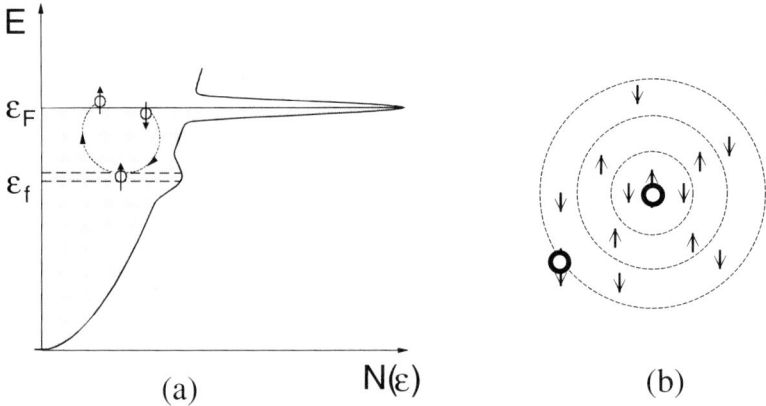

(a) $N(\epsilon)$ (b)

Fig. 14.1. (a) Schematic representation of the quasiparticle DOS caused by the Kondo fluctuations of the spin of localized f-electrons via states near the Fermi level. (b) Interaction between the magnetic impurities (solid circles) caused by RKKY spin-density oscillations.

the electronic specific heat, enhanced Pauli-like susceptibility, clear-cut T^2 dependence of the specific resistivity at low temperatures, and unusual properties in the SC state.

Local magnetic moments are formed at the f-atom sites at weak hybridization, and the demagnetization of these moments takes place at temperature lowering through the "cloud" of the conduction electrons. This is known as Kondo compensation of magnetic moments. This results in effective enhancement of the band electron mass at $T \ll T_\text{K}$, where

$$T_\text{K} \sim (\epsilon_\text{F}/k_\text{B}) \exp(-1/N(\epsilon_\text{F})\,J)$$

is Kondo temperature [see also (6.6)], and J is the exchange integral between the localized spin and the conduction-electron spin. Kondo temperature delineates the crossover from the high-temperature local magnetic moment behavior to the low-temperature nonmagnetic state. This is seen, e.g., in the transport measurements. At $T \gg T_K$, the logarithmic increase of $\rho(T)$ with lowering T, characteristic for isolated Kondo impurity, is clearly seen for the HF samples. A transition occurs from isolated Kondo impurity to the many-body coherent dynamical screening of the local magnetic moments by further temperature lowering. For dilute Kondo systems, this leads to resistivity saturation at $T \to 0$ (so-called unitary limit), whereas a steep decrease of the resistivity occurs in the case of regular lattice of magnetic ions (so-called Kondo lattice). This is because in such a periodic magnetic moment lattice, coherent nature of the ground state results in the Fermi liquid behavior with $\rho(T) = \rho_0 + AT^2$ at $T \to 0$ and gigantic value of the A coefficient. As A is proportional to the square of the effective mass of quasiparticles m^*, this indicates a narrow band formation. Correspondingly, the Sommerfeld coefficient γ in the electronic specific heat and Pauli spin susceptibility χ, both proportional to m^*, have huge values at low temperatures compared with room temperature. Apart from Kondo screening, at low temperatures, an indirect interaction between localized magnetic moments caused by Ruderman–Kittel–Kasuya–Yosida (RKKY) spin-density oscillations occurs (Fig. 14.1(b)). This leads to an antiferromagnetic ordering of the magnetic moments below the characteristic temperature

$$T_{\mathrm{RKKY}} \sim N(\epsilon_{\mathrm{F}})\,J^2.$$

The competition between RKKY and Kondo processes is governed by the exchange constant J. Following the Doniach (1977) approach, the crossover from the magnetic ground state at $T_{\mathrm{RKKY}} \gg T_K$ to the nonmagnetic one at $T_{\mathrm{RKKY}} \ll T_K$ takes place at the critical value J_c, below which the magnetically ordered state is formed, and above J_c, the effective suppression (screening) of the magnetic moments are realized, preventing their ordering.

Normal state properties of the HF compounds are believed to play an important role in the formation of the SC state. Sure, that transition to the SC state in the materials with such a high concentration of Ce or U local magnetic moments is very surprising. Moreover, the size of the specific heat jump at T_c, which is of comparable magnitude as the specific heat in the normal state, as well as the extremely large slope of the upper critical field at T_c indicates that Cooper pairs are formed by the strongly hybridized, itinerant band of the heavy-mass quasiparticles. As might be expected in this case, the SC properties of the HF compounds differ radically from that in the conventional BCS superconductors. Well-established power law of temperature dependence of the specific heat and the magnetic field penetration depth in the SC state of the HF compounds gives a strong evidence of zeros of the SC energy gap along the lines or at points on the Fermi surface for some crystallographic directions. Therefore, like for high-T_c materials, different symmetries of the SC wave function as well as the non-electron–phonon mediating pairing mechanism are considered by explanation of the experimental observations. Additionally, the SC state of the HF systems is very sensitive to the small concentration of impurities. For example, even a few percent of nonmagnetic impurities can totally suppress the superconductivity. All mentioned features bear evidence about the unconventional anisotropic order parameter $\Delta(\mathbf{k})$ in the HF materials.

14.2 Normal state

14.2.1 PC spectra interpretation

Moser et al. (1986a,b) investigated point contacts with $CeAl_3$, $CeCu_6$, and UPt_3. The main peculiarity they found in $dV/dI(V)$ was a minimum at zero bias with a width comparable with the expected narrow maximum width in DOS at ϵ_F for these materials. By analogy with the tunneling approach (see Section 2.4), Moser et al. (1985a,b) stated that the dI/dV differential conductance of contacts is proportional to the electronic DOS $N(\epsilon)$. However, it is seen from (3.2) that DOS does not enter directly to the current through the constriction or to its resistance. Another explanation of the $dV/dI(V)$ behavior of the point contact with the HF compounds is proposed by Naidyuk et al. (1985a) and Paulus and Voss (1985). A huge residual resistivity of the HF materials at low temperatures was taken into account, and a heating model was examined (see Sections 3.5, 7.3, and 7.4).

First of all, lets consider the $LaCu_2Si_2$ compound isostructural to the well-known HF system $CeCu_2Si_2$. The $d^2V/dI^2(V)$ curve of $LaCu_2Si_2$ point contact (Fig. 14.2) shows a number of features, which correspond well to the phonon density of states. On the contrary, the $d^2V/dI^2(V)$ of $CeCu_2Si_2$ differs radically from that of $LaCu_2Si_2$. The lack of peculiarities caused by the electron–phonon interaction points to the nonballistic regime in this case. On the other hand, it may indicate weak electron–phonon contribution to the resistivity compared, e. g., with a magnetic one. Note that the behavior of the dV/dI (V) curve is similar to the $\rho(T)$ dependence of the bulk material (Fig. 14.3), as it is expected for the thermal regime. This is also clearly seen for the other investigated HF compounds presented in Fig. 14.3. Moreover, the asymmetric part of dV/dI (V) for the heavy-fermion heterocontacts is usually in a qualitative agreement with temperature dependence of the Seebeck coefficient $S(T)$ (see, e. g., Fig. 7.11) in accordance with (3.28) for the thermal regime. As was found for UPt_3 homocontacts by Lysykh et al. (1988), in some cases, dV/dI (V) even coincides with dV/dI $(V = 0, T) \propto \rho(T)$ curve, if (3.21) is used for connection between the V and T scale, taking into account the temperature-dependent Lorenz number. Computing the $I - V$ characteristics according to (3.23) for $UPt_x(x = 1, 2, 3, 5)$ compounds by Jansen et al. (1987), for $UCu_{4+x}Al_{8-x}$ by Naidyuk et al. (1993), and for UPd_2Al_3 and UNi_2Al_3 by Kvitnitskaya et al. (1999) further proves the thermal regime in the point contacts with these high-resistivity compounds.

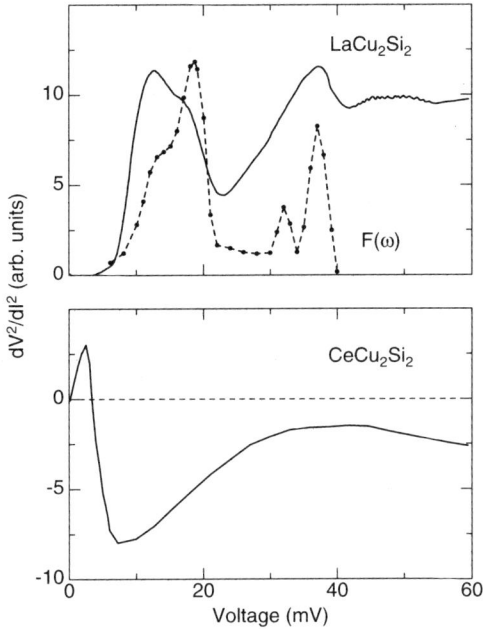

Fig. 14.2. $d^2V/dI^2(V)$ of a LaCu$_2$Si$_2$ point contact ($R_0 = 13.5\,\Omega$) at 4.2 K (solid curve) and the phonon density of states $F(\omega)$ (dashed curve) at 8 K [data taken from Gompf et al. (1987)] along with the $d^2V/dI^2(V)$ characteristic for a CeCu$_2$Si$_2$ point contact ($R_0 = 30\,\Omega$) at 4.2 K from Naidyuk et al. (1985b).

As often encountered for the HF point contacts in the thermal limit, the relative change in $dV/dI\,(V, T=\text{const})$ with voltage increasing is smaller compared with the $\rho(T)$ change versus T. In other words, calculated by (3.23), $dV/dI\,(V)$ has larger amplitude compared with the measured one, especially with increasing the contact resistance. This observation was explained by including an additional constant contribution to the contact resistance caused by the following reasons: (1) increased residual resistivity of the metal inside the constriction [Naidyuk et al. (1993)], (2) spoiled, poorly conducting interface layer [Nowack et al. (1992), Gloos et al. (1998)], (3) low transparent barrier at the boundary [Naidyuk et al. (1991)], (4) strong reflection of the electrons by mismatch of the effective masses or Fermi momenta of contacting metals [Gloos et al. (1995a)]. It means that it would be more logical to use the temperature dependence of the point-contact resistance R_0 instead of the bulk $\rho(T)$ by calculation according to (3.23).

The magnetic field acts on the heavy-fermion ground state giving rise, e. g., to an essential reduction of their resistivity as in the case of CeCu$_6$ [Onuki et al. (1985)] or CeB$_6$ [Peysson et al. (1986)]. Then one would expect a weakening of the inelastic scattering and an overcoming of the heating ef-

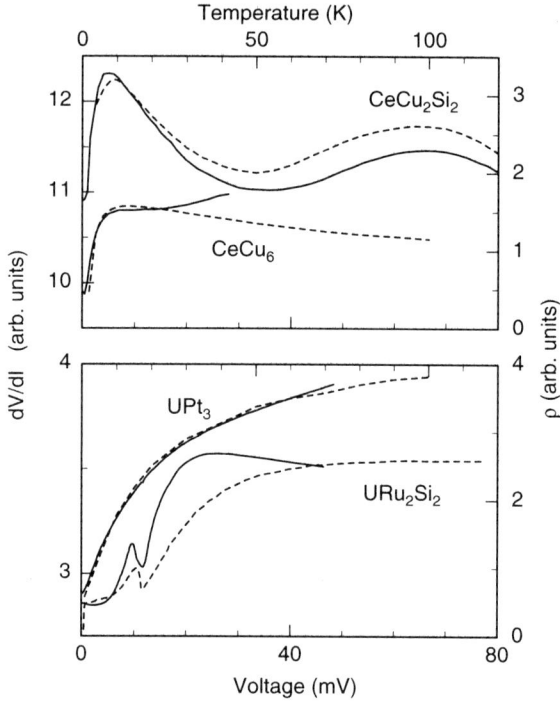

Fig. 14.3. $dV/dI\,(V)$ of point contacts (solid curves, the bottom x-axis) along with the specific resistivity $\rho(T)$ (dashed curves, the upper x-axis) of some heavy-fermion compounds. The $\rho(T)$ curves are from Stewart (1984) – CeCu$_6$ and UPt$_3$, Brandt and Moshchalkov (1984) – CeCu$_2$Si$_2$, and Mydosh (1987) – URu$_2$Si$_2$.

fects in strong magnetic fields. Nevertheless, the $dV/dI\,(V)$ curves for these compounds behave in the high magnetic field similar to the bulk resistivity as shown by Naidyuk et al. (1991), which again supports the resistive nature of the measured nonlinearities. Albeit Kunii et al. (1987) noticed a remarkable difference between $dV/dI\,(V)$ and bulk $\rho(T)$ for CeB$_6$, additional measurements of the temperature dependence of the point-contact resistance are desirable in this case to ensure that the latter behaves for these contacts like the bulk $\rho(T)$.

14.2.2 Scrutiny of URu$_2$Si$_2$

Naidyuk et al. (1995) found that the behavior of $dV/dI\,(V)$ for some URu$_2$Si$_2$ contacts contradicts (even qualitatively) the $\rho(T)$ dependence. $dV/dI\,(V)$ shows the maximum at $V = 0$; that is, at first dV/dI decreases with a voltage around zero bias [Fig. 14.4(b)]. Contrary to that zero-bias differential resistance $dV/dI\,(V = 0, T)$ increases with temperature [Fig. 14.4(a)].

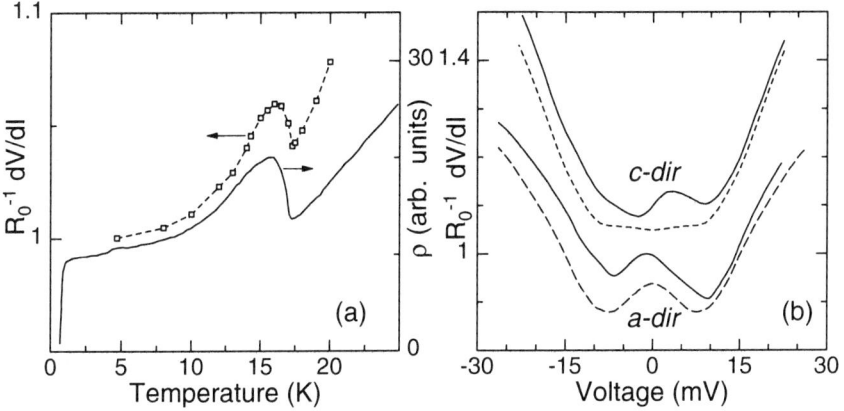

Fig. 14.4. (a) Temperature dependence of the normalized point-contact resistance at $V = 0$ of a URu_2Si_2 homocontact ($R_0 = 3.2\ \Omega$) along the ab plane (symbols) compared with the bulk resistivity from Mydosh (1987) (solid curve). (b) Reduced dV/dI (V) (solid curves) and their symmetric parts (dashed curves) for contacts established along the c (two upper curves) and in the ab direction (two bottom curves). R_0 are 3.3 and 3.2 Ω, respectively, $T = 4.2\,K$. Data taken from Naidyuk et al. (1995).

dV/dI $(V = 0, T)$ possesses also a maximum at T_N, which corresponds well to $\rho(T)$ [Fig. 14.4.(a)]. This correspondence is very important because it testifies to the consistency of transport properties of the material in the constriction to the bulk. Contradiction to the thermal model, at least at small biases, allows us to suppose the spectroscopic nature of the maximum at $V = 0$ in dV/dI (V). Partially gapping of the Fermi surface in the antiferromagnetic (AFM) state below the Néel temperature T_N, caused by a spin-density wave formation, was considered as a reason [Naidyuk et al. (1995, 1996a)]. As shown in Fig. 5.10(b), dV/dI (V) curves for the archetypical spin-density wave metal Cr exhibit analogous maximum as well. The origin of the maximum in simple phrase is the following. Sharvin resistance (3.3) is governed by the Fermi surface area S_F, because $R_{Sh} \propto 1/k_F^2 \propto 1/S_F$. Hence, R_{Sh} should increase at voltages eV smaller than the gap Δ_{AF} because of removing of the part of the Fermi surface. This leads to the appearance of the maximum at zero bias. Note that the Fermi surface gapping is different from the renormalization of the electronic DOS $N(\epsilon)$ at the Fermi surface because of the many-body interaction, which does not influence the Sharvin resistance, as mentioned in Section 3.1. Further support of the model provides observations that the analogous zero-bias maximum was absent by partially replacing Ru by Re in the $URu_{2-x}Re_xSi_2$ compounds in paramagnetic or ferromagnetic states [Steiner et al. (1996)], i.e., in the absence the of spin-density wave state.

Subsequently, it turned out that the zero-bias maximum in the symmetric part of dV/dI: $(dV/dI)^{sym} = [dV/dI(V > 0)+dV/dI(V < 0)]/2$ was well pronounced for the contacts along the ab plane of the tetragonal lattice, but it was mainly suppressed or absent at measurements along the c direction, whereas the asymmetric maximum was present for both curves (Fig. 14.4). This observation is in consensus with the STM experiments of Aliev et al. (1991) and Aarts et al. (1994), showing that the gap at the Fermi surface in URu_2Si_2 is strongly anisotropic and opens in the ab plane.

Figure 14.5 shows the behavior versus temperature for a few contacts of the width of the zero-bias dV/dI maximum symmetric part related to the gap. The gap value estimated from the width of the zero-bias maximum in the symmetric part of dV/dI is of the order of 10 meV, almost the same as it was obtained from tunneling, resistivity, specific heat, and neutron scattering measurements (see Aarts et al. (1994) and references therein). The zero-bias maximum decreases gradually with the temperature increasing (see Fig. 14.5), but its width decreases much slowly persisting practically up to the temperature close to $T_N = 17.5$ K. The last behavior looks like a first-order transition near T_N and deviates from the expected one in the mean-field theory. Remember that the thermodynamics of the transition to the spin-density wave state in the weak-coupling limit is described in the same way as the condensation in the BCS superconducting state. In both cases, a gap develops in the single-particle excitation spectrum with well-known mean-field (or BCS for superconductivity) temperature dependence.

Escudero et al. (1994) discussed the structure with double minimum near zero bias in the $dV/dI(V)$ characteristics of URu_2Si_2 contacts. They attracted for explanation the Andreev-like scattering electrons on the spin-density wave condensate in the AFM state below T_N by analogy as in the case of superconductors (see Section 3.7). Correspondingly, the distance between the minima in the dV/dI characteristics was taken as a gap width. However, only the zero-bias maximum for $T > T_N$ disappears, but the overall dV/dI minimum does not still vanish. Hence, the latter does not relate to the AFM state and, in this fashion, to the spin-density wave gap. It is easy to imagine that the double minimum develops because of the combination of the zero-bias maximum and increasing with a voltage (temperature) like $\rho(T)$ Maxwell contribution [see (3.18)] to the contact resistance.

Hasselbach et al. (1992) proposed that the change in the scattering amplitude of the hybridized heavy quasiparticles, as they undergo a magnetic transition, can be the possible mechanism of the "gap-like" feature formation in $dV/dI(V)$. Unfortunately, the details of the mechanism were not discussed and the qualitative estimation was not carried out. On the other hand, Rodrigo et al. (1997) analyzed the $dI/dV(V)$ curves of URu_2Si_2-Pt contacts or URu_2Si_2 homocontacts in terms of conductance spectra considering the effect of resonance in DOS at the Fermi level that corresponds to the Kondo system. They argued that the results are consistent with the existence of a Kondo-

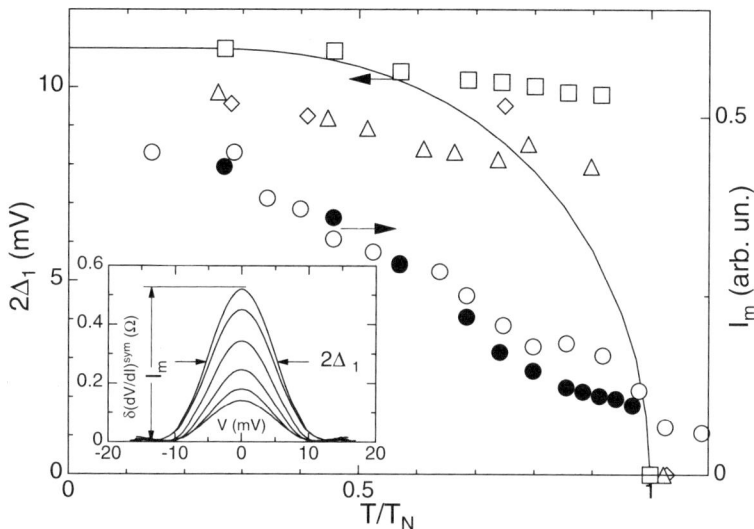

Fig. 14.5. Temperature dependence of the width of the zero-bias maximum for three URu$_2$Si$_2$ homocontacts (squares, triangles, and diamonds) along the ab plane. Solid curve indicates a mean-field BCS temperature dependence. Solid circles show the temperature dependence of the zero-bias maximum intensity (see inset), and the open ones show intensity of the AFM Braggs peaks describing behavior of the magnetic order parameter [see, e. g., van Dijk et al. (1997)]. Inset shows the determination of the width Δ and the intensity I_m of zero-bias maximum $\delta\,(\mathrm{d}V/\mathrm{d}I)^{\mathrm{sym}}$ extracted from the symmetric part of $\mathrm{d}V/\mathrm{d}I\,(V)$. The temperature decreases from 4.2 K (upper curve) to 17 K (bottom curve). Data taken from Naidyuk et al. (1995, 1996a).

like bound state of the U^{4+} ionic configuration and conduction electrons. In spite of the used STM technique, the authors dealt with true metallic point contacts. Their zero-bias resistance and $\mathrm{d}I/\mathrm{d}V\,(V)$ behavior were similar to $(\mathrm{d}V/\mathrm{d}I)^{-1}$ from the above discussed papers, e. g., Nowack et al. (1992), Hasselbach et al. (1992), Escudero et al. (1994), and so on. In this case, for a contact with metallic conductivity, direct reflection of DOS in point-contact spectra is questionable, as mentioned in the Section 3.1.

So far the nature of the order parameter in the AFM state of URu$_2$Si$_2$ remains under discussion in the literature. Transport measurements in the magnetic field lead to the estimation of a critical field of the AFM gap about 40 T [Mentink et al. (1996)], whereas the tiny staggered magnetic moments vanish nearly 14 T according to the neutron scattering [Mason et al. (1995)]. This leads to the speculation about some additional "hidden" magnetic order parameter, which challenged point-contact measurements as well.

$\mathrm{d}V/\mathrm{d}I(V)$ curves of URu$_2$Si$_2$ point contacts reveal two kinds of structure: the asymmetric zero-bias maximum (ZBM) [Fig. 14.4(b)] and a kink at the higher energy (see dV/dI curve for URu_2Si_2 in Fig. 14.3), both varying with temperature rise and vanishing above the Néel temperature $T_N = 17.5$ K. It was pointed out that the kink is determined by T_N and is connected with the transition of the point-contact region from AFM to the paramagnetic state by increasing an applied voltage V most likely because of constriction heating. On the contrary, ZBM is more pronounced for curves with gentle or not resolved kink features [Fig. 14.4(b)]; therefore, ZBM has a nonthermal origin. According to the foregoing discussion, ZBM is caused by the gap development in the excitation spectra of electrons by spin-density wave formation in the AFM state. Remember that PCS is sensitive to the magnetic type of in-

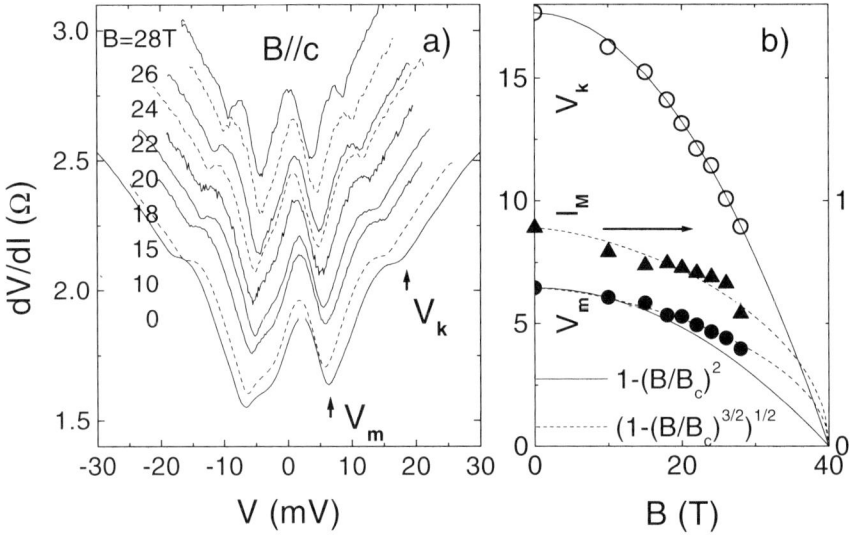

Fig. 14.6. (a) $dV/dI\,(V)$ curves for URu_2Si_2-Cu heterocontact in magnetic fields along the c axis of the sample. The solid curves correspond to the field sweep up, and the dashed one corresponds to the field sweep down. The arrows show position of the kink V_k and the minimum V_m. (b) The dependence of V_k, V_m and zero-bias maximum intensity I_m versus magnetic field. The solid lines represent the dependence characteristic for $T_N(B)$ behavior and $\Delta(B)$, and the dashed line is the dependence taken by Mason et al. (1995) for staggered magnetic moments versus magnetic field. Data taken from Naidyuk et al. (2001).

teractions (see Section 6.1). It is not improbable that ZBM is connected also with the magnetic order parameter in some way. The intensity of ZBM decreases with increasing of temperature, analogously to the intensity of AFM Bragg peaks describing magnetic order parameter behavior (see Fig. 14.5 and figure caption). On the other hand, the temperature dependence of V_{m} (see Fig. 14.6(a) for definition) and the ZBM width [Steiner et al. (1996)] for some contacts is close to the mean-field BCS-like dependence characteristic for the order parameter. Recent experiments by Naidyuk et al. (2001) in the magnetic field yielded that both V_{m} and zero-bias maximum intensity I_{m} behave as described for the magnetic moments [Fig. 14.6(b)]. On the contrary, the position of the kink V_{k} [Fig. 14.6(b)] follows the magnetic field dependence similar to $T_{\mathrm{N}}(B)$ [Mentink et al. (1996)]. The latter dependence is also found for the width of ZBM related to the spin-density wave gap. Eventually, all mentioned features in $\mathrm{d}V/\mathrm{d}I(V)$ – kink, ZBM width, and V_{m} are described by dependencies versus magnetic field [Fig. 14.6(b)] specific for the behavior of the transition temperature T_{N}, the spin-density wave gap width, and the magnetic order parameter, correspondingly. Moreover, it is seen, that for all features, the critical field has been estimated about 40 T [Fig. 14.6(b)], independent of the type of behavior. Therefore, point-contact measurements show the presence of one order parameter vanishing at $T_{\mathrm{N}} = 17.5\,\mathrm{K}$ and $B_{\mathrm{c}} \simeq$ 40 T. This correlates with the statements of van Dijk et al. (1997) in that the results of inelastic neutron scattering measurements at 12 T are in disagreement with the prediction of a decoupling of the ordered moment and the energy gap in the high magnetic field.

A pronounced feature in the $\mathrm{d}V/\mathrm{d}I\,(V)$ characteristics of heterocontacts with the HF compounds is an asymmetry (see Fig. 7.11), which is mainly governed by the large thermopower of the HF materials. In the case of $\mathrm{URu_2Si_2}$ contacts, the asymmetric part of $\mathrm{d}V/\mathrm{d}I\,(V)$ has a sign corresponding to that of $S(T)$. However, as mentioned by Naidyuk et al. (1995), its temperature dependence contradicts at low biases to the heating model. This is an additional confirmation of nonthermal origin of the above-mentioned zero-bias maximum structure. A possible source of the asymmetry in the frame of the ballistic point-contact theory was proposed by Nowack and Klug (1992). Exploiting an artificial supposition about energy-independent inelastic processes in the ballistic limit, they showed that the $\mathrm{d}V/\mathrm{d}I$ characteristics still might depict the electronic DOS (see Fig. 14.7). They consider Lorentz-shaped maximum in the background electronic DOS $N(0)$ at the Fermi energy:

$$N(\epsilon) = N(0)\left(1 + \sigma\frac{\Gamma^2}{\Gamma^2 + (\epsilon - \epsilon_0)^2}\right). \tag{14.1}$$

If this takes place, then the $\mathrm{d}V/\mathrm{d}I$ asymmetry in the heterocontact with $\mathrm{URu_2Si_2}$ can be partly caused by (at least at low biases) the asymmetric peak in DOS at ϵ_{F} for the latter compound.

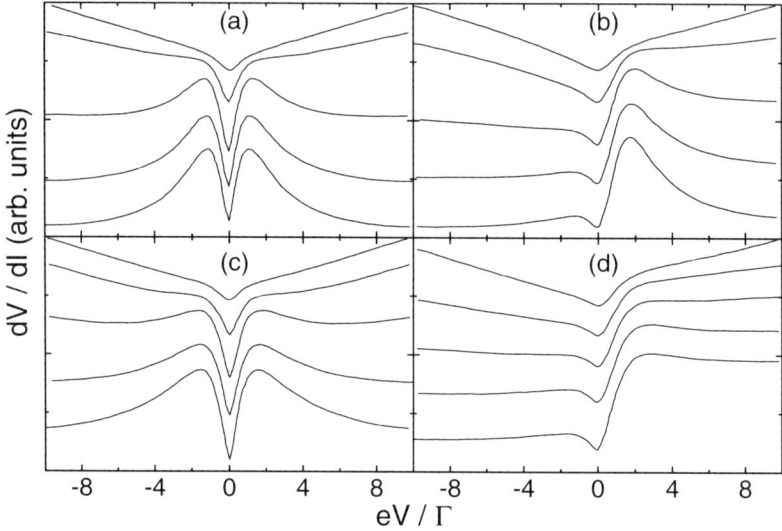

Fig. 14.7. Point-contact differential resistance calculated with DOS given by (14.1) within the "one band" (a), (b) and the "two band" (with "light" and "heavy electrons") (c), (d) for a symmetric ($\epsilon_0 = 0$) (a, c) and an asymmetric ($\epsilon_0 = \Gamma$) position (b, d) of the DOS peak with respect to the Fermi energy. For (a) and (b), σ values are 1, 3, 10, 30, and 100 from the top curve to the bottom one. For (c) and (d), σ values are 3, 10, 30, 100, and 300 from the top curve to the bottom one. The nonlinearities are smaller by factor σ in the "two band" model. After Nowack and Klug (1992).

14.3 Superconducting state

14.3.1 S-c-N contacts

Study of S-c-N contacts with HF superconductors, namely, UBe$_{13}$, was began by Nowack et al. (1987a). They measured a pronounced minimum in $dV/dI(V)$ at $V = 0$ below T_c with the width roughly corresponding to the expected SC gap value. The magnitude of the minimum was up to 50% of R_N for UBe$_{13}$ in accordance with the BTK prediction (see Section 3.7). In contrast, the $dV/dI(V)$ curves for UPt$_3$ showed only a shallow zero-bias minimum about a few percent of the normal state resistance as it was found by Goll et al. (1993), de Wilde et al. (1994), Naidyuk et al. (1996b), and Obermair et al. (1998). The latter authors reported that the weak SC signal in UPt$_3$ (presumably caused by the suppression of the order parameter at the surface) cannot be enhanced even by different surface treatments.

Goll et al. (1993) observed a minimum (quite often even a double-minimum) in dV/dI for the c direction of hexagonal UPt$_3$ crystal, but such features were absent for the perpendicular direction. The authors suggested an anisotropic order parameter in UPt$_3$ with the gap vanishing in the basal plane. Furthermore, as mentioned by Goll et al. (1995), dV/dI (V) cannot be fitted reasonably by the BTK model with an isotropic gap (Fig. 14.8). Their analysis favors two-dimensional representation of the order parameter, with the orbit part having a line of nodes in the basal plane and point nodes along the c axis.

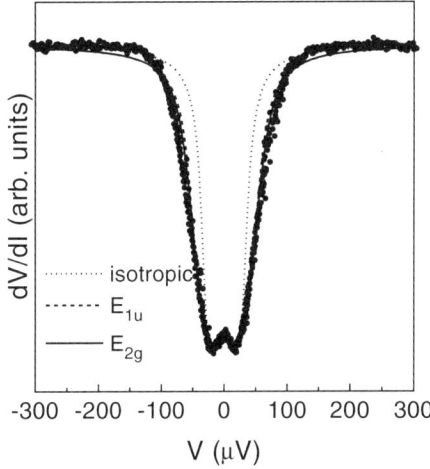

Fig. 14.8. dV/dI curve of a UPt$_3$-Pt point contact along the c direction showing a double-minimum structure (symbols) along with calculated ones using different order parameter representation: $\Delta(\mathbf{k}) = \eta_1|k_z|k_x + \eta_2|k_z|k_y$ for E_{1u} and $\Delta(\mathbf{k}) = \eta_1|k_z|(k_x^2 - k_y^2) + 2\eta_2|k_z|k_xk_y$ for E_{2g}. By courtesy of Goll et al. (1995).

de Wilde et al. (1996) also claimed that dV/dI (V) behavior of UPt$_3$ point contact cannot be described by the BTK model, even including lifetime electronic DOS broadening (that is, at $\Gamma \neq 0$). Instead of it, an anisotropic SC order parameter with a broadening effect leads to fairly good agreement with the experimental data. At the same time, the results for CeCu$_2$Si$_2$ and URu$_2$Si$_2$ according to de Wilde et al. (1994) point to the absence of zeros in $\Delta(\mathbf{k})$. On the contrary, Hasselbach et al. (1992) and Samuely et al. (1995) stated that an anisotropic d-wave order parameter formally included in the BTK model better describes the dV/dI (V) characteristics of URu$_2$Si$_2$. However, in the latter cases, the SC minimum at $V = 0$ was very weak and smooth, which does not allow us to draw an unequivocal conclusion concerning the gap symmetry by using fit procedure. As stated by Anders and Gloos (1997) a strongly reduced or even missing Andreev reflection signal could be also explained by the short lifetime of heavy quasiparticles.

Naidyuk et al. (1996c) observed a clear double-minimum structure in the dV/dI (V) curves for a few (about 5% of the total amount) contacts between URu$_2$Si$_2$ single crystal and Pt. Fitting the dV/dI curves by virtue of the modified BTK model, (3.30) and (3.32) yielded a large reduced SC gap ratio $2\Delta(0)/k_BT_c \sim 10$ for the c axis and a few times smaller gap for the a axis (see Table 14.3.1). Such a strong anisotropy of the SC gap in URu_2Si_2 can be due by suppression of superconductivity in the ab plane by the exchange field of the ferromagnetically ordered layers of the U magnetic moments in the basal plane. Kulic et al. (1991) predicted that the reduced gap is a few times larger than the standard BCS value 3.5 in this case, which is in line with the measurements. Another point is noted that a double-minimum structure was found for contacts with suppressed AFM features in dV/dI; i.e., the broad asymmetric zero-bias maximum in normal state [see Fig. 14.4(b)] was absent.

Wälti et al. (2000) reported on measurements of the differential conductivity $dI/dV(V)$ of UBe_{13}-Au contacts. They found a huge zero-bias conductance peak that was connected with the existence of low-energy Andreev surface bound states. These bound states may form only in superconductors with nontrivial energy-gap functions. From the width of the zero-bias structure in dI/dV at $T < T_c$, they established a lower limit of the normalized energy gap, such that $2\Delta(0)/k_BT_c >6.7$, much in excess of the weak coupling BCS value of 3.5, and indicating strong coupling effects in superconducting UBe_{13}

In the comments on the Wälti et al. (2000) paper, Gloos (2000) noticed that the large SC anomalies (zero-bias resistance minimum or, in other words, zero-bias conductance peak) usually observed for contacts with UBe_{13} are mainly caused by the Maxwell resistance vanishing below $T < T_c$ of UBe_{13}. Analyzing contribution of the Maxwell term [see (2.3)] in the conductivity of S-c-N contacts, Gloos et al. (1996a,b,c) found that the diminishing of the zero-bias resistance δR below T_c scales inversely with the contact radius instead of the inverse constriction area as expected for Andreev reflection. Furthermore, the magnitude of δR corresponds well to the Maxwell contribution in the contact resistance with ρ as in the bulk: $\delta R \simeq \rho/d$. This gives strong evidence that the zero-bias minimum in the dV/dI curves at $V = 0$ is mainly caused by the Maxwell resistance vanishing in the SC state. On the other hand, the zero-bias resistance at low temperatures $R(V = 0, T \ll T_c)$ in the SC state scales fairly well with the inverse contact area, albeit being approximately two orders of magnitude higher than that estimated from the Sharvin formula (3.3). This means that the Andreev-reflection hole current leading normally to the zero-bias resistance decreasing is suppressed in these contacts (analogous results of the same authors will be discussed in the next Section 14.3.3 in details; see also Figs. 14.13, 14.14, and 14.15, where the results on homocontacts, $viz.$, break junctions are presented).

For S-c-N contact in a diffusive regime, Maxwell contribution to the resistance appears at $eV \geq \Delta$ (see Fig. 3.14) when normal quasiparticle current through the S–N interface is possible. One would expect the abrupt increase of the contact resistance (at least on the R_M value) at this energy. Usually, maxima or kinks in dV/dI are often observed in the energy range of about

Table 14.1. Superconducting gap of the HF compounds determined from the $dV/dI\,(V)$ curves of point contacts using BTK or modified BTK fit both with isotropic and anisotropic order parameter, from full width (FW) of the minimum at zero-bias, otherwise by peak-to-peak distance (PP), or just by estimation. Here T_c is the temperature at which the SC features vanish in dV/dI.

Compound	$2\Delta(0)$, μeV	$2\Delta(0)/k_B T_c$	Method	Refs.
CeCu$_2$Si$_2$	≥ 160	≥ 2	estim.	de Wilde et al. (1994)
UBe$_{13}$	260	3.4	FW	Nowack et al. (1987a)
	290	4.2	from $I-V$	Moreland et al. (1994)
	–	6.7	FW	Wälti et al. (2000)
UPt$_3$	140	3.5	FW	Nowack et al. (1987a)
	58–78	1.3–1.8	isotr. BTK	Goll et al. (1993)
	150	3.96	anis. BTK	de Wilde et al. (1994)
	51–107	2.6-4.4	anis. BTK	Goll et al. (1995)
	100	~ 2.6	modif. BTK	Naidyuk et al. (1996b)
U$_2$PtC$_2$	440	3.4	isotr. BTK	Naidyuk et al. (1988)
URu$_2$Si$_2$	$\sim$1000	~ 9	PP	Nowack et al. (1992)
	420	~ 4	anis. BTK	Hasselbach et al. (1992)
	700	~ 6	anis. BTK	Samuely et al. (1995)
	≥ 340	≥ 3	estim.	de Wilde et al. (1994)
a-dir	~ 500	~ 5	modif. BTK	Naidyuk et al. (1996c)
c-dir	~ 1400	~ 12	modif. BTK	Naidyuk et al. (1996c)

Δ and they are taken as a manifestation of the SC gap. However, it should be kept in mind that the transition to the normal state could be caused by the ordinary local heating of the point-contact region, accounting for large specific resistivity of the HF compounds and its steep increase with the temperature even at very low temperatures. The latter testifies to strong inelastic scattering, which favors the transition to the thermal regime. This occurs at critical voltage $eV_{cr} \simeq 3.63 k_B T_c$[1] (3.22); that is, the thermal features at eV_{cr} could be close to the spectral gap features defined by the BCS relation $2\Delta \simeq 3.52 k_B T_c$. In this case, also the temperature dependence of the thermal peculiarities (3.21) may even simulate BCS dependence. This can lead to misinterpretation of the thermal-regime features in dV/dI as spectroscopic ones and wrong determination of the spatial and temperature dependencies of the SC order parameter.

Superconductivity suppression in the point contact can also be caused by the current density exceeding the critical value or by the influence of the self

---

[1] Coefficient 3.63 corresponds to the standard Lorentz number L_0. For most of the HF systems, Lorentz number $L > L_0$ and the coefficient at T will be $\sqrt{L/L_0}$ times larger.

magnetic field. For all of these causes, as well as for the heating, according to Iwanyshyn and Smith (1972), an excess current should disappear at higher biases compared with the gap position. Experimentally, the excess current is often observable up to the high voltage biases, especially for the high-ohmic junctions. There are also other experimental contradictions to the heating model. As was shown for URu$_2$Si$_2$ in the previous section, nonheating effects may define the $I - V$ characteristic at low biases for some contacts. Another example of absence of remarkable heating is the observation of Andreev reflection for the UPt$_3$-Zn point contact by Naidyuk et al. (1996b). Here the Andreev reflection features of Zn were fitted well at different temperatures, with a supposition $T = T_{bath}$; i.e., no noticeable temperature rise occurs in point contact at $eV \leq \Delta$. Nevertheless, the trivial heating should be taken into consideration in each case, before dealing with more exotic phenomena like Andreev reflection.

The ratio $2\Delta(0)/k_BT_c$ for the HF superconductors reported by different authors scatters around the BCS value with a few exceptions and is shown in Table 14.3.1.

14.3.2 S-c-S contacts

The first measurements of the S-c-S contacts between ordinary singlet and the HF superconductors were carried out by Poppe (1985) and Han et al. (1985). The former author measured $I - V$ characteristics of CeCu$_2$Si$_2$-Al contact at different temperatures and magnetic fields (Fig. 14.9). The $I - V$ curves in the SC state display a supercurrent at $V = 0$ and an excess current at a finite voltage. The absolute magnitude of the supercurrent I_c was estimated to be 15–80% of the theoretical value (3.37) for ordinary superconductors, which leads to the conclusion that CeCu$_2$Si$_2$ behaves like an s-wave superconductor. It is worth noting that the excess current vanishes at temperature 0.7 K (Fig. 14.9), noticeably lower than T_c of Al, or in the magnetic field being negligible with respect to the critical field of CeCu$_2$Si$_2$; i.e., at least one of the electrodes remains superconducting in both cases. In this case of S-c-N contact, I_{exc} should be visible, but its value decreases about two times compared with the situation when both electrodes are superconducting (see Section 3.7). This observation questions the formation of the weak link between two contacted superconductors; instead of this, it forces us to assume that the real weak link is apparently located somewhere inside in Al.

Poppe (1985) carried out vacuum tunneling experiments below T_c of UPt$_3$ using both Pt and UPt$_3$ counterelectrodes. He did not find neither Josephson coupling nor any other SC gap features.

Han et al. (1985) have observed the ac Josephson effect and the Fraunhofer-type pattern of the supercurrent in a magnetic field for contact between Nb and HF superconductors CeCu$_2$Si$_2$, UBe$_{13}$, and ordinary LaBe$_{13}$ surprisingly well above the critical temperature of the HF samples. The authors explained the results as an anomalous proximity induced by Nb in the HF side s-wave

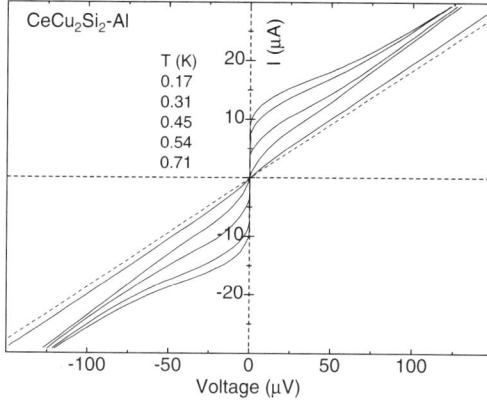

Fig. 14.9. $I - V$ characteristics of a CeCu$_2$Si$_2$-Al Josephson contact at different temperatures. Dashed curve was measured in the external magnetic field at 0.14 K. Data taken from Poppe (1985).

superconductivity. The decreasing of I_c by about 10% below T_c of UBe$_{13}$, in contrast to I_c increase by using Mo as a counterelectrode, was interpreted by Han et al. (1986) as an occurrence of a negative s-wave proximity effect. They stated that these observations favors an unconventional SC ground state in HF compounds. However, Kadin (1990) found the model of the proximity-induced Josephson effect fundamentally flawed and proposed an alternative explanation based on the presence of a phase-slip center near the tip of a superconducting electrode in an S-c-N contact.

dc Josephson effect between UPd$_2$Al$_3$ and Nb was reported by He et al. (1992). They observed a persistent current in a composite Nb ring bridged by small UPd$_2$Al$_3$ pointed rod and trapped flux, which was in discrete quantum states separated by the flux quantum $h/2e$. However, the supercurrent in the $I - V$ curves had a very small product $I_c R_N$, only a few tenths of a percent of the theoretical value given by (3.37). The measurements confirmed the bulk superconducting state in UPd$_2$Al$_3$, but they did not clarify the nature of the SC ground state in this material.

The $dV/dI\,(V)$ curves of point contacts between URu$_2$Si$_2$ and Zn investigated by Nowack et al. (1995) behaved like a superconductor–normal metal–superconductor junction with a thick normal layer at the HF part of the contact. $dV/dI\,(V)$ showed the Andreev-reflection minima, typical as for the S-c-N contact, and related both to the heavy fermion and conventional superconductor. On the contrary, contacts between URu$_2$Si$_2$ and NbTi are superconducting at low bias current (i.e., they exhibit the supercurrent at $V = 0$), although the corresponding critical voltage $I_c R_N \approx 10\,\mu$V is about one order of magnitude below the theoretical value (3.37). Nowack et al. (1995) succeeded in fabrication of closed-loop setup with two NbTi contacts on a

URu_2Si_2 sample, which even showed oscillations in weak magnetic fields vanishing above the critical temperature of the HF superconductor (Fig. 14.10). The last observation proves that the superconducting state of the HF material

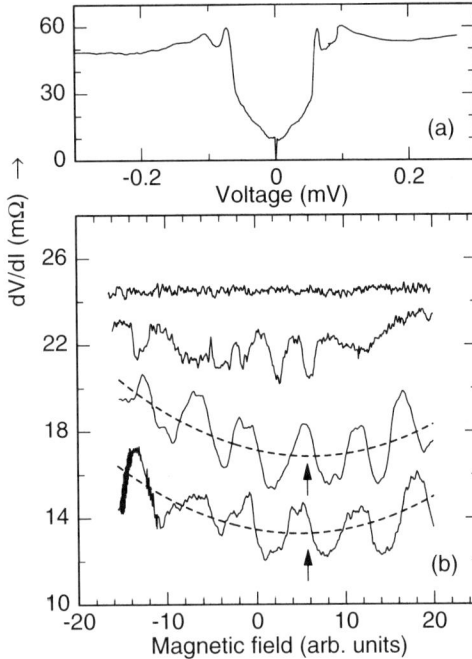

Fig. 14.10. (a) dV/dI (V) of a URu_2Si_2 - NbTi contact with zero resistance at $V = 0$. (b) Oscillations of the differential resistance dV/dI of this contact at constant bias voltage $V_{bias} \neq 0$ versus magnetic field. Temperature from the top to the bottom curve is 1.3, 1.0, 0.6, and 0.2 K. The arrows show the position of the background minimum (dashed lines) taken as the position of the zero-field point. Data taken from Nowack et al. (1995).

plays a role here. Josephson coupling between a conventional superconductor and the HF superconductor URu_2Si_2 was succeeded, in spite of the fact that the coupling between HF and other superconductors is hindered by formation of the mentioned normal layer at the surface of the HF material. If the proximity-induced superconductivity in this thick normal-conducting zone is strong enough, it will make the effective layer of the normal zone thinner [Fig. 14.11(a)], leading to the increase of the coupling. The authors suggested that the strong-coupling conventional superconductors with high T_c should be used to obtain Josephson contacts and, in particular, to carry out the Sigrist and Rice (1992) type experiment designed [Fig. 14.11(b)] to establish the symmetry of the order parameter in unconventional superconductors.

Nowack et al. (1995) detected a finite phase difference at zero field (i. e., the minimum rather than the maximum of critical current, Fig. 14.10) for fabricated two contact SQUID, as expected for SQUID established in a special way [Fig. 14.11(b)] between a conventional and a d-wave superconductor.

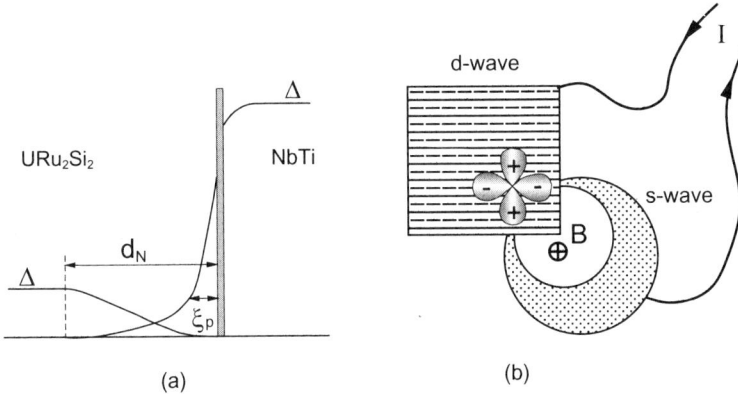

(a)   (b)

Fig. 14.11. (a) Schematic view of the S-c-S contact with a thick nonsuperconducting N layer with a width d_N and proximity-induced superconductivity with characteristic coherence length $\xi_p = (\hbar v_F l_{el}/2\pi k_B T)^{1/2}$. To succeed in Josephson coupling, the proximity-induced order parameter in N zone should overlap sufficiently with the gradually suppressed at the surface HF order parameter. After Nowack et al. (1995). (b) Model of two-contact SQUID on the base of d-wave and conventional superconductors proposed by Sigrist and Rice (1992) to determine symmetry of the order parameter. For junctions with an equal critical current in the limit of zero loop inductance, the maximal supercurrent is $I_c \sim |\cos(\pi\Phi/\Phi_0 + \phi)|$, where Φ is the external magnetic flux, Φ_0 is the flux quantum, and ϕ is the intrinsic phase shift between two orthogonal direction. ϕ is equal to zero for the s-wave superconductor and equal to π for the d-wave superconductor with $d_{x^2-y^2}$ symmetry. As a result, in the last case, at zero external field, I_c will be a minimum instead of a maximum for the s-wave superconductor [see van Harlingen (1995)].

By further investigation, Wasser et al. (1998) observed an interference pattern for a single URu_2Si_2-Nb contact aligned in the basal plane. In contrast, the contact established along the perpendicular to the basal plane (i. e., along the c axis) showed much weaker SC features and absence of zero resistance or critical current. This strong anisotropy was attributed by Wasser et al. (1998) to an unconventional order parameter, with the symmetry leading to the destructive interference of Josephson current along the c direction, or in other words, the gap should vanish in the c direction. This is in contrast with measurements of Naidyuk et al. (1996c) for S-c-N URu_2Si_2 - Ag contacts described above, which yielded a maximal gap along the c axis.

Tachibana et al. (2003) observed the Josephson effect between URu_2Si_2 and Nb planar junctions for both c- and a- axis directions. The critical current values of various SS' or SNS' (where N marks thin 1–10-μm Cu film) junctions were of the same order of magnitude independent of current direction or surface roughness of the (001) plane. These results indicate that the c axis Josephson coupling, which was unresolved by Wasser et al. (1998), is still possible between URu$_2$Si$_2$ and Nb. However, the strength of the Josephson coupling estimated from $I_c R_N \approx 0.1\,\mu$V was much smaller, as mentioned above for the point contacts. The same group also measured the Josephson effect by the same type of the junctions for CeCu$_2$Si$_2$, UBe$_{13}$, and UPt$_3$ as well [see Sumiyama et al. (2000)]. The small product $I_c R_N$ (below 0.1 μV) is specific for all compounds. Tachibana et al. (2003) attributed the small $I_c R_N$ value to the reduction of the condensation amplitude at the URu$_2$Si$_2$-Nb interface caused by the proximity effect.

The presence of the normal layer or the suppression of superconductivity at the contact interface could be just caused by a mechanical stress in the contact region in connection with the sensitivity of the HF superconducting state to impurities or distortions. Nowack et al. (1995) speculated that the development of a normal layer relates to the magnetic properties of URu$_2$Si$_2$, which orders antiferromagnetically.

Naidyuk et al. (1996b) carried out experiments with the S-c-S contacts on the base of UPt$_3$. dV/dI (V) of UPt$_3$-Zn contacts well below T_c for both materials showed a pronounced zero-bias double-minimum structure caused by the superconducting Zn. The fit of the dV/dI curves on the basis of the modified BTK theory yields for Zn an energy gap $\Delta(T)$ perfectly following the BCS theory. Using a small magnetic field to suppress superconductivity in Zn Naidyuk et al. (1996b) estimated also Δ_0 in UPt$_3$ (see Table 14.3.1), which is comparable with the values obtained by other authors. It should be pointed out that the barrier-strength parameter Z was found in the range of 0.45 ± 0.02; i.e., it is the same as for Zn-Ag contacts and for many others contacts with both conventional and high-T_c superconductors (see Chapters 12 and 13). This supports the theory of Deutscher and Nozieres (1994) that the boundary condition at the interface with the HF compounds involves the Fermi velocities without a large mass enhancement factor. Otherwise, the large Fermi velocity mismatch between HF samples and an ordinary metal would lead according to (3.31) to the Z value characteristic for a tunnel junction. On the other hand, the same Z parameter of about 0.5 for different systems (including the contacts with high-T_c superconductors) hints that all of these contacts might be in the dirty limit as follows from Fig. 9 in the well-known BTK paper (see also discussion on the page 203).

Naidyuk et al. (1996b) did not observe oscillations of the supercurrent in small magnetic fields for UPt$_3$-NbTi contacts, as described above for URu$_2$Si$_2$-NbTi contacts. It looks like the surface layer of UPt$_3$ has a more strong pair-breaking, taking into account the high Γ value (3.32), increasing with the contact resistance (see Fig. 12.14). The latter figure shows that the Γ parameter is about one order of magnitude larger than that for Zn-Ag contacts. This provides clear evidence that the interface with HF compounds, in the case of UPt$_3$, produces a much stronger pair-breaking effect than normal metal like Ag.

14.3.3 Break junctions

Mechanically controllable break junctions (MCBJ) (see Section 4.1.5) prepared *in situ* at low temperatures and at ultra-high vacuum conditions seem to be the most effective tool to get clean, nondegraded interfaces. Identical metals on both sides of the junction permit us to avoid normal quasiparticle reflection and discard a contribution from the second dissimilar electrode both in the normal and in the SC state, neglecting undesirable proximity effects. Moreover, MCBJ enables fine control of the contact resistance by many orders of magnitude and to measure $I - V$ characteristics from the low-ohmic metallic contacts to the high-resistance tunnel junctions.

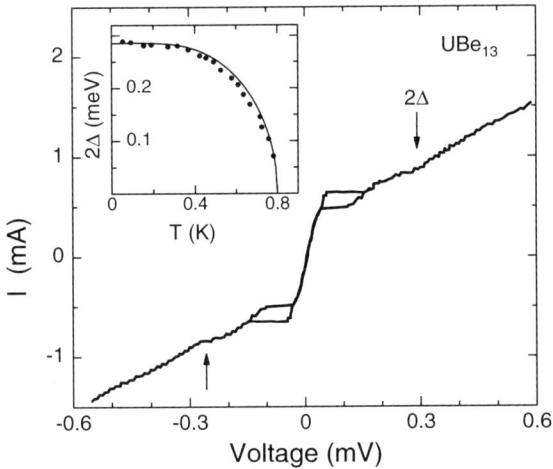

Fig. 14.12. $I - V$ characteristic of a UBe$_{13}$ break junction. The arrows show position of the gap-related feature placed symmetrically around $V = 0$. Inset: temperature dependence of the gap-feature position. Data taken from Moreland et al. (1994).

The first break junction study of the HF superconductors was carried out by Moreland et al. (1994). They obtained distinct SC features in the $I - V$ characteristic of UBe$_{13}$ break contact, namely, large slope at zero voltage, hysteresis at the higher voltages, and shallow dip at the voltage close to the expected $2\Delta/e$ (Fig. 14.12). The authors took the latter feature as a gap position, which gives a reasonable reduced gap value $2\Delta(0)/k_BT_c = 4.21$ with $\Delta(T)$ temperature dependence close to the BCS one (Fig. 14.12). It should be noted that the junction exhibits a finite conductance at zero bias; that is, no Josephson-like current is seen. As it is also mentioned by the authors, Shapiro steps were absent under radio frequency irradiation. Moreover, the SC gap features in the $I - V$ characteristics in a vacuum-tunneling regime were not resolved.

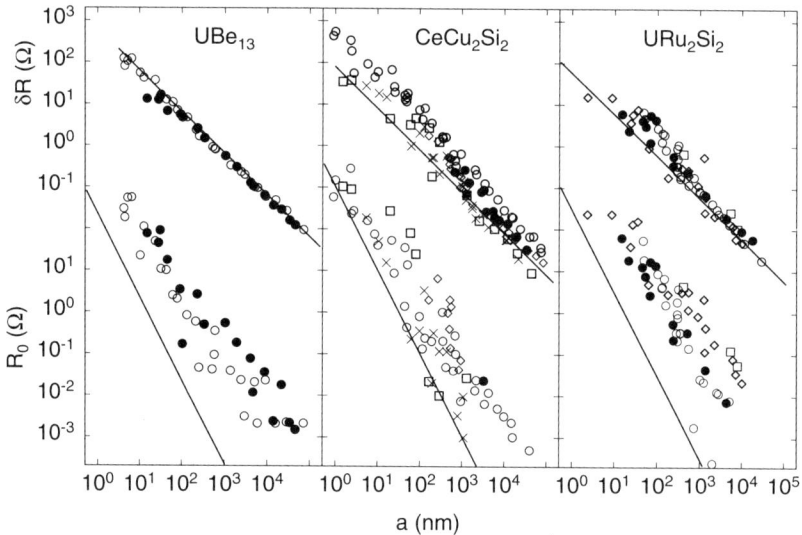

Fig. 14.13. Resistance drop δR at zero bias by transition into the SC state and zero-bias resistance R_0 (at $V = 0$ and $T \ll T_c$) versus the contact radius determined by (3.19) for MCBJ with three typical HF superconductors. Different symbols denote different samples, both polycrystals and single crystals. Solid lines are $\delta R \propto d^{-1}$ and $R_0 \propto d^{-2}$. By courtesy of Gloos et al. (1998).

The break junction experiments on the various HF systems by Gloos et al. (1997, 1998) confirmed in general their previous findings with the HF heterocontacts (see Section 14.3.1). The decreasing (or drop) of the zero-bias resistance δR below T_c scales inversely with the radius of the contacts (Figs. 14.13 and 14.14) at the same time the magnitude of δR corresponds well to the Maxwell contribution in the contact resistance with ρ as in the bulk. The correspondence is even better if the residual resistivity of the material in the point contact is taken (Fig. 14.15). The zero-bias resistance R_0 (at $V = 0$ and $T \ll T_c$) scales well with the inverse contact area (Figs. 14.13 and 14.14), albeit being enhanced about two orders of magnitude as compared with the estimation using the Sharvin formula (2.1). Thereby the R_M contribution to the SC minimum at $V = 0$ was clearly resolved, and no clear-cut signatures of the Andreev reflection or Josephson effects were found.

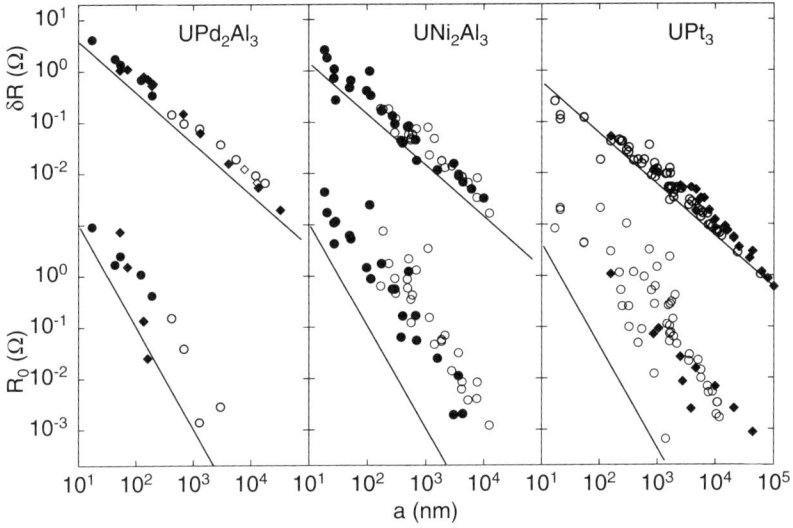

Fig. 14.14. The same as in Fig. 14.13 for three other HF compounds. By courtesy of Gloos et al. (1998).

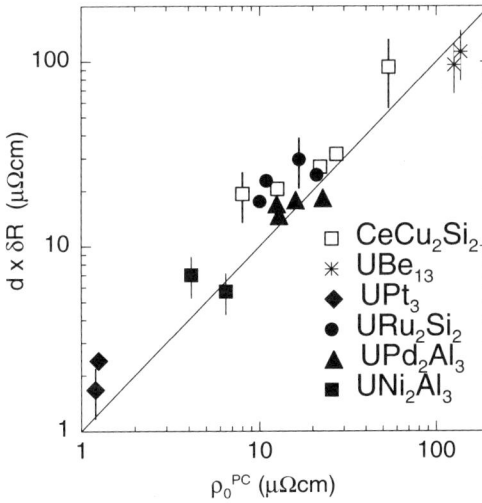

Fig. 14.15. Average product $d \times \delta R$ versus residual resistivity of corresponding compound for junctions with HF superconductors shown in Figs. 14.13 and 14.14. A few symbols for one compound denote different samples. The solid line is the expected Maxwell behavior. By courtesy of Gloos et al. (1998).

Naidyuk et al. (1997) analyzed structures in $dV/dI\,(V)$ of MCBJ with URu_2Si_2 single crystal. Peaks in $dV/d\,I(V)$ often observed for HF compounds at some critical voltage V_c (Fig. 14.16) cannot be caused by Andreev reflection in the frame of the BTK or OBTK theory (see Section 3.7), and they were attributed to the destruction of superconductivity in the constriction supposedly by the local heating. The correspondence between V_c and T_c by virtue of (3.21) was found, supposing that $L \simeq L_0$ below 1 K. Additionally, the temperature dependence of $V_c(T)$ obeys well (3.21) (Fig. 14.16, insert) in support of the thermal regime picture.

$dV/dI\,(V)$ in Fig. 14.16 have additionally a gentle double-minimum zero-bias structure at voltages smaller than V_c, which is usually viewed as Andreev reflection processes at the N–S interface. Kondo scattering at localized magnetic moments (supposedly uranium) in the contact region (see Section 6.1) yields also a maximum at zero bias and can create similar features. The authors claimed that it was not simple to distinguish between the Andreev reflection or the Kondo nature of the zero-bias maximum. It seems to be reasonable to include Kondo scattering in the BTK theory (Section 3.7). Conceivably, the elastic spin-flip scattering can be formally described by the Z parameter if the Andreev reflection and the Kondo effect do not blockade each other completely.

Gloos et al. (1997, 1998) and Naidyuk et al. (1997) investigated the Josephson-like $I - V$ characteristics of MCBJ with zero resistance at $V = 0$. These contacts usually had low normal resistance – $R_N \leq 0.2\,\Omega$ in the case of URu_2Si_2. No oscillatory pattern of the critical current was found in a magnetic field. This may be caused by the lack of a clear-cut normal area penetrated by magnetic flux like, for example, in Josephson junction with the circular interface. Another point worth noting is that ballistic contribution to the resistance of these contacts is extremely small. For $R_{PC} \leq 0.2\,\Omega$ or $d \geq 1\mu m$ (see Fig. 3.9), Sharvin resistance is less than $1/100$ of the total normal-state resistance, assuming the bulk value of the residual resistivity and typical metallic electron density in URu_2Si_2. This means that the vanishing of Maxwell resistance in the SC state can simulate appearance of a supercurrent, whereas $R_{PC} \simeq R_{Max} \to 0$ because of $R_{Max} \gg R_{Sh}$.

The absence of the Josephson effect could be explained by the normal interface layer, which also manifests itself by the enhanced residual point-contact resistance R_0 compared with the Sharvin one (Figs. 14.13 and 14.14). For the S-c-S type of weak links, the Josephson effect is expected to survive as long as a width of the region with the reduced order parameter (e.g., the width of the normal layer) is less than 3ξ [Likharev (1979)]. Therefore, already a thin (about 30 nm) normal layer may suppress Josephson coupling in the case of the HF superconductors, the coherence length of which is of the order of 10 nm. This layer, which is eventually of magnetic origin, appears to hinder also the observation of the SC features by the vacuum tunneling

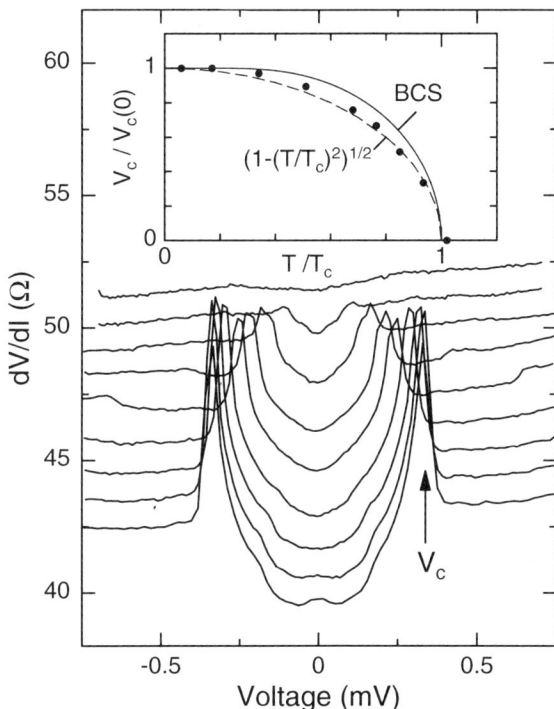

Fig. 14.16. $dV/dI\,(V)$ along a direction of URu_2Si_2 MCBJ with $R_0 \simeq 40\,\Omega$ at temperatures $T\,(K)$: 1.2, 1.1, 1.0, 0.9, 0.8, 0.6, 0.4, 0.2 (from the top curve to the bottom one). Inset: Position of the dV/dI maximum V_c (symbols) versus reduced temperature T/T_c for the same contact ($T_c \simeq 1.1\,K$). Solid and dashed lines show BCS and thermal regime behavior, correspondingly. Data taken from Naidyuk et al. (1997).

[Poppe (1985), Moreland et al. (1994), and Goschke et al. (1996)], which senses the topmost atomic layer.

Naidyuk et al. (1997) measured simultaneously the transition from the metallic to the vacuum tunneling regime for URu_2Si_2 MCBJ. The latter regime was proved by the perfect exponential dependence of the current on the vacuum gap width [Fig. 14.17(a)]. The authors reported about the lack of both clear SC features and supercurrent in the tunnel regime, which testifies that it takes place also for the atomary clean surfaces. Only broad asymmetric minimum in $dI/dV\,(V)$ was resolved with a width more than one order of magnitude larger than the expected SC energy gap [Fig. 14.17(b)]. More expressive minimum in $dI/dV\,(V)$ is seen for c direction in contrast with observation of Aliev et al. (1991) and Aarts et al. (1994). This questioned the formation of the above-mentioned minimum because of the Fermi surface gapping in the AFM state. It is worth noting that as a consequence of a small

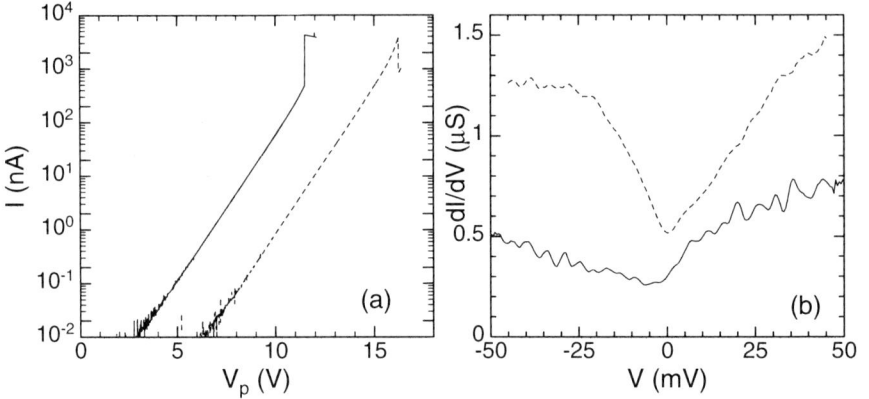

Fig. 14.17. (a) Tunnel current versus change of the piezovoltage V_p at $T \leq 0.1$ K for URu$_2$Si$_2$ break junction along a (solid line) and c (dashed line) direction. Constant bias voltage is 0.1 V in both cases. (b) Numerical derivatives of the $I - V$ curves in the tunnel regime at $T \leq 0.1$ K for the a direction (solid curve) and c direction (dashed curve). The zero-bias resistance is above 1 MΩ in both cases. Data taken from Naidyuk et al. (1997).

capacitance C of MCBJ, Coulomb blockade energy $e^2/2C$ can be larger than the thermal energy k_BT, which can produce similar zero-bias minimum in dI/dV [van Bentum et al. (1988)]. On the other hand, Gloos et al. (1998) pointed out that the small capacitance (~ 1 fF) of MCBJ results also in zero-point energy $E_0 = \sqrt{e\hbar I_c/2C}$ of Josephson plasma oscillations, which can easily exceed Josephson coupling energy $E_J = hI_c/2e$ and suppress the supercurrent I_c.

The SC features in the $I - V$ characteristics of MCBJ on the HF superconductors already vanish, as mentioned by Naidyuk et al. (1997) and Gloos et al. (1998), for the contact resistance above 1 kΩ. This value is well below a transition from the metallic regime to the tunneling one, which occurs at $R_k = h/2e^2 = 12.9$ kΩ. This fact points additionally to the nonsuperconducting layer on the heavy-fermion surface.

When studying URu$_2$Si$_2$ by MCBJ in the tunneling regime, Naidyuk et al. (1997) found a nearly free-electron mass for the tunneling charge carriers, as well as a typical metallic carrier density or an ordinary Fermi wave number in URu$_2$Si$_2$. The curious thing is that the temperature dependence of the tunneling current can be used to determine the behavior of the thermal expansion coefficient and its anisotropy in URu$_2$Si$_2$ in a simple way (Fig. 14.18). This was possible because HF materials have a relatively large thermal expansion coefficient even at very low temperatures, and the tunnel current or resistance exponentially depends (2.7) on the vacuum gap width.

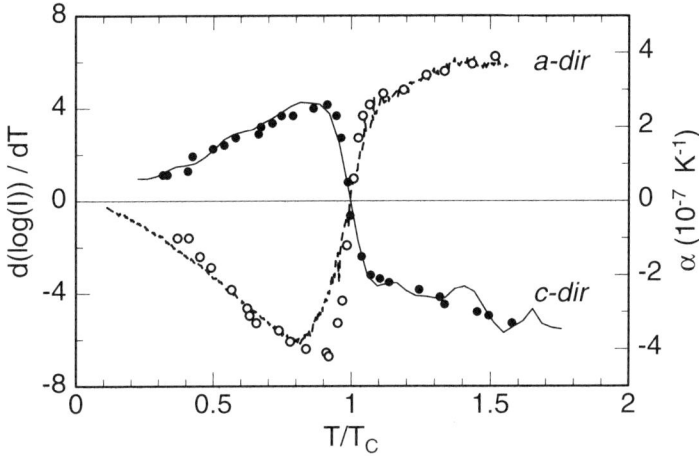

Fig. 14.18. Logarithmic derivatives of the tunnel current versus reduced temperature T/T_c along c and a direction (curves) at the constant bias voltage 100 mV and constant piezovoltage after Naidyuk et al. (1997). Open and closed circles show the temperature dependence of the thermal-expansion coefficients α determined by the conventional method for two corresponding directions [data taken from van Dijk et al. (1995)].

References

Aarts J., Volodin A. P., Menovsky A. A., Nieuwenhuys C. J. and Mydosh J. A. (1994) Europhys. Lett. **26** 203.

Aliev A., Agraït N., Vieira S., Villar R., Kovachik V., Moshchalkov V. V., Pryadun V. V., Alekseevskii N. E. and Mitin A. V. (1991) J. Low Temp. Phys. **85** 359.

Anders F. B. and Gloos K. (1997) Physica B **230-232** 230.

Brandt N. B. and Moshchalkov V. V. (1984) Adv. in Phys. **33** 373.

Deutscher G. and Nozieres P. (1994) Phys. Rev. B **50** 13557.

de Wilde Y., Heil J., Jansen A. G. M., Wyder P., Deltour R., Assmuss W., Menovsky A. A., Sun W. and Taillefer L. (1994) Phys. Rev. Lett. **72** 2278.

de Wilde Y., Klapwijk T. M., Jansen A. G. M., Heil J. and Wyder P. (1996) Physica B **218** 165.

Doniach S. (1977) Physica B **91** 231.

Escudero R., Morales F. and Lejay P. (1994) Phys.Rev. B **49** 15271.

Fisk Z., Hess D. W., Pethick C. J., Pines D., Smith J. L., Thompson J. D. and Willis J. O. (1988) Science **239** 33.

Gloos K. (2000) Phys. Rev. Lett. **85** 5257.

Gloos K., Anders F. B., Assmus W., Buschinger B., Geibel C., Kim J. S., Menovsky A. A., Müller-Reisener R., Neuttgens S., Schank C., Stewart G. R. and Naidyuk Yu. G. (1998) J. Low Temp. Phys. **110** 873.

Gloos K., Anders F. B., Buschinger B. and Geibel C. (1997) Physica B **230-232** 391.

Gloos K., Anders F. B., Buschinger B., Geibel C., Heuser K., Järling F., Kim J. S., Klemens R., Müller-Reisener R., Schank C. and Stewart G. R. (1996a) J. Low Temp. Phys. **105** 37.

Gloos K., Geibel C., Müller-Reisener R. and Schank C. (1996b) Physica B **218** 169.

Gloos K., Kim S. J. and Stewart G. R. (1996c) J. Low Temp. Phys. **102** 325.

Gloos K., Martin F., Schank C., Geibel C. and Steglich F. (1995a) Physica B **206** & **207** 279.

Goll G., Bruder C. and von Löhneysen H. (1995) Phys. Rev. B**52** 6801; (1995) Physica B **206** & **207** 609.

Goll G., von Löhneysen H., Yanson I. K. and Taillefer L. (1993) Phys. Rev. Lett. **70** 2008; (1994) Physica B **199** & **200** 110.

Gompf F., Gering E., Renker B., Rietschel H., Rauchschwalbe U. and Steglich F. (1987) J. Magn. and Magn. Mat. **63** & **64** 344.

Goschke R. A., Gloos K., Geibel C., Ekino T. and Steglich F. (1996) Czech. J. Phys. **46** 797.

Grewe N. and Steglich F. (1991) in *Handbook on the Physics and Chemistry of Rare Earths*, ed. K. A. Gschneidner, Jr. and L. Eyring (North-Holland, Amsterdam) Vol. 14, p. 343.

Hasselbach K., Kirtley J. R. and Lejay P. (1992) Phys. Rev. B **46** 5826.

Han S., Ng K. W., Wolf E. L., Braun H. F., Tanner L., Fisk Z., Smith J. L. and Beasley M. R. (1985) Phys. Rev. B **32** 7567.

Han S., Ng K. W., Wolf E. L., Millis A., Smith J. L. and Fisk Z. (1986) Phys. Rev. Lett. **B 57** 238.

He Y., Muirhead C., Bradshaw A., Abell J. S., Schank C., Geibel C. and Steglich F. (1992) Nature **357** 227.

Hess D. W., Riseborough P. S. and Smith J. L. (1993) Encycl. Appl. Phys. **7** 435.

Iwanyshyn O. and Smith H. J. T. (1972) Phys. Rev. **6** 120.

Jansen A. G. M., de Visser A., Duif A. M., France J. J. M. and Perenboom J. A. A. J. (1987) J. Magn. and Magn. Mat. **63** & **64** 670.

Kadin A. M. (1990) Phys. Rev. B **41** 4072.

Kulic M. L., Keller J. and Schotte K. D. (1991) Solid State Commun. **80** 345.

Kunii S. (1987) J. Magn. and Magn. Mat. **63** & **64** 673; (1992) Sov. J. Low Temp. Phys. **18** 343; (1996) Physica B **218** 181.

Kvitnitskaya O. E., Naidyuk Yu. G., Nowack A., Gloos K., Geibel C., Jansen A. G. M. and Wyder P. (1999) Physica B **259-261** 638.

Likharev K. K. (1979) Rev. Mod. Phys. **51** 101.

Lysykh A. A., Duif A. M., Jansen A. G. M., Wyder P. and de Visser A. (1988) Phys. Rev. B **38** 1067.

Mason T. E., Buyer S. W. J. L., Petersen T., Menovsky A. A. and Garrett J. D. (1995) J. Phys.: Condens. Matter **7** 5089.

Mentink S. A. M., Mason T. E., Süllow S., Nieuwenhuys G. J., Menovsky A. A., Mydosh J. A. and Perenboom J. A. A. J. (1996) Phys. Rev. B **53** 6014.

Moreland J., Clark A. F., Soulen R. J. Jr. and Smith J. L. (1994) Physica B **194-196** 1727.

Moser M., Hulliger F. and Wachter P. (1985a) Physica B **130** 21.

Moser M., Wachter P. and France J. J. M. (1986a) Sol. State Commun. **58** 515.

Moser M., Wachter P., France J. J. M., Meisner J. P. and Walker E. (1986b) J. Magn. and Magn. Mat. **54-57** 373.

Moser M., Wachter P., Hulliger F. and Etourneau J. R. (1985b) Sol. State Commun. **54** 241.

Mydosh J. A. (1987) Physica Scripta **19** 260.

Naidyuk Yu. G., Gloos K. and Menovsky A. A. (1997) J. Phys.: Cond. Matter **9** 6279.

Naidyuk Yu. G., Gribov N. N., Lysykh A. A., Yanson I. K., Brandt N. B. and Moshchalkov V. V. (1985a) JETP Lett. **41** 399.

Naidyuk Yu. G., Gribov N. N., Lysykh A. A., Yanson I. K., Brandt N. B. and Moshchalkov V. V. (1985b) Sov. Phys. Solid State **27** 2153.

Naidyuk Yu. G., Kvitnitskaya O. E., Jansen A. G. M., Wyder P., Geibel C. and Menovsky A. A. (2001) Low Temp. Phys. **27** 493.

Naidyuk Yu. G., Kvitnitskaya O. E., Nowack A., Yanson I. K. and Menovsky A. A. (1995) Low Temp. Phys. **21** 236.

Naidyuk Yu. G., Kvitnitskaya O. E., Nowack A., Yanson I. K. and Menovsky A. A. (1996a) Physica B **218** 157.

Naidyuk Yu. G., Kvitnitskaya O. E., Yanson I. K., Suski W. and Folchik L. (1993) Low Temp. Phys. **19** 204.

Naidyuk Yu. G., Reiffers M., Jansen A. G. M., Wyder P. and Yanson I. K. (1991) Z. Phys. B - Cond. Matter **82** 221.

Naidyuk Yu. G., von Löhneysen H., Goll G., Paschke C., Yanson I. K. and Menovsky A. A. (1996b) Physica B **218** 161.

Naidyuk Yu. G., von Löhneysen H., Goll G., Yanson I. K. and Menovsky A. A. (1996c) Europhys. Lett. **33** 557.

Naidyuk Yu. G., Yanson I. K., Chubov P. N., Kirzhner V. M., Panova G. Kh. and Alekseeva Z. M. (1988) Sov. Phys.-Sol. State **30** 1343.

Nowack A., Heinz A., Oster F., Wohlleben D., Güntherodt D., Fisk Z. and Menovsky A. A. (1987a) Phys. Rev. B **36** 2436.

Nowack A. and Klug J. (1992) Low Temp. Phys. **18** 367.

Nowack A., Naidyuk Yu. G., Chubov P. N., Yanson I. K. and Menovsky A. A. (1992) Z. Phys. B - Cond. Matter **88** 295.

Nowack A., Naidyuk Yu. G., Ulbrich E., Freimuth A., Schlabitz W., Yanson I. K. and Menovsky A. A. (1995) Z. Phys. B - Cond. Matter **97** 77; (1995) Low Temp. Phys. **21** 259.

Obermair C., Goll G., von Löhneysen H., Yanson I. K. and Taillefer L. (1998) Phys. Rev. B **57** 7506.

Onuki Y., Shimizu Y. and Komatsubara T. (1985) J. Phys. Soc. Jap. **54** 304.

Paulus E. and Voss G. (1985) J. Magn. and Magn. Mat. **47 & 48** 539.

Peysson Y., Ayache C., Salce B., Kunii S. and Kasuya T. (1986) J. Magn. and Magn. Mat. **59** 33.

Poppe U. (1985) J. Magn. and Magn. Mater. **52** 157.

Rodrigo J. G., Guinea F., Vieira S. and Aliev F. G. (1997) Phys. Rev. B **55** 14318.

Samuely P., Szabo P., Flachbart K., Mihalik M. and Menovsky A. A. (1995) Physica B **206-207** 612.

Sigrist M. and Rice T. M. (1992) J. Phys. Soc. Japan **61** 4283.

Steglich F., Aarts J., Bredl C. D., Lieke W., Meschede D., Franz W. and Schäfer H. (1979) Phys. Rev. Lett **43** 1892.

Steiner P., Degiorgi L., Maple M. B. and Wachter P. (1996) Physica B **218** 173.

Stewart G. R. (1984) Rev. Mod. Phys. **56** 755.

Sumiyama A., Shibata S., Oda Y., Kimura N., Yamamoto E., Haga Y. and Onuki Y. (2000) Physica B **281 & 282** 1010.

Tachibana R., Sumiyama A., Oda Y., Yamamoto E., Haga Y., Honma T., and Onuki Y. (2003) J. Phys. Soc. Japan **56** 364.

van Bentum P. J. M., van Kempen H., van de Leemput L. E. C. and Teunissen P. A. A. (1988) Phys. Rev. Lett. **60** 369.

van Dijk N. H., Bourdarot F., Fåk B., Lapierre F., Regnault L. P., Burlet P., Bossy J., Pyka N. and Menovsky A. A. (1997) Physica B **234-236** 693.

van Dijk N. H., de Visser H., Franse J. J. M. and Menovsky A. A. (1995) Physica B **206** 583.

van Harlingen D. J. (1995) Rev. Mod. Phys. **67** 515.

Wälti C., Ott H. R., Fisk Z. and Smith J. L. (2000) Phys. Rev. Lett. **84** 5616.

Wasser S., Nowack A., Schlabitz W., Freimuth A., Kvitnitskaya O. E., Menovsky A. A. and Bruder C. (1998) Phys. Rev. Lett. **81** 898.

15 New trends in research

At the end of the last chapter, a few issues were discussed that are not connected directly with the metallic point contacts; the main topic of this book, however, is the logical development of the point-contact technique. In this chapter, we describe additionally some new trends of research that have been growing by further developing or application of the point contacts.

15.1 Conductance quantization

The conductivity of a smooth narrow constriction can be represented in the quantum-mechanical approach by a sum of a finite number of the quantized modes. According to the Landauer formalism, each propagating one-dimensional electron modes constitute a conductance channel with a conductance quantum $G_0 = 2e^2/h$. Therefore, the conductance of a ballistic contact can be written as NG_0, where N is the number of channels. By decreasing the lateral constriction size to the value of the order of atomic size, this quantized conductivity should be clearly visible. Indeed a step-like decrease of conductance through the mechanically controlled break junction was observed [see Fig. 15.1(a)] by gradual bending force increasing, which pulls the electrodes apart. Such conductance steps[1] were observed by decreasing the lateral constriction size by a few experimental groups using an MCBJ or STM technique both at room and liquid helium temperatures (see Ruitenbeek (2000) and references therein). However, the conductance quantization should be distinguished from rearrangements of the atomic structure of matter by formation of a point contact, the size and conductance of which cannot be changed continuously in atomic scale. It turns out that the degeneracy of the conduction modes of the smooth cylindrical symmetry point contact results in a characteristic sequence of the conductance values where some integer multiples of G_0 are excluded (e. g., $N \neq 2$, 4, 7). Observation by Krans et al. (1995) of the conductance quantization with $N=1$, 3, 5, 6 .... [see

---

[1] The real steps occur in the case of nearly one-dimensional constriction geometry. The increase of the aperture angle of the contact from zero for the case of one-dimensional model to π in the case of orifice model results in gradually smearing and vanishing of the steps [Fig. 15.1(c)].

Fig. 15.1(b)] for atomic size sodium point contacts gives the real evidence for the presence of quantized atomic channels.

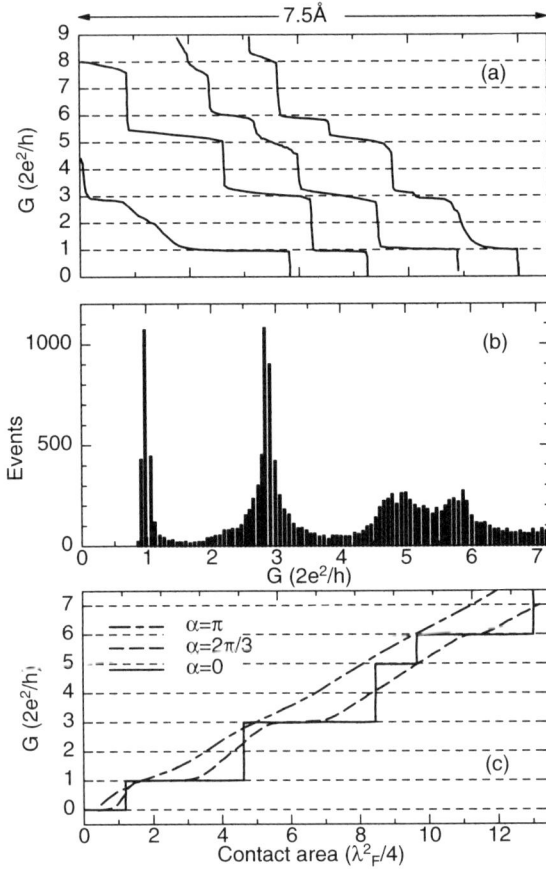

Fig. 15.1. (a) A few records of a sodium MCBJ conductance by increasing the distance between the electrodes. (b) Histogram of conductance values constructed for sodium MCBJ from more than 100 curves as shown in the upper panel. Data taken from Krans et al. (1995). (c) Calculated by Bogachek et al. (1990) conductance for a free electron gas and cylindrically symmetric contact, as a function of the contact area in units $(\lambda_F/2)^2$ for three different aperture angels: $\alpha = 0, 2\pi/3$, and π.

It is necessary to account that each conductance channel (i) is characterized by a transmission coefficient T_i. The transmission probability for a mode caused by scattering may be smaller than 1, leading to a noninteger in the value of G_0 conductance of constriction. The method for determination of the individual transmission coefficient was developed by Scheer et al. (1998)

who investigated atomic-size contacts in the superconducting state. The determination was based on the comparison of the subharmonic gap structure in $I - V$ characteristic of one-atom superconducting contact caused by the multiple Andreev reflection with the prediction of the theory (Fig. 15.2). It

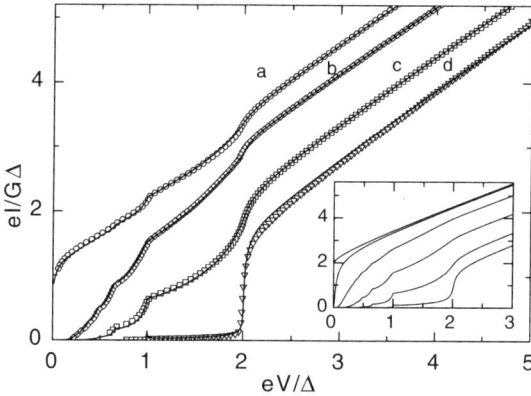

Fig. 15.2. $I - V$ characteristics of four thin-film aluminum MCBJ at $30\,\mathrm{mK}$ with clear structure caused by multiple Andreev reflection. The amount of individual channels (first number) and their transmission for each curve determined by fitting procedure are (a) 3, 0.997, 0.46, and 0.29; (b) 2, 0.74, and 0.11; (c) 3, 0.46, 0.35, and 0.07; (d) 1 and 0.025. The last curve (d) corresponds to the tunnel regime. The superconducting gap for Al equal to $182.5 \pm 2.0\,\mu\mathrm{eV}$ was measured. Inset shows calculated $I - V$ curves for a single-channel superconducting contact with different values of transmission coefficient: 0.99, 0.9, 0.7, 0.4, and 0.1 from the top curve to the bottom one. By courtesy of Scheer et al. (1998).

was shown that more than one channel will contribute to the conductivity of the one-atom point contact even if the total conductance is lower than G_0. Moreover, the authors mentioned that they had never found a contact that can be described by a single channel. This leads to the conclusion that the number of the conduction channels of an one-atom contact is determined by the number of available valence orbitals. This means that chemical valence of atoms in solid determines the conduction properties of one-atom contact between two conductors.

15.2 From one-atom contact to one-atom wire

The MCBJ or STM technique caused by a very stable lateral position of two electrodes and extremely precise control of the electrode separation allows us to form a nanosized constriction or even one-dimension wires. Untiedt et al. (1997) showed that the shape of metallic constriction produced by STM depends on the fabrication procedure. Repeated cycles of plastic deformation by elongation and contraction procedure make it possible to obtain neck or nanowires with an almost crystalline structure. Striking repetitiveness of the conductance steps during elongation-contraction cycles along with the clear EPI spectra testify correspondingly to reversible rearrangements of atoms formed the constriction and their crystalline structure. The last plateau in the conductivity with the conductance equal to G_0 corresponds to one atom contact. During the further elongation procedure, the transition to the tunnel regime occurs. Yanson et al. (1998) brought to notice the fact that the last plateau with conductance, which never exceeds G_0, has a different length while repeating the procedure of elongation and contraction (Fig. 15.3). As-

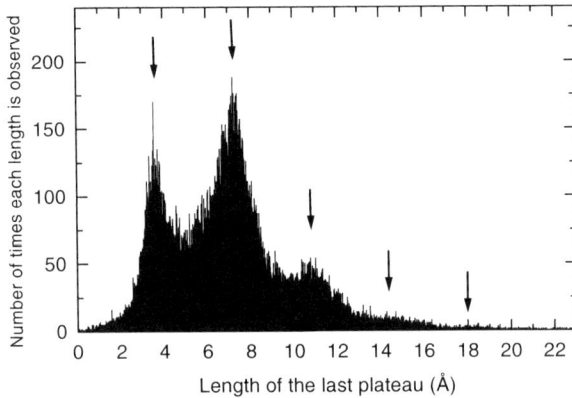

Fig. 15.3. Histogram of distribution for the last plateau length obtained at about 100,000 scans at 4.2 K for gold junction. The arrows, indicated by the maxima in the histogram, are positioned at multiples of 3.6 Å, which is somewhat longer than the nearest-neighbor spacing 2.88 Å of gold atoms in the crystal. By courtesy of Yanson et al. (1998).

suming that the length of the last plateau is proportional to the number of single atoms forming a chain, the establishing of one-atom-thick wire with the length up to 5 atoms was considered. This leads to potentialities of study of both electrical and mechanical properties of truly one-dimensional objects as well as investigation of real one-dimensional excitations.

Direct imaging of an Au nanowire was done by Ohnishi et al. (1998) using STM for the atomic wire formation simultaneously with an ultra-high vacuum electron microscope (Fig. 15.4). They observed strands of gold atoms that were about one nanometer long, transforming into one single atomic chain suspended between electrodes while withdrawing the tip. Measuring the conductivity of this strand verifies that the conductivity of the single strand of atoms is G_0 and that the conductance of the double strand is twice as large.

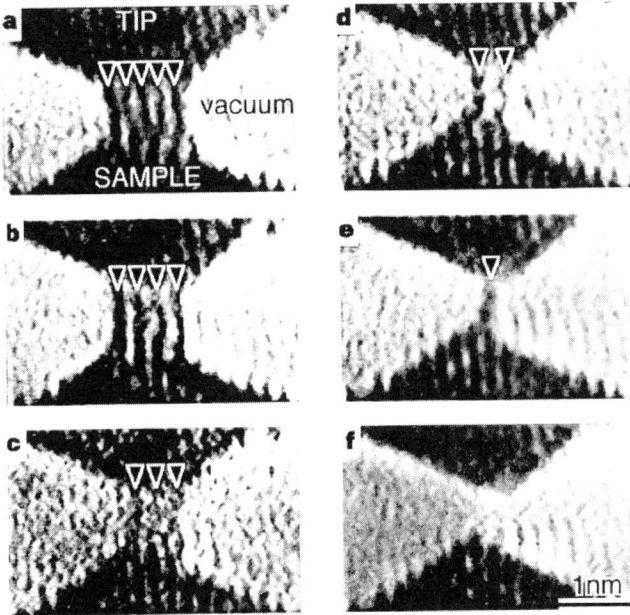

Fig. 15.4. Electron microscope imaging of a gold contact (dark region) while withdrawing the upper electrode using the STM piezodriver. A gold bridge is formed between the top and the bottom electrode, which thinned from **a** to **e** and breaks at **f**. Dark vertical lines indicated by arrowheads are rows of gold atoms. After Ohnishi et al. (1998).

Erts et al. (2000) investigated conductance of gold point contacts using the STM setup inside a transmission electron microscope. They measured the conductance of these point contacts, which size was between a single atom and 20 nm, as a function of radius. Using Wexler interpolation formula (3.18), a mean free path of 4 nm, ten times shorter than the room-temperature bulk value, was obtained. This was explained by an enhanced scattering on a large number of scattering centers created during the point-contact formation process. Erts et al. (2000) also observed that in real-time, thicker contacts looked semiliquid and changed shape on a time scale of seconds.

Agraït et al. (2002) studied by STM electronic transport in gold atomic chains up to seven atoms in length. They found nondissipative current in these atomic wires up to the voltage of the order of several mV or maximum dissipationless current of about $1\,\mu$A corresponding to the current density $\sim 10^9\,$A/cm^2. By further voltage rise, the inelastic scattering of electrons occurs because of the excitation of atomic-chain ion vibrations. The position and amplitude of the phonon peak in the spectra were very sensitive to its state of strain, so that wire stretching causes the phonon frequency softening and an increase of the one-dimensional electron–phonon interaction.

The MCBJ technique can be applied not only to the study of the one-atom contacts or atomic wires, but also to the investigation of the electronic transport through an individual molecule as shown by Smit et al. (2002). They succeeded in fabrication of a bridge between two platinum electrodes consisting of a hydrogen molecule H_2. It is amazing that such a bridge has a nearly perfect conductance of one quantum unit carried by a single channel. The authors found a resonance at the energy about 63.5 meV (higher than typical phonon energies of Pt) at measuring the differential conductivity and its derivative of the bridge. By repeating the experiments with isotopes D_2 (or DH), the resonance is shifted by value $\sqrt{m_{D_2(DH)}/m_{H_2}}$, further confirming the interpretation of molecule bridge establishing.

Many other directions in the field of atomic-sized conductors study can be found in the neoteric comprehensive review by Agraït et al. (2003).

15.3 Carbon nanotubes

The discovery of carbon nanotubes by Iijima (1991) has led to enormous growth of interest to their transport and mechanical properties [see, e. g., Dresselhaus et al. (1996), Saito et al. (1998), and Dekker (1999)], bearing in mind their potential possibilities in nanoscale science and applications. Carbon nanotube has been formed by carbon atoms and can be imagined as a single honeycomb graphite layer wrapped into a cylinder with a diameter of about 1 nm (Fig. 15.5). Meanwhile, the length of nanotubes can reach as high as a few micron; thus, the length-diameter ratio is typically greater than 100 and can be up to 10,000 and higher; therefore, they represent an excellent example for study of electronic properties of low-dimensional conductors. Small-diameter nanotubes (~ 1 nm) represent a single-walled nanotube, and nanotubes with a larger diameter (~ 10 nm) consist of multiple shells; that is, a number of tubes are arranged in a coaxial fashion. The conductivity of nanotubes strongly depends on both the diameter and chirality or twist (see also Fig. 15.5 caption). Their electrical properties can be modified in the presence of certain substances, topological defects, or as a result of mechanical deformation (stress).

A single-walled carbon nanotube possesses conductivity of two conductance quantum $2G = 4e^2/h$ with electrons traveling ballistically through the tube without being scattered. By electrical investigations, the coupling of the

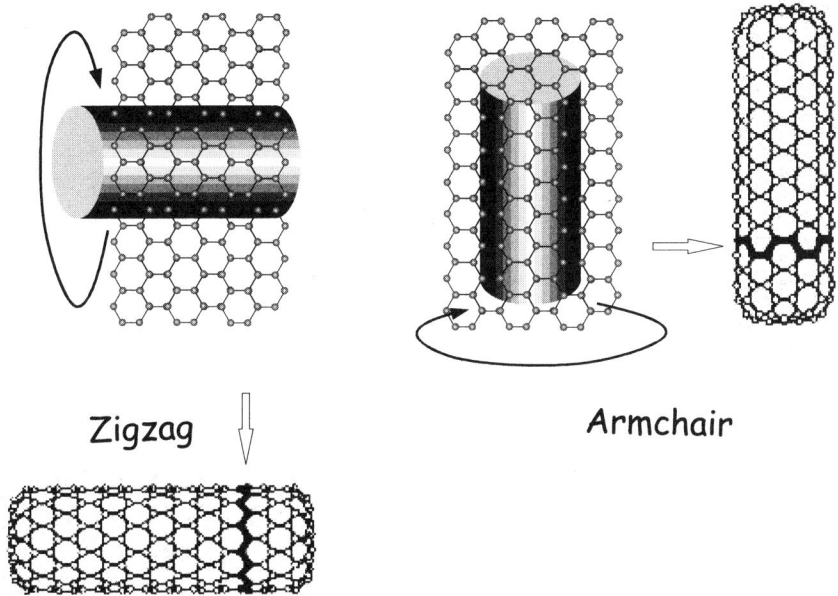

Fig. 15.5. Carbon nanotubes can be formed by wrapping sheets of graphite into a tube as shown in this figure. Wrapping the sheets along hexagon rows result in *zigzag* nicknamed tubes, whereas in the perpendicular direction – in *armchair* tubes. In another direction of wrapping, the winding of hexagon rows along tubes takes place and tubes are characterized by some chiral angle Φ. A slight change of Φ, in other words the helicity of the graphene sheet, can transform the tube from metallic to semiconducting and vice versa.

tube to metallic leads (pads) is of importance because creation of transparent contacts to single-walled nanotube requires a clever piece of work.

Search for superconductivity in the carbon nanotubes is in the spotlight as well. Kasumov et al. (1999) showed that it is possible to create superconducting (below 1 K) junctions with carbon nanotubes embedded between superconducting leads. Kociak et al. (2001) observed superconductivity in ropes of single-walled carbon nanotubes (SWNT) in low-resistance contacts to non superconducting (normal) metallic pads at low voltage and at temperature below 0.55 K, which was destroyed by a magnetic field of the order of 1 T, or by a dc current greater than 2.5 μA.

Nanotubes can even be stuffed with other materials, e. g., the quasi-one-dimensional (1D) phase of C_{60} (buckyball) molecules embedded in carbon nanotubes (in other words, buckyball-filled nanotubes) nicknamed "peas in a pod (peapods)". This opens new capabilities for carbon nanotubes, e. g., as a reservoir for material (gas) storage.

Since the discovery of carbon nanotubes, there have been reports of nanotubes produced from other elements, taking carbon nanotubes and porous membranes as templates. Recently, Goldberger et al. (2003) reported growth of single-crystal semiconducting (gallium nitride) hollow nanotubes, which should be applicable to many other semiconducting systems. So, perspective for nanotubes is very bright.

This brings up another point. How all mentioned above are related to PCS, because underlying physics in conductivity of carbon nanotubes and point contacts is different. Here we put attention on the possibility of creating of a nanotube filled by some magnetic or superconducting metals. In this case, nanotubes represent a geometrically well-defined object that can be described by a long channel model [see Fig. 3.1(b)]. Thus, we can apply much of what is written in the book for study of a such system and to investigate peculiarities in their conductivity, both in the magnetic, normal, and superconducting state. Certainly, electron–phonon interaction in carbon nanotubes takes place, and analyzing of their nonlinear conductivity can provide a valuable information about peculiarity of electron–phonon coupling in low-dimension systems at specific geometry.

15.4 Study of mesoscopic particles

The nanofabrication technique can be applied for production of mesoscopic devices in a controlled way. By such a method both homo- and heterocontacts can be formed [Gribov et al. (1996)], the crystalline structure of which can be tested, e. g., by measuring point-contact electron–phonon spectra. Some experiments concerning the study of transport properties of lithographically prepared nanocontacts are given in the *Proceedings of the PCS Conference* [Physica B **218** (1996)].

The STM technique can be applied to investigate mesoscopic particles. Evaporation of a metal on a surface under special condition can produce very small size particles. STM allows us to carry out both the topography of the particles and the spectroscopic analysis of the metal properties in definite points. Poza et al. (1996) studied superconducting lead clusters with maximal length as high as 50 nm. The superconducting state was found for clusters with a dimension smaller than the bulk coherence length in Pb. However, the superconducting features in the $I - V$ characteristics were smeared compared with the bulk.

Ralph et al. (1995, 1996) fabricated a point-contact tunnel junction with a single Al particle with the diameter below 10 nm by the electron beam lithography. This allows them to measure the discrete spectrum of the electronic energy level, which can lead to the detailed understanding of forces acting on the nucleons and atomic electrons.

15.5 Light-induced electron focusing

Real space imaging of ballistic carrier propagation can be studied by the method called light-induced electron focusing developed by Heil et al. (1995). Here nonequilibrium carriers are injected into crystal by the hot spot produced by illumination of a small area on the sample surface by the laser light[2] [see Fig. 15.6(a)].

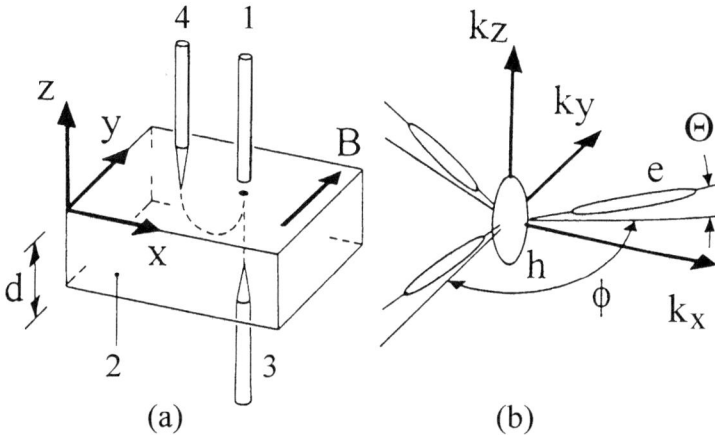

Fig. 15.6. (a) Experimental setup: 1 – moveable optical fiber, 2 – sample, 3 and 4 – metallic point contacts. (b) Sketch of the Fermi surface of Bi (e and h denote electron and hole ellipsoids, correspondingly). After Heil et al. (1995).

The exited quasiparticles propagate under a ballistic condition through the bulk crystal and can be detected by a metallic point contact on the opposite surface. The experimental idea here is very close to Sharvin's electron focusing [see Fig. 2.2(a)]; however, the physics of processes is different. It is known that the direction of group velocity of quasiparticles is perpendicular to the Fermi surface. As the Fermi surface of many metals differs considerably from the sphere, the electron flow concentrates in the direction of the Fermi surface curvature vanishing. In other words, the flat parts of the Fermi surface produce an electron beam; therefore, the electron current density exhibits singularity in these directions, which results in an enhanced potential on the

---

[2] Other carries sources like electron beam, ultrasonic foci, thin-film heaters, or tunneling contacts could be used as well [Knauth et al. (1999)].

sample surface registered by metallic point contact. The hot spot produced by optic fiber can be scanned on the sample surface resulting in the real space resolved measurements, even at a fixed position of a point-contact detector. Signals produced by electrons, holes, and phonons can be distinguished by applying the magnetic field.

The collector voltage in gray scale such that the electron signal appears bright as a function of the fiber position for two Bi slabs with different thickness is shown in Fig. 15.7. Three bright lines at zero magnetic field result from

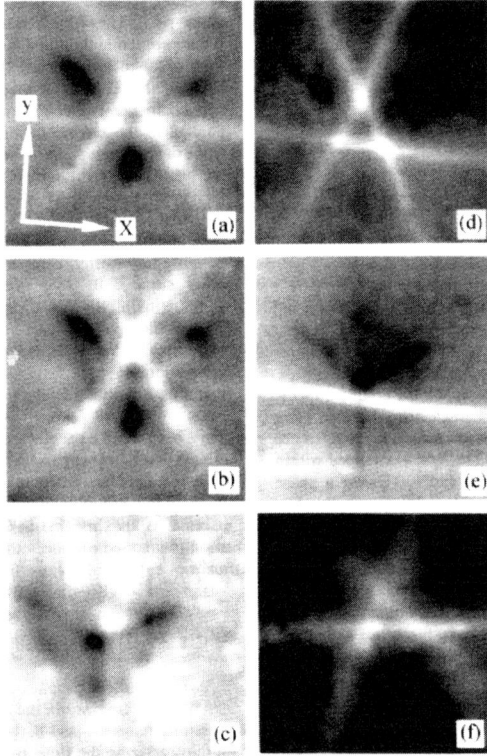

Fig. 15.7. Electron focusing image represents a collector voltage in a gray scale as a function of a fiber position for Bi crystal with the thickness of 0.2 mm (a), (b), and (c) and 0.5 mm (d), (e). The magnetic field is 0 (a), 0.3 mT (b), 1.1 mT (c), 0 (d), and 0.4 mT (e). (f) shows experimental results with a scanning point contact as emitter. After Primke et al. (1996).

electrons of three Fermi ellipsoids shown in Fig. 15.6(b). These ellipsoids extremely stretched in one direction do not strictly meet the requirements of zero curvature. However, they can be modeled by cylinders, which results in an enhanced electron flux perpendicular to the main axis. The three focusing lines observed do not intersect in one point, owing to the tilt of the main ellipsoid axis by angle θ against the base plane. In the magnetic field, electron trajectories bend in accordance with the Fermi surface topology and field direction with respect to the electron velocity. Therefore, the field increase suppresses the contribution to the focusing signal from different ellipsoids in a different way, as seen from Fig. 15.7, depending on the orientation of the field with respect to the ellipsoids axis.

If the collector point contact is placed on the same surface near the hot spot and the magnetic field is applied to bent electron trajectories back to the upper surface, such geometry will be analogous to the conventional transverse electron focusing (see Fig. 2.2). The difference is that in this case, the transverse focusing signal can be registered not only along one direction but for all scanned by the fiber surface as well. That is, the creation of the spatial resolved image of transverse focusing is possible.

It should be noted that the nonequilibrium quasiparticles are created by light apparently because of heating of the hot spot by irradiation. Therefore, the nonequilibrium phonon transport should also be necessarily taken into account. The phonons can produce the thermal gradient, drag the carriers, and generate a thermoelectric signal at the crystal-point contact interface. Correspondingly, these effects produce an additional potential at the collector, which have to be accounted for or separated. However, the whole above-mentioned phenomena can be represented as a new technique for the investigation of the Fermi surface.

References

Agraït N., Untiedt C., Rubio-Bollinger G. and Vieira S. (2002) Phys. Rev. Lett. **88** 216803.

Agraït N., Yeyati A. L. and van Ruitenbeek J. M. (2003) Physics Reports **377** 81.

Bogachek E. N., Zagoskin A. N. and Kulik I. O. (1990) Sov. J. Low Temp. Phys. **16** 796.

Dekker C. (1999) Physics Today, **52(5)** 22.

Dresselhaus M. S., Dresselhaus G. and Eklund P. C. (1996) *Science of Fullerenes and Carbon Nanotubes* (Academic, San Diego).

Erts D., Olin H., Ryen L., Olsson E. and Thölén A. (2000) Phys. Rev. B **61** 12725.

Goldberger J., He R., Zhang Y., Lee S., Yan H., Choi H.-J. and Yang P. (2003) Nature **422**, 599.

Gribov N. N., Caro J. and Radelaar S. (1996) Physica B **218** 97,

Heil J., Primke M., Würz M. and Wyder P. (1995) Phys. Rev. Lett. **74** 146.

Iijima S. (1991) Nature **354**, 56.

Kasumov A.Yu., Deblock R., Kociak M., Reulet B., Bouchiat H., Khodos I. I., Gorbatov Yu. B., Volkov V. T., Journet C. and Burghard M. (1999) Science **284** 1508.

Kociak M., Kasumov A.Yu., Guron S., Reulet B., Khodos I. I., Gorbatov Yu. B., Volkov V. T., Vaccarini L. and Bouchiat H. (2001) Phys. Rev. Lett. **86** 2416.

Knauth S., Lenzner J., Herrnberger H., Grill W., Böhm A., Gröder A., Heil J., Primke M. and Wyder P. (1999) J. Appl. Phys. **85**(1) 1.

Krans J. M., van Ruitenbeek J. M., Fisun V. V., Yanson I. K. and de Jongh L. J. (1995) Nature **375** 767.

Ohnishi H., Kondo Y. and Takayanagi K. (1998) Nature **395** 780.

Poza M., Rodrigo J. G. and Vieira S. (1996) Physica B **218** 265.

Ralph D. C., Black C. T. and Tinkham M. (1995) Phys. Rev. Lett. **74** 3241.

Ralph D. C., Black C. T. and Tinkham M. (1996) Physica B **218** 258.

Primke M., Heil J. and Wyder P. (1996) Physica B **218** 26.

Saito R., Dresselhaus G. and Dresselhaus M. S. (1998) *Physical Properties of Carbon Nanotubes*, (Imperial College Press, London).

Scheer E., Agraït N., Joyez P., Cuevas J. C., Yeyati A. L., Ludoph B., Martin-Rodero A., Bollinger G. R., van Ruitenbeek J. M. and Urbina C. (1998) Nature (1998) **394** 154.

Smit R. H. M., Noat Y., Untiedt C., Lang N. D., van Hemert M. and van Ruitenbeek J. M. (2002) Nature **419** 906.

Untiedt C., Rubio G., Vieira S. and Agraït N. (1997) Phys. Rev. B **56** 2154.

van Ruitenbeek J. M. (2000) in: *Metal Clusters on Surfaces: Structure, Quantum Properties, Physical Chemistry* ed. K. H. Meiwes-Broer (Springer-Verlag, Heidelberg) 175.

Yanson A. I., Jr., Rubio-Bollinger G., van den Brom H. E., Agraït N. and van Ruitenbeek J. M. (1998) Nature **395** 783.

Index

Springer Series in
SOLID-STATE SCIENCES

Series Editors:
M. Cardona   P. Fulde   K. von Klitzing   R. Merlin   H.-J. Queisser   H. Störmer

Springer Series in
SOLID-STATE SCIENCES

Series Editors:
M. Cardona   P. Fulde   K. von Klitzing   R. Merlin   H.-J. Queisser   H. Störmer